Exponential Fitting

Mathematics and Its Applications

Managing Editor:

M. HAZEWINKEL

Centre for Mathematics and Computer Science, Amsterdam, The Netherlands

Exponential Fitting

Edited by

Liviu Gr. Ixaru

National Institute for Research and Development for Physics and Nuclear Engineering, "Horia Hulubei", Department of Theoretical Physics, Bucharest, Romania

and

Guido Vanden Berghe

University of Gent,
Department of Applied Mathematics and Computer Science, Gent, Belgium

KLUWER ACADEMIC PUBLISHERS
DORDRECHT / BOSTON / LONDON

A C.I.P. Catalogue record for this book is available from the Library of Congress.

ISBN 978-90-481-6590-2 (PB)
ISBN 978-1-4020-2100-8 (e-book)

Published by Kluwer Academic Publishers,
P.O. Box 17, 3300 AA Dordrecht, The Netherlands.

Sold and distributed in North, Central and South America
by Kluwer Academic Publishers,
101 Philip Drive, Norwell, MA 02061, U.S.A.

In all other countries, sold and distributed
by Kluwer Academic Publishers,
P.O. Box 322, 3300 AH Dordrecht, The Netherlands.

Printed on acid-free paper

Contents

*This book is dedicated to our
wives, children and
grandchildren.*

Preface

Exponential fitting is a procedure for an efficient numerical approach of functions consisting of weighted sums of exponential, trigonometric or hyperbolic functions with slowly varying weight functions. Operations on such functions like numerical differentiation, quadrature, interpolation or solving ordinary differential equations whose solution is of this type are of a real interest nowadays in many phenomena as oscillations, vibrations, rotations or wave propagation. The behaviour of quantum particles is also described by this type of functions.

We witnessed the field for many years and contributed in it. Since the total number of papers accumulated so far in this field is over 200, and these papers are spread over journals with various profiles, to mention only those of applied mathematics, of computer science or of computational physics and chemistry, we thought that the time has come for a compact and systematic presentation of this vast material. It is hoped that in this way many persons who are faced with such problems in their own activity would be better helped than searching for the needed information in such an abundant literature.

This is what we do in this book which covers a series of aspects, ranging from the theory of the procedure up to direct applications and sometimes including ready-to-use programs.

The writing of this book was decided two years ago. Since our working places are not close to each other, we agreed to share the effort: the first author has taken upon him Chapters 1 to 5, while the second author was responsible for Chapter 6. However, we both share equal responsibility for the whole text.

We are indebted to a number of colleagues for discussions and the collaboration in several periods of our research:
S. Berceanu and M. Rizea (Bucharest, Romania), R. Cools (Leuven, Belgium), H. De Meyer, M. Van Daele and H. Vande Vyver (Universiteit Gent, Belgium),

T. Van Hecke (Hogeschool Gent, Belgium), K. J. Kim (Seoul National University, South-Korea), B. Paternoster (Universitá di Salerno, Italy).

We like to thank W. Dewolf, Universiteit Gent, for technical assistance.

We want to thank especially Prof. J. P. Coleman, University of Durham, U.K., who helped us a lot by his pertinent and extremely valuable suggestions.

The authors

Liviu Gr. Ixaru   Guido Vanden Berghe

ixaru@theory.nipne.ro   guido.vandenberghe@UGent.be

January 2004

Acknowledgments

The authors like to thank the following institutions for financial support during the period of research and the preparation of the manuscript:

- Fund for Scientific Research (FWO-Flanders), Belgium.

- Research funding of the University Research Fund, Universiteit Gent, Belgium.

- National Institute of Physics and Numerical Engineering "Horia Hulubei", Bucharest, Romania.

- Katholieke Universiteit Leuven, Belgium.

- Ministry of the Flemish Community, Science, Innovation and Media Department, International Scientific and Technological Cooperation, Belgium.

We acknowledge permission for reprinted material from Elservier, Copyright (2003) from the following publications:

- Computer Physics Communications:

  - Volume number **105**, (1997), 1–19:
    Ixaru, L. Gr.,
    Operations on oscillatory functions.
  - Volume number **133**, (2001), 177–188:
    Ixaru, L. Gr. and Paternoster, B.,
    A Gauss quadrature rule for oscillatory integrands.
  - Volume number **140**, (2001), 346–357:
    Vanden Berghe, G. , Ixaru, L. Gr. and Van Daele, M.,
    Optimal implicit exponentially-fitted Runge-Kutta methods.

– Volume number **150**, (2003), 116–128:
Ixaru, L. Gr. , Vanden Berghe, G. and De Meyer, H. ,
Exponentially fitted variable two-step BDF algorithm for first order
ODE.

■ Computational Biology & Chemistry (formerly known as Computers and
Chemistry):

– Volume number **25**, (2001), 39–53:
Ixaru, L. Gr.,
Numerical operations on oscillatory functions.

■ Journal of Computational and Applied Mathematics:

– Volume number **132**, (2001), 95–105:
Vanden Berghe, G. , Ixaru, L. Gr. and De Meyer, H. ,
Frequency determination and step–lenght control for exponentially-
fitted Runge–Kutta methods.

– Volume number **140**, (2002), 423–434:
Ixaru, L. Gr. , Vanden Berghe, G. and De Meyer, H.,
Frequency evaluation in exponential mutistep algorithms for ODEs.

– Volume number **140**, (2002), 479–497:
Kim, K. J. , Cools, R. and Ixaru, L. Gr.,
Quadrature rules using first derivatives for oscillatory integrands.

– Volume number **149**, (2002), 407–414:
Franco, J .M. ,
An embedded pair of exponentially fitted explicit Runge-Kutta methods.

– Volume number **159**, (2003), 217–239:
Vanden Berghe G. , Van Daele, M. and Vande Vyver H.,
Exponential fitted Runge–Kutta methods of collocation type: fixed or
variable knot points?

■ Applied Numerical Mathematics:

– Volume number **46**, (2003), 59–73:
Kim, K. J. , Cools, R. and Ixaru, L. Gr.,
Extended quadrature rules for oscillatory integrands.

Chapter 1

INTRODUCTION

The simple approximate formula for the computation of the first derivative of a function $y(x)$,

$$y'(x) \approx \frac{1}{2h}[y(x+h) - y(x-h)], \qquad (1.1)$$

is known to work well when $y(x)$ is smooth enough. However, if $y(x)$ is an oscillatory function of the form

$$y(x) = f_1(x)\sin(\omega x) + f_2(x)\cos(\omega x) \qquad (1.2)$$

with smooth $f_1(x)$ and $f_2(x)$, the slightly modified formula

$$y'(x) \approx \frac{1}{2h} \cdot \frac{\theta}{\sin(\theta)} \cdot [y(x+h) - y(x-h)], \qquad (1.3)$$

where $\theta = \omega h$, becomes appropriate.

Likewise, the well known trapezoidal rule for the quadrature,

$$\int_{x-h}^{x+h} y(z)dz \approx h[y(x+h) + y(x-h)], \qquad (1.4)$$

gives good results for smooth $y(x)$ but for functions of the form (1.2) its modified version

$$\int_{x-h}^{x+h} y(z)dz \approx h \cdot \frac{\sin(\theta)}{\theta\cos(\theta)} \cdot [y(x+h) + y(x-h)], \qquad (1.5)$$

where again $\theta = \omega h$, has to be used.

Formulae like (1.1) and (1.4) are classical formulae and it is often believed that the only way of improving the accuracy of the results would consist in taking

1

small values for h and/or in replacing them by formulae with an increased number of points. The alternative way of keeping the form of the formula unchanged while modifying the coefficients in terms of the current behaviour of $y(x)$ is comparatively less popular among the users although this also has a long history.

Particularly interesting for many applications are functions of the forms (1.2) or

$$y(x) = f_1(x)\sinh(\lambda x) + f_2(x)\cosh(\lambda x), \tag{1.6}$$

and also linear combinations of such forms,

$$
\begin{aligned}
y(x) &= \sum_{j=1}^{J_1} f_1^j(x)\sin(\omega_j x) + f_2^j(x)\cos(\omega_j x) \\
&+ \sum_{j=1}^{J_2} g_1^j(x)\sinh(\lambda_j x) + g_2^j(x)\cosh(\lambda_j x)
\end{aligned}
\tag{1.7}
$$

where the coefficients of the trigonometric and of the hyperbolic functions are slowly varying. In such applications only the values of the whole y are currently available, not those of the individual coefficient functions. A situation of this type is met in quantum mechanics: the wavefunctions produced by the coupled-channel treatment of the Schrödinger equation are piecewise of the type (1.7), see, e. g., Ixaru [21].

The numerical treatment of oscillatory functions received a special attention along the time. In the field of the numerical quadrature several methods have been developed, see, e.g., the books of Davis and Rabinowitz [10] and of Evans [15].

A substantial set of such methods has been devised for the calculation of the finite Fourier integrals, and the oldest one is perhaps that of Filon. To compute the integrals

$$I_s = \int_a^b y_s(x)dx \text{ and } I_c = \int_a^b y_c(x)dx,$$

where the functions $y_s(x) = f(x)\sin(\omega x)$ and $y_c(x) = f(x)\cos(\omega x)$ depend on the parameter ω, and f is assumed smooth, the interval $[a, b]$ is divided into $2N$ subintervals of equal length $h = (b-a)/(2N)$. With $x_k = a + kh$, $\theta = \omega h$ and

$$\alpha(\theta) = 1/\theta + \cos(\theta)\sin(\theta)/\theta^2 - 2\sin^2(\theta)/\theta^3,$$

$$\beta(\theta) = 2\{[1 + \cos^2(\theta)]/\theta^2 - \sin(2\theta)/\theta^3\},$$

$$\gamma(\theta) = 4[\sin(\theta)/\theta^3 - \cos(\theta)/\theta^2],$$

$$S_{2N} = \sum_{r=0}^{N} y_s(x_{2r}) - [y_s(a) + y_s(b)]/2, \quad S_{2N-1} = \sum_{r=1}^{n} y_s(x_{2r-1}),$$

and

$$C_{2N} = \sum_{r=0}^{N} y_c(x_{2r}) - [y_c(a) + y_c(b)]/2, \; C_{2N-1} = \sum_{r=1}^{n} y_c(x_{2r-1}),$$

Filon's formulae are

$$I_s \approx h\{\alpha(\theta)[y_c(a) - y_c(b)] + \beta(\theta)S_{2N} + \gamma(\theta)S_{2N-1}\},$$

and

$$I_c \approx h\{\alpha(\theta)[y_s(b) - y_s(a)] + \beta(\theta)C_{2N} + \gamma(\theta)C_{2N-1}\}.$$

To obtain these expressions a technique based on the assumption that $f(x)$ is well approximated by a parabola on each double subinterval $[x_{2r}, x_{2r+2}]$ has been used.

Rules based on different assumptions and/or techniques have been also derived, to quote only the papers of Bakhvalov and Vasil'eva [3], Gautschi [18], Piessens [27], Levin [23], Patterson [26], Alaylioglu et al. [1], and Evans and Webster [16]. For example, Levin, [23], considered the numerical evaluation of the integral

$$I = \int_a^b f(x) \exp[i\tau q(x)]dx,$$

for given functions $f(x)$, $q(x)$ and constant τ; the functions f and q are assumed slowly varying on $[a, b]$. The idea of this approach is that if $f(x)$ had the form

$$f(x) = i\tau q'(x)p(x) + p'(x)$$

then the integral I would integrated exactly,

$$\int_a^b f(x) \exp[i\tau q(x)]dx = p(x) \exp[i\tau q(x)]|_a^b.$$

The evaluation of the integral is then reduced to solving the written differential equation for $p(x)$, despite there being no boundary conditions to use. However, a slowly oscillating solution does exist, which can be highlighted by the basis set method for ordinary differential equations.

This type of rules is not suited for integrands of the form which we are interested in; it is sufficient to notice that they rely on the numerical values of the smooth component of the integrand, which in our case are unknown. However, rules for the integration of functions of the form (1.7) where only the trigonometric terms are considered do exist. Examples include Ehrenmark's version [14] of the Simpson rule suited for the form (1.2) (this is actually exact if f_1 and f_2 are constants), a version due to Vanden Berghe et al., [32], of the N-point Newton–Cotes formula suited for

$$y(x) = f_1(x)\sin(\omega x) + f_2(x)\cos(\omega x) + \phi(x), \tag{1.8}$$

(it is exact if f_1 and f_2 are constants and $\phi(x) \in \mathcal{P}_{N-3}$), and a version of the same Newton–Cotes formula suited for

$$y(x) = \sum_{j=1}^{2} [f_1^j(x) \sin(\omega_j x) + f_2^j(x) \cos(\omega_j x)] + \phi(x),$$

due to Van Daele et al., see [31]. The latter is exact if f_1^j, f_2^j are constants and $\phi(x) \in \mathcal{P}_{N-5}$. Two different techniques were used to determine the coefficients of the new rules. While in [14] a direct technique to implement the requirement that the formula be exact for $y(x) = 1$, $\sin(\omega x)$ and $\cos(\omega x)$ was used, the mixed interpolation technique of De Meyer et al. [11] was applied in [31, 32]. The latter has been also used in [4] to derive a version of the Gregory rule which makes this suited for (1.8).

Rules of this type are of use not only for the evaluation of integrals whose integrands are of the specified behaviour but also for the solution of integral or integro-differential equations, see, e. g., Bocher et al. [5], and Brunner et al. [7].

In the field of the numerical solution of ordinary differential equations (ODE for short) a series of methods have been devised for the case when the solution is known to exhibit a specific behaviour. Special attention was paid to the linear multistep methods whose classical versions, as described by Henrici [20], are designed to be exact when the solution is a polynomial of sufficiently low degree. The idea of using a basis of functions other than polynomials has a long history, going back at least to the papers of Greenwood [19], Brock and Murray [6] and Dennis [13], where sets of exponential functions were used to derive the coefficients of the methods for the first order ODE

$$y' = f(x, y). \tag{1.9}$$

Methods using trigonometric polynomials have also been considered; for theoretical aspects see Gautschi [17]. Salzer [28] assumed that the solution is a linear combination of trigonometric functions, of the form

$$y(x) = \sum_{j=0}^{J} [a_j \sin(jx) + b_j \cos(jx)], \tag{1.10}$$

with arbitrary constant coefficients a_j and b_j, to obtained predictor-corrector methods which are exact for this form; expressions of the coefficients of these methods are given in that paper for small values of J.

Sheffield [29] and Stiefel and Bettis [30] considered the second order ODE of the form

$$y'' = f(x, y), \tag{1.11}$$

for the orbit problem in celestial mechanics. They constructed multistep methods which are exact if the solution is of the form

$$y(x) = \sum_{j=1}^{J} [f_1^j(x) \sin(\omega_j x) + f_2^j(x) \cos(\omega_j x)], \qquad (1.12)$$

where f_1^j and f_2^j are low degree polynomials. A simple two-step method which is exact for the form (1.12) with $J = 1$ and constant f_1^1 and f_2^1 has been derived by Denk [12] by means of a principle of coherence.

Coleman [8] considered a special family of methods, hybrid versions included, by means of a technique based on rational approximations of the cosine. This technique is able to cover both the construction of such methods and the description of their stability properties; see also [9].

Also in this context are the papers of Andrew and Paine [2] and of Vanden Berghe et al. [34] on higher eigenvalues of the Sturm–Liouville problems; in these papers an asymptotic correction technique is used.

Methods for the solution of

$$y^{(r)} = f(x, y), \; r = 1, 2, \ldots \qquad (1.13)$$

were derived by Vanthournout et al. [35] and Vanden Berghe et al. [33] as an application of the mixed interpolation technique; they are exact if $y(x)$ is of the form (1.8) where f_1 and f_2 are constants and ϕ is a polynomial of low degree.

The existence of a large variety of techniques, which sometimes seem to have distinct areas of applicability, though this is not always true, is rather discouraging for a user who, though not directly working on numerical techniques, is interested in getting the pertinent information for his or her own problem. Such a user would be better served if one and the same technique, hopefully transparent and simple enough, would be available for as many cases of interest as possible.

The exponential fitting technique is the best suited in this context. It is aimed at deriving linear approximation formulae for various operations on functions of the form

$$y(x) = \sum_{i=1}^{I} f_i(x) \exp(\mu_i x), \qquad (1.14)$$

where μ_i are constants (complex in general) whose exact values, or some reasonable approximations of these, are assumed known. In the literature on exponential fitting these constants are called frequencies in spite that this term does not overlap with its usual acceptance, which refers to the real coefficient ω in a trigonometric function like $\sin(\omega x)$. However, they are correlated. Since $\sin(\omega x)$ is a linear combination of $\exp(i\omega x)$ and $\exp(-i\omega x)$ then, with the new acceptance in mind, this trigonometric function is characterized by a pair of two imaginary frequencies, $\pm i\omega$. It is also assumed that the functions $f_i(x)$ are smooth enough but only the numerical values of the whole $y(x)$ are available. It is obvious that any of the forms (1.2), (1.6) and (1.7) are of this type; for example, (1.2) is the particular case of (1.14) when $I = 2$ and $\mu_1 = -\mu_2 = i\omega$.

The idea of the approach consists in constructing the coefficients of the formula by asking it be exact for each of the following M functions:

$$x^k \exp(\mu_i x),\ k = 0,\ 1,\ 2,\ \ldots,\ m_i - 1,\ i = 1,\ 2,\ \ldots,\ I\,, \qquad (1.15)$$

where m_i satisfy $m_1 + m_2 + \cdots + m_I = M$. The value of M depends of the number N of coefficients to be evaluated. As a rule one has $M = N$ but exceptions do also exist. The condition $m_1 + m_2 + \cdots + m_I = M$ restricts the values to be accepted for I and for m_i but, once I has been fixed, there is still some flexibility in the choice of m_i. The value of the latter is meant to reflect the behaviour of the smooth function $f_i(x)$. If, for example, $f_1(x)$ is expected to exhibit a stronger variation than $f_2(x)$ then we should normally take $m_1 > m_2$.

As for the theoretical background of the procedure, this is inspired by the generalization of Lyche [25] of the approach of Henrici [20] on multistep methods for ODEs. This perhaps explains why for a long period of time the exponential fitting procedure was thought to cover only this field. As a matter of fact, the expression *exponential fitting* with the stated meaning seems to have been first used also in the context of solving ODEs, by Liniger and Willoughby [24]. However, as shown by Ixaru [22], the area of applicability of this procedure is much broader; it covers operations such as numerical differentiation, quadrature or interpolation, as well.

In addition to the facts that the technique for the calculation of the coefficients is simple and that so many different numerical operations are gathered under the same umbrella, the exponential fitting procedure is also advantageous from a theoretical perspective. Thus it guarantees that each classical linear approximation formula can be extended by exponential fitting and also that the latter tends to the former when all frequencies tend to zero. Also the error formula, in particular the leading term of this, has a sharply defined expression which can be easily particularized for the formula of interest. Last, but not least, the exponential fitting procedure may be used as a basis for a subsequent generalization in which the set (1.15) is replaced by a set of other functions chosen in terms of the needs of the current problem.

A short description of the book follows. In Chapter 2 some mathematical concepts are introduced and several important properties of the exponential fitting approach are derived. A computational problem which is typical in the practical applications of the approach is also discussed and the corresponding ready-to-use program is presented. As a matter of fact, the set of subroutines developed for this book, all written in FORTRAN–95 and collected in the attached CD, are aimed at helping the reader in his or her own application. Each of these subroutines has been successfully tested on a series of cases but, of course, no general guarantee is given for any of them.

While the theoretical material presented in Chapter 2 refers to functions of the form (1.14), in Chapter 3 we consider the case when these are of the forms (1.2), (1.6) and (1.7). After discussing the Schrödinger equation as a typical environment where functions of this form do appear, we present the special ingredients needed for a direct approach with such functions. There are two such ingredients, Ixaru's set of functions $\eta_s(Z)$, $s = -1, 0, 1, \ldots$, for which we also give four subroutines, and the algorithm-like six-step flow chart. Both will be used repeatedly in the forthcoming chapters of the book.

Chapters 4–6 are devoted to applications.

In Chapter 4 a series of formulae are presented for the numerical differentiation, quadrature and interpolation. The presentation is definitely more detailed and perhaps also more rigid in the first section than in the other sections. The reason is that we wanted to first make the reader familiar with the use of the two ingredients and for this purpose a classroom exercise-like style appeared to us as the most appropriate.

Many of the exponential fitting-based formulae of this chapter are extensions of traditional formulae; this is the case, among others, for the Newton–Cotes and Gauss–Legendre rules for quadrature. Some less familiar formulae are also considered, to mention only a quadrature rule based not only on the values of the integrand at the nodes of the rule but also on the values of a number of derivatives of it. The section on interpolation is more limited since it deals with the two-point interpolation only, but at that moment the reader is supposed to have acquired sufficient expertize in how to use the two ingredients for developing a version corresponding to his or her own interest.

Chapter 5 is devoted to the application of the exponential fitting procedure for the generation of multistep methods for ODEs. The literature is extremely vast on this subject though, surprisingly enough, a number of important issues seem to have been disregarded. For reasons of brevity and coherence we had to adopt a severe selection in the existing material, on one hand, and to address some new items, on the other hand. Specifically, only first and second order ODEs are considered and, for each of the two cases, only one family of multistep algorithms is investigated in detail. This is the two-step backwards differentiation (bdf) formula for first order ODEs and the two-step Numerov algorithm for second order ODEs. For the former we discuss the problem of the automatic determination of the frequencies when the solution is propagated from one step to another, the variable step version of this algorithm and the steplength control. Among the newly addressed items we quote the investigation, though incomplete, of the stability of the multistep methods for the first order ODEs, and a function fitting version of the two-step bdf algorithm. A set of subroutines is also included.

Chapter 6 presents the Runge–Kutta solvers for ODEs. The special feature here is that such algorithms are nonlinear and then some suited adaptation of the exponential fitting procedure is needed. After a short survey on the formalism of the classical explicit Runge–Kutta methods we describe exponential fitted extensions of them, an embedded pair of such methods, the determination of the suited frequencies and the steplength control. We also present the implicit Runge–Kutta methods paying particular attention to the exponential-fitted two-stage versions of the second, third and fourth order. The discussion covers both theoretical aspects and technical details for their implementation. Some Runge–Kutta-Nyström methods for ODEs of the second order are considered, as well.

References

[1] Alaylioglu, G., Evans, G. A. and Hyslop, J. (1976). The use of Chebyshev series for the evaluation of oscillatory integrals. *Comput. J.*, 19: 258–267.

[2] Andrew, A. L. and Paine, J. W. (1985). Correction of Numerov's eigenvalue estimates. *Numer. Math,*, 47: 289–300.

[3] Bakhvalov, N. S. and Vasil'eva, L. G. (1969). Evaluation of the integrals of oscillating functions by interpolation at nodes of Gaussian quadratures. *USSR Comput. Math and Math. Phys.*, 8: 241–249.

[4] Bocher, P., De Meyer H. and Vanden Berghe G. (1994). Modified Gregory formulae based on mixed interpolation. *Intern. Computer Math.*, 52: 109–122.

[5] Bocher, P., De Meyer H. and Vanden Berghe G. (1994). Numerical solution of Volterra equations based on mixed interpolation. *Computers Math. Applic.*, 27: 1–11.

[6] Bock, P. and Murray, F. J. (1952). The use of exponential sums in step by step integration. *Mathematical Tables and other Aids to Computation*, 6: 63–78 and 138–150.

[7] Brunner, H., Makroglou, A. and Miller, R. K. (1997). Mixed interpolation collocation methods for first and second order Volterra integro-differential equations with periodic solution. *Appl. Num. Math.*. 23: 381–402.

[8] Coleman, J. P. (1989). Numerical methods for $y'' = f(x, y)$ via rational approximations for the cosine. *IMA J. Numer. Anal.*, 9: 145–165.

[9] Coleman, J. P. and Ixaru, L. Gr. (1996). P-stability and exponential-fitting methods for $y'' = f(x, y)$. *IMA J. Numer. Anal.*, 16: 179–199.

[10] Davis, P. J. and Rabinowitz, P. (1984). *Methods of Numerical Integration.* Academic Press, New York.

[11] De Meyer, H., Vanthournout, J. and Vanden Berghe, G. (1990). On a new type of mixed interpolation. *J. Comput. Appl. Math.*, 30: 55–69.

[12] Denk, G. (1993). A new numerical method for the integration of highly oscillatory second-order ordinary differential equations. *Appl. Numer. Math.*, 13: 57–67.

[13] Dennis, S. C. R. (1960). The numerical integration of ordinary differential equations possesing exponential type solutions. *Proc. Cambridge Phil. Soc.*, 65: 240-246.

[14] Ehrenmark, U. T. (1988). A three–point formula for numerical quadrature of oscillatory integrals with variable frequency. *J. Comput. Appl. Math.*, 21: 87–99.

[15] Evans, G. (1993). *Practical Numerical Integration*. John Wiley & Sons Ltd., Chichester.

[16] Evans, G. A. and Webster, J. R. (1997). A high order, progressive method for the evaluation of irregular oscillatory integrals. *Appl. Num. Math.*, 23: 205-218.

[17] Gautschi, W. (1961). Numerical integration of ordinary differential equations based on trigonometric polynomials. *Numer. Math.*, 3: 381–397.

[18] Gautschi, W. (1970). Tables of Gaussian quadrature rules for the calculation of Fourier coefficients. *Math. Comp.*, 24: microfiche.

[19] Greenwood, R. E. (1949). Numerical integration of linear sums of exponential functions. *Ann. Math. Stat.*, 20: 608–611.

[20] Henrici, P. (1962). *Discrete Variable Methods in Ordinary Differential Equations*. Wiley, New York.

[21] Ixaru, L. Gr. (1984). *Numerical Methods for Differential Equations and Applications*. Reidel, Dordrecht - Boston - Lancaster.

[22] Ixaru, L. Gr. (1997). Operations on oscillatory functions. *Comput. Phys. Commun.*, 105: 1–19.

[23] Levin, D. (1982). Procedures for computing one and two dimensional integrals of functions with rapid irregular oscillations. *Math. Comp*, 38: 531–538.

[24] Liniger, W. , Willoughby, R. A. (1970). Efficient integration methods for stiff systems of ordinary differential equations. *SIAM J. Numer. Anal.*, 7: 47–66.

[25] Lyche, T. (1974). Chebyshevian multistep methods for ordinary differential equations. *Numer. Math.*, 19: 65–75.

[26] Patterson, T. N. L. (1976). On high precision methods for the evaluation of Fourier integrals with finite and infinite limits. *Numer. Math.*, 24: 41–52.

[27] Piessens, R. (1970). Gaussian quadrature formulae for the integration of oscillating functions. *ZAMM*, 50: 698–700.

[28] Salzer, H. E. (1962). Trigonometric interpolation and predictor–corrector formulas for numerical integration. *ZAMM*, 9: 403–412.

[29] Sheffield, C. (1969). Generalized multi-step methods with an application to orbit computation. *Celestial Mech.*, 1: 46–58.

[30] Stiefel, E. and Bettis, D. G. (1969). Stabilization of Cowell's method. *Numer. Math.*, 13: 154–175.

[31] Van Daele, M., De Meyer, H. and Vanden Berghe, G. (1992). Modified Newton–Cotes formulae for numerical quadrature of oscillatory integrals with two independent frequencies. *Intern. J. Comput. Math.*, 42: 83–97.

[32] Vanden Berghe, G., De Meyer, H. and Vanthournout, J. (1990). On a class of modified Newton–Cotes quadrature formulae based upon mixed-type interpolation. *J. Comput. Appl. Math.*, 31: 331–349.

[33] Vanden Berghe, G., De Meyer, H. and Vanthournout, J. (1990). A modified Numerov integration method for second order periodic initial–value problems. *Intern. J. Computer Math.*, 32: 233–242.

[34] Vanden Berghe, G. and De Meyer, H. (1991). Accurate computation of higher Sturm–Liouville eigenvalues. *Numer. Math.*, 59: 243–254.

[35] Vanthournout, J., Vanden Berghe, G. and De Meyer, H. (1990). Families of backward differentiation methods based on a new type of mixed interpolation. *Computers Math. Applic.*, 11: 19–30.

Chapter 2

MATHEMATICAL PROPERTIES

In this chapter we present the main mathematical elements of the exponential fitting procedure. It will be seen that this procedure is rather general. However, later on in this book the procedure will be mainly applied in the restricted area of the generation of formulae and algorithms for functions with oscillatory or hyperbolic variation.

The first two sections introduce some specific mathematical issues to be repeatedly invoked in various contexts throughout the whole book. Thus, in Sections 2.1 and 2.2 we examine the properties of the solutions of the linear and homogeneous differential equation with constant coefficients, and present a regularization procedure for solving linear systems of algebraic equations of a special form, respectively.

In subsequent sections we gradually construct the main elements of the procedure. For clarity reasons, in Section 2.3 we treat a simple case in some detail. Here we explicitly place ourselves in the position of a reader interested in getting a first glimpse on the procedure, explain how the problem is approached in this frame, introduce some specific terms and notations, and also mention what kind of advantages may result in practice.

In Section 2.4 the procedure is examined from a theoretical point of view. A series of mathematical aspects are considered, such as the $\mathcal{L}$ operator and its moments, the rules for constructing the algebraic equations for the coefficients of the formula of interest, and the error formulae. We also show that the classical formulae represent a particular case of the new ones.

1. A reference differential equation

The construction of the exponential fitting procedure has a differential equation of a special form as one of its main ingredients. This is the linear and homogeneous differential equation with constant coefficients. The M-th order equation of this form reads:

$$y^{(M)} + c_1 y^{(M-1)} + c_2 y^{(M-2)} + \cdots + c_{M-1} y' + c_M y = 0, \; x \in \Re \quad (2.1)$$

where $c_1, c_2, \ldots, c_n$ are complex constants. In the following, the equation (2.1) will be called the reference differential equation and abbreviated as RDE. It admits M linearly independent solutions and any linear combination of these solutions is also a solution of (2.1).

To determine the linearly independent solutions it is convenient to introduce the following M-th degree polynomial

$$P_M(\mu) := \mu^M + c_1 \mu^{M-1} + c_2 \mu^{M-2} + \ldots + c_{M-1} \mu + c_M \quad (2.2)$$

and to consider two kinds of arguments. If μ is a complex variable, then $P_M(\mu)$ is an algebraic polynomial, called the characteristic polynomial of the equation (2.1). Alternatively, μ can be replaced by an operator D acting in the function space. In this case, $P_M(D)$ is also an operator in the same space. As usual, D^m means that D is applied m times on the considered function.

Particularly useful for our problem is to take $D := \dfrac{d}{dx}$ (and therefore $D^m = \dfrac{d^m}{dx^m}$) because this allows writing the equation (2.1) in compact form:

$$P_M(D)y = 0. \quad (2.3)$$

On applying $P_M(D)$ on the function $f_1(x; \mu) := \exp(\mu x)$ we get

$$P_M(D)f_1(x; \mu) = P_M(\mu) \cdot f_1(x; \mu), \quad (2.4)$$

where a dot was placed in the right-hand side to stress that this is a product of two algebraic expressions, not the application of an operator on a function, as it is in the left-hand side.

The characteristic polynomial $P_M(\mu)$ has M roots. We distinguish two situations:

(i) The M roots are distinct. If they are denoted $\mu_1, \mu_2, \ldots, \mu_M$ the right-hand side of (2.4) vanishes when $\mu = \mu_i$, $i = 1, 2, \ldots, M$ and therefore the M linearly independent solutions of (2.1) are $y_i(x) = \exp(\mu_i x)$, $i = 1, 2, \ldots, M$.

(ii) Multiple roots are present. Let μ_i, $i = 1, 2, \ldots, I < M$ be the distinct roots and let m_i be the multiplicity of μ_i. [Some of the roots can be simple, i.e. of multiplicity one. Anyway, the relation $m_1 + m_2 + \cdots + m_I = M$ must be fulfilled because the total number of roots is M.] This allows writing the characteristic polynomial $P_M(\mu)$ in the product form, viz:

$$P_M(\mu) = (\mu - \mu_1)^{m_1}(\mu - \mu_2)^{m_2} \cdots (\mu - \mu_I)^{m_I}. \quad (2.5)$$

A simple way to construct the required set of solutions in the latter case consists in taking the partial derivatives of the equation (2.4) with respect to μ. We use the fact that, if $\bar{\mu}$ is one of the roots of $P_M(\mu)$ with multiplicity $m > 1$, then this $\bar{\mu}$ is a common root of $P_M(\mu)$, $P_M'(\mu)$, $P_M''(\mu)$, ..., $P_M^{(m-1)}(\mu)$.

We define the function

$$f_{k+1}(x; \mu) := \frac{\partial^k f_1(x; \mu)}{\partial \mu^k} = x^k \exp(\mu x), \quad k = 1, 2, \ldots \qquad (2.6)$$

and take the k-th derivative of (2.4) with respect to μ:

$$P_M(D)f_{k+1}(x; \mu) = \sum_{j=0}^{k} \binom{k}{j} P_M^{(j)}(\mu) f_{k-j+1}(x; \mu), \qquad (2.7)$$

where the binomial coefficient is defined by

$$\binom{k}{j} := \frac{k!}{j!(k-j)!}, \quad 0 \le j \le k. \qquad (2.8)$$

If $\mu = \bar{\mu}$ the right-hand side of (2.7) vanishes for all $k = 1, 2, \ldots, m-1$. Then, not only $f_1(x; \bar{\mu})$ but also $f_{k+1}(x; \bar{\mu})$, $k = 1, 2, \ldots, m-1$ are solutions of the equation (2.1). Their expressions show that these solutions are linearly independent, indeed. On successively taking $\bar{\mu} = \mu_1, \mu_2, \ldots, \mu_I$ we have the final result:

The M linearly independent solutions of RDE are fixed by the roots of the characteristic polynomial $P_M(\mu)$. If the root μ_i has the multiplicity m_i, $i = 1, 2, \ldots, I$ ($m_1 + m_2 + \cdots + m_I = M$) then the set of M linearly independent solutions is

$$x^k \exp(\mu_i x), \quad k = 0, 1, 2, \ldots, m_i - 1 \ (i = 1, 2, \ldots, I). \qquad (2.9)$$

The case of M distinct roots corresponds to $I = M$ and $m_i = 1$, $i = 1, 2, \ldots, M$.

This result is useful not only for the direct problem, that is to construct the solutions of the given RDE, but also for the inverse problem, i. e. to construct the differential equation if the set (2.9) is known. In this case the variable μ in $P_M(\mu)$ is formally replaced by the operator D. The required equation then is $P_M(D)y = 0$. If the form (2.5) is used this reads:

$$(D - \mu_1)^{m_1}(D - \mu_2)^{m_2} \cdots (D - \mu_I)^{m_I} y = 0. \qquad (2.10)$$

The order of the factors is arbitrary because the parameters μ_i, $i = 1, 2, \ldots, I$ are constants.

For example, the equation which has the functions

$$1, \; x, \; x^2, \; \exp(\lambda_1 x), \; x \exp(\lambda_1 x) \; \text{and} \; \exp(\lambda_2 x) \qquad (2.11)$$

as a complete set of linearly independent solutions is $D^3(D-\lambda_1)^2(D-\lambda_2)y = 0$ or, in expanded form,

$$y^{(6)} - (2\lambda_1 + \lambda_2)y^{(5)} + (\lambda_1^2 + 2\lambda_1\lambda_2)y^{(4)} - \lambda_1^2\lambda_2 y^{(3)} = 0 \qquad (2.12)$$

because the set (2.11) is of form (2.9) with $M = 6$ and because its characteristic polynomial has three distinct roots: $\mu_1 = 0$ with multiplicity $m_1 = 3$, $\mu_2 = \lambda_1$ with multiplicity $m_2 = 2$, and $\mu_3 = \lambda_2$ with multiplicity $m_3 = 1$.

In general, if μ_i, $i = 1, 2, \dots, M$ are the M roots (distinct or not) then the following well known relations between the roots and the coefficients hold:

$$
\begin{aligned}
c_1 &= -(\mu_1 + \mu_2 + \cdots + \mu_M), \\
c_2 &= \mu_1\mu_2 + \mu_1\mu_3 + \cdots + \mu_{M-1}\mu_M \qquad (2.13) \\
&\quad \cdot \quad \cdot \quad \cdot \\
&\quad \cdot \quad \cdot \quad \cdot \\
c_M &= (-1)^M \mu_1\mu_2 \cdots \mu_M \, .
\end{aligned}
$$

The reference to the RDE when deriving numerical formulae is useful in several contexts.

1 It helps becoming aware that there is full flexibility in choosing the origin of the x axis. In fact the set of M linearly independent solutions of the equation (2.1) is closed with respect to a translation in x: if $y_j(x)$, $j = 1, 2 \dots, M$ are such solutions, then this is also true for the set $y_j(x + \Delta x)$, $j = 1, 2 \dots, M$. The origin can then be chosen in terms of practical convenience. Another useful property is that the family of solutions is closed with respect to differentiation: if $y(x)$ is a solution of the equation (2.1), then $Dy(x)$ is also a (nontrivial or trivial) solution.

   In general the family of solutions is not closed with respect to reflection: if $y(x)$ does satisfy (2.1), then this is not automatically true for $y(-x)$. A notable exception is when all the coefficients c_i with odd i are vanishing, and this happens when the set of nonvanishing roots of the characteristic polynomial consists of pairs with opposite values: $\mu_1 = -\mu_2$, $\mu_3 = -\mu_4$ and so on. This will be the case for most of the problems discussed in this book.

2 The RDE allows re-expressing some standard formulations in a convenient way. To describe the accuracy of numerical formulae for differentiation, integration or interpolation of some function $y(x)$, sentences like "this formula is exact if $y(x)$ is of class $\mathcal{P}_{M-1}$" are often used in the literature, with

the meaning that it is exact if that function is a polynomial of degree smaller than or equal to $M - 1$. Now, the simplest RDE of order M is $y^{(M)} = 0$. Its M linearly independent solutions are the power functions

$$1, \, x, \, x^2, \, \ldots, \, x^{M-1} \tag{2.14}$$

because the characteristic polynomial has only one distinct root, $\mu_1 = 0$, and its multiplicity is $m_1 = M$. Their linear combinations, also solutions of $y^{(M)} = 0$, are polynomials of degree $M - 1$ at most and then the above statement can be re-expressed as "this formula is exact if the function $y(x)$ satisfies $y^{(M)} = 0$ ".

An advantage of the latter formulation is that it suggests directly the structure of the error formula. Indeed, as it will be detailed later on, to each approximation formula $\mathcal{O}y(X) \approx A[y(X); h, \mathbf{a}]$ where X is chosen by the user (for example $\mathcal{O} = D$ and $A[y(X); h, \mathbf{a}] = (a_0 y(X - h) + a_1 y(X) + a_2 y(X + h))/h$ for a three point approximation of the first derivative at X) an operator $\mathcal{L}[h, \mathbf{a}]$ acting on $y(x)$ is introduced by $\mathcal{L}[h, \mathbf{a}]y(x) := \mathcal{O}y(x) - A[y(x); h, \mathbf{a}]$. It will be assumed that $\mathcal{L}[h, \mathbf{a}]$ can be expanded in powers of D. If the set $\mathbf{a}$ is chosen upon the condition that $\mathcal{L}[h, \mathbf{a}]y(x)$ vanishes identically in x and in h if and only if $y(x)$ satisfies $y^{(M)} = 0$, then $\mathcal{L}[h, \mathbf{a}]y(x)$ can be written as a formal series, viz.:

$$\mathcal{L}[h, \mathbf{a}]y(x) = \sum_{j=0}^{\infty} t_j D^j \, y^{(M)}(x), \tag{2.15}$$

where t_j, $j = 0, 1 \ldots$ ($t_0 \neq 0$) are some suited coefficients which depend on h. As for the error of the approximation formula, this is obtained by taking $x = X$.

The specific feature is that, since for this case we have $P_M(D) = y^{(M)}$, equation (2.15) can be written equivalently as

$$\mathcal{L}[h, \mathbf{a}]y(x) = \sum_{j=0}^{\infty} t_j D^j \, P_M(D)y(x).$$

3 The use of the RDE enables a natural generalization. Not only power function based (or, equivalently, polynomial based) formulae can be derived but also formulae which are exact if $y(x)$ is a linear combination of functions (2.9). The technique will be explained later on but what is important is that, since now $P_M(D)$ is as in (2.10), the series expansion of $\mathcal{L}[h, \mathbf{a}]y(x)$ reads

$$\mathcal{L}[h, \mathbf{a}]y(x) = \sum_{j=0}^{\infty} t_j D^j \, (D - \mu_1)^{m_1}(D - \mu_2)^{m_2} \cdots (D - \mu_I)^{m_I} y(x),$$

$$\tag{2.16}$$

where the coefficients $t_0 \neq 0$, t_1, t_2, ... depend on the discussed formula, h, and also on μ_i and m_i. Again, the error will be given by taking $x = X$ in the right-hand side of this equation.

Quite interesting is the case when all μ_i tend to zero. The equation (2.10) tends to $y^{(M)} = 0$, i.e. the power function case represents a particular, limiting case of the general one. This is not visible directly if the sets (2.9) and (2.14) are compared as they stand but one can always construct linear combinations of elements of the former to tend to any desired element in the latter. For example, if the set (2.11) corresponding to equation (2.12) is used to obtain the elements of the associated simplest equation $y^{(6)} = 0$, the element x^3 in the power function set is the limit when $\lambda_1 \to 0$ of

$$y(x; \lambda_1) := \frac{2}{\lambda_1^2}(x \exp(\lambda_1 x) - x) - \frac{2}{\lambda_1}x^2.$$

As a matter of fact, this function is not the only one that enjoys the stated property and it would be a good exercise to try constructing an alternative expression, and also to find expressions which tend to x^4 and to x^5.

Finally we note that when all c coefficients in the RDE are real, then the roots of the characteristic polynomial are real and/or pairs of complex conjugate values.

2.   A regularization procedure

The determination of the coefficients of various formulae will often imply the solution of linear systems of algebraic equations with a particular structure of the coefficients. Such a system reads

$$\mathbf{A}\mathbf{x} = \mathbf{b}, \tag{2.17}$$

in which $\mathbf{A}$ and $\mathbf{b}$ are an n by n matrix and a column vector with n components, respectively, whose elements are generated in terms of the values of $n+1$ given functions, viz. $f_j(z)$, $j = 1, 2, \ldots, n$ and $g(z)$, by the rule $A_{ij} := f_j(z_i)$ and $b_i := g(z_i)$, $(i = 1, 2, \ldots, n)$. Their form then is

$$\mathbf{A} = \begin{bmatrix} f_1(z_1) & f_2(z_1) & \cdots & f_n(z_1) \\ f_1(z_2) & f_2(z_2) & \cdots & f_n(z_2) \\ \vdots & \vdots & & \vdots \\ f_1(z_n) & f_2(z_n) & \cdots & f_n(z_n) \end{bmatrix}, \tag{2.18}$$

and

$$\mathbf{b} = [g(z_1), g(z_2), \cdots, g(z_n)]^T. \tag{2.19}$$

The functions $f_1, f_2, \ldots, f_n$ and g which typically occur in our problems are real functions of a real variable. They have relatively simple forms, currently consisting of low degree polynomials multiplied by exponential, trigonometric or hyperbolic functions.

For small n analytic solutions can be obtained just by hand, while when n is increased suited packages (as MATHEMATICA, see [11]) are of help. However, the volume of calculation and the length of the final expressions increases dramatically with n such that, for big n, only the numerical solution must be considered. As for the relevance of these solutions for subsequent considerations, the knowledge of the analytic expressions is important mainly when a better theoretical insight in the behaviour of these solutions is needed. In applications, however, only the numerical solutions are required.

A major difficulty in solving systems of the quoted form numerically comes from the fact that quite often the solution is needed for tightly clustered values of the arguments $z_1, z_2, \ldots, z_n$. A highly accurate direct solution of system (2.17) is not possible in such a case because the equations tend to coincide to each other. If analytic solutions are available this is reflected in the appearance in their expressions of a $0/0$ type of indeterminacy. A separate transformation of these expressions is then needed to eliminate the indeterminacy and in general this can be done by some elementary manipulations. However, when the solution can be accessed only numerically such a way of improving the situation is impossible.

The following example will illustrate the things and it will also suggest a suitable way for the numerical treatment. The system

$$\begin{cases} \sin(z_1)x_1 + \cos(z_1)x_2 = \sin(2z_1), \\ \sin(z_2)x_1 + \cos(z_2)x_2 = \sin(2z_2), \end{cases} \tag{2.20}$$

is of the form (2.17)–(2.19) with $n = 2$, $f_1(z) = \sin(z)$, $f_2(z) = \cos(z)$ and $g(z) = \sin(2z)$. Its determinant is $d = \sin(z_1 - z_2)$ and then the system becomes singular when $z_1 \to z_2$. Indeed, the expression of the analytic solution

$$x_1 = \frac{2\cos(z_1)\cos(z_2)(\sin(z_1) - \sin(z_2))}{\sin(z_1 - z_2)},$$

$$x_2 = \frac{2\sin(z_1)\sin(z_2)(\cos(z_2) - \cos(z_1))}{\sin(z_1 - z_2)},$$

shows that both x_1 and x_2 exhibit a $0/0$ indeterminacy at $z_1 = z_2$, but these expressions can be easily regularized by means of standard trigonometric formulae, to obtain

$$x_1 = \frac{2\cos(z_1)\cos(z_2)\cos[(z_1 + z_2)/2]}{\cos[(z_1 - z_2)/2]},$$

$$(2.21)$$

$$x_2 = \frac{2\sin(z_1)\sin(z_2)\sin[(z_1+z_2)/2]}{\cos[(z_1-z_2)/2]}.$$

such that, if these are used, the numerical computation of x_1 and x_2 becomes free of difficulties also when z_1 and z_2 are close together, as wished.

The instructive point is that the same effect can be obtained in an alternative way, that is by first regularizing the given system and then by solving the new system. If we subtract the first equation from the second and divide by $z_2 - z_1$ we get

$$\frac{\sin(z_2)-\sin(z_1)}{z_2-z_1}x_1 + \frac{\cos(z_2)-\cos(z_1)}{z_2-z_1}x_2 = \frac{\sin(2z_2)-\sin(2z_1)}{z_2-z_1}. \quad (2.22)$$

The computation of the coefficients of x_1 and x_2 and of the free term in equation (2.22) becomes inaccurate when z_1 and z_2 are close to each other, due to the near cancellation of like terms, but this difficulty is easily avoided by using the Taylor series, for instance

$$\frac{\sin(z_2)-\sin(z_1)}{z_2-z_1} = \cos(z_1)-\frac{1}{2!}(z_2-z_1)\sin(z_1)-\frac{1}{3!}(z_2-z_1)^2\cos(z_1)+\cdots$$

$$(2.23)$$

This suggests introducing some threshold value $\Delta > 0$ to compute the three z-dependent quantities in (2.22) by their definition formulae when $|z_2 - z_1| > \Delta$ and by truncated Taylor series when $|z_2 - z_1| \leq \Delta$. In this way the computation of the coefficients in equation (2.22) will become free of problems irrespective of how mutually close are z_1 and z_2. In particular, when $z_1 = z_2$ the equation (2.22) becomes

$$\cos(z_1)x_1 - \sin(z_1)x_2 = 2\cos(2z_1).$$

The new system of two equations, in which the second equation of (2.20) is replaced by (2.22), is equivalent to the whole original system (2.20) but it has the advantage of being regularized when z_1 and z_2 are close together. In particular, it can be verified without difficulty that in the worst case, that is $z_1 = z_2$, its determinant equals -1, hence not zero, as it was with the original system.

The extension of this idea to the general problem (2.17)–(2.19) was considered in [6]. Let us denote generically by $f(z)$ any of the functions $f_1(z)$, $f_2(z)$, ..., $f_n(z)$ and $g(z)$. We assume that $f(z)$ is a real function of a real variable z and that it is differentiable indefinitely many times.

We construct the successive divided differences of $f(z)$ by the usual rule :

$$f^1(z_m, z_{m+1}) := \frac{f(z_{m+1})-f(z_m)}{z_{m+1}-z_m}, \quad m = 1, 2, \ldots, n-1 \quad (2.24)$$

Table 2.1. A table of divided differences.

$$f(z_1)$$

$$f^1(z_1, z_2)$$

$$f(z_2) \qquad \qquad f^2(z_1, z_2, z_3)$$

$$f^1(z_2, z_3)$$

$$f(z_3)$$

$$\vdots$$

$$f^{n-1}(z_1, z_2, \ldots, z_n)$$

$$\vdots$$

$$f(z_{n-2})$$

$$f^1(z_{n-2}, z_{n-1})$$

$$f(z_{n-1}) \qquad \qquad f^2(z_{n-2}, z_{n-1}, z_n)$$

$$f^1(z_{n-1}, z_n)$$

$$f(z_n)$$

and

$$f^q(z_m, \ldots, z_{m+q}) := \frac{1}{z_{m+q} - z_m}$$
$$\times \left(f^{q-1}(z_{m+1}, z_{m+2}, \ldots, z_{m+q}) - f^{q-1}(z_m, z_{m+1}, \ldots, z_{m+q-1}) \right),$$
$$m = 1, 2, \ldots, n - q, \qquad (2.25)$$

for $q = 2, 3, \ldots, n-1$, see Table 2.1. The total number of divided differences is $n(n - 1)/2$.

Each divided difference f^q depends on $q + 1$ arguments with labels in the ascending order, and each column in the table consists of $n - q$ components, uniquely identified by the label of their first argument. It is also worth mentioning that f^q is invariant upon permutation of its arguments. This can be easily seen if f^q is written in terms of the original $f(z_1), f(z_2), \ldots, f(z_n)$. For instance, the invariance is obvious for $q = 1$, that is $f^1(z_m, z_{m+1}) = f^1(z_{m+1}, z_m)$, while for $q = 2$ we have :

$$
\begin{aligned}
f^2(z_1, z_2, z_3) &= -\frac{(z_3 - z_2)f(z_1) + (z_2 - z_1)f(z_3) + (z_1 - z_3)f(z_2)}{(z_1 - z_3)(z_3 - z_2)(z_2 - z_1)} \\
&= f^2(z_1, z_3, z_2) = f^2(z_3, z_1, z_2) = f^2(z_3, z_2, z_1) \\
&= f^2(z_2, z_3, z_1) = f^2(z_2, z_1, z_3).
\end{aligned}
$$

The scheme of Table 2.1 is applied to each of the functions $f_1(z)$, $f_2(z)$,..., $f_n(z)$ and $g(z)$ to obtain the values of $f_i^q(z_m, \ldots, z_{m+q})$, $i = 1, 2, \ldots, n$ and of $g^q(z_m, \ldots, z_{m+q})$.

Particularly important is the case $m = 1$. By the very construction, the divided difference $f^q(z_1, \ldots, z_{q+1})$ is some linear combination of $f(z_1)$, $f(z_2)$, ..., $f(z_{q+1})$ and the coefficients are the same in all $n + 1$ functions used for the generation of system (2.17). Replacing the elements of the $(q + 1)$-th row in matrix $\mathbf{A}$ by $f_i^q(z_1, \ldots, z_{q+1})$, $i = 1, 2, \ldots, n$ and the $(q + 1)$-th component of vector $\mathbf{b}$ by $g^q(z_1, \ldots, z_{q+1})$ is equivalent to replacing the $(q+1)$-th equation by a linear combination of the first $(q + 1)$ equations of the original system, an operation which obviously leaves the solution unchanged. On doing this on all admissible q-s the original system is replaced by

$$\mathbf{A}_R \mathbf{x} = \mathbf{b}_R, \tag{2.26}$$

where

$$\mathbf{A}_R = \begin{bmatrix} f_1(z_1) & f_2(z_1) & \cdots & f_n(z_1) \\ f_1^1(z_1, z_2) & f_2^1(z_1, z_2) & \cdots & f_n^1(z_1, z_2) \\ f_1^2(z_1, z_2, z_3) & f_2^2(z_1, z_2, z_3) & \cdots & f_n^2(z_1, z_2, z_3) \\ \vdots & \vdots & & \vdots \\ f_1^{n-1}(z_1, \ldots, z_n) & f_2^{n-1}(z_1, \ldots, z_n) & \cdots & f_n^{n-1}(z_1, \ldots, z_n) \end{bmatrix} \tag{2.27}$$

and

$$\mathbf{b}_R = [g(z_1), g^1(z_1, z_2), g^2(z_1, z_2, z_3), \cdots, g^{n-1}(z_1, \ldots, z_n)]^T. \tag{2.28}$$

The new system has the same solution as (2.17) but it has the advantage of being regularized and then it is convenient to treat the case when some or all values of the arguments $z_1, z_2, \ldots, z_n$ are close together. To see this, we have to check that

(i) a safe procedure can be formulated for the determination of the components of $\mathbf{A}_R$ and of $\mathbf{b}_R$ irrespective of the mutual separation distance of the arguments,

(ii) the system (2.26) does not necessarily tend to a system with coinciding equations when the arguments merge into each other, in contrast with what happens with (2.17).

The following theorem, taken from [6], is of help for both issues because it shows how a divided difference has to be computed when the arguments are close together, and it also gives the limit value of f^q when all arguments tend towards one and the same value.

THEOREM 2.1 *The function $f^q(z_m, z_{m+1}, \ldots, z_{m+q})$ has the following series representation in which the derivatives of the original function f are taken at one and the same point z_j, $(m \leq j \leq m+q)$:*

$$f^q(z_m, z_{m+1}, \ldots, z_{m+q}) = \sum_{k=0}^{\infty} \frac{1}{(k+q)!} f^{(k+q)}(z_j)$$

$$\times C_k^q[z_m - z_j, z_{m+1} - z_j, \ldots, z_{j-1} - z_j, z_{j+1} - z_j, \ldots, z_{m+q} - z_j].$$

(2.29)

The coefficients C_k^q, which depend on q variables, satisfy the recurrence

$$C_k^q[x_1, x_2, \ldots, x_q] = C_k^{q-1}[x_1, x_2, \ldots, x_{q-1}] + x_q C_{k-1}^q[x_1, x_2, \ldots, x_q],$$

$$(k \geq 1) \qquad (2.30)$$

with $C_0^q[x_1, \ldots, x_q] = 1$ and $C_k^1[x] = x^k$.

Proof By definition

$$f^1(z_m, z_{m+1}) = \frac{1}{z_{m+1} - z_m} (f(z_{m+1}) - f(z_m)).$$

By means of Taylor series expansion around one of the two points, say z_m, one obtains

$$f^1(z_m, z_{m+1}) = \sum_{k=0}^{\infty} \frac{(z_{m+1} - z_m)^k}{(k+1)!} f^{(k+1)}(z_k),$$

which is of the form (2.29) with $C_k^1[x] = x^k$. Substituting in the right hand side of

$$f^2(z_m, z_{m+1}, z_{m+2}) = \frac{1}{z_{m+2} - z_m} \left(f^1(z_{m+1}, z_{m+2}) - f^1(z_m, z_{m+1}) \right)$$

twice an expansion of the form (2.29), with $q = 1$ and $z_j = z_{m+1}$, one finds

$$f^2(z_m, z_{m+1}, z_{m+2}) = \frac{1}{z_{m+2} - z_m}$$

$$\sum_{k=0}^{\infty} \left(C_{k+1}^1[z_{m+2} - z_{m+1}] - C_{k+1}^1[z_m - z_{m+1}] \right) \frac{f^{(k+2)}(z_{m+1})}{(k+2)!}.$$

Since $f^2(z_m, z_{m+1}, z_{m+2})$ is invariant under any permutation of its arguments, this equality is of the form (2.29), if one defines :

$$C_k^2[x_1, x_2] := \frac{1}{x_2 - x_1} \left(C_{k+1}^1[x_2] - C_{k+1}^1[x_1] \right).$$

From this definition, it follows that

$$
\begin{aligned}
C_k^2[x_1, x_2] &= \frac{x_2^{k+1} - x_1^{k+1}}{x_2 - x_1} = x_1^k + \frac{x_2(x_2^k - x_1^k)}{x_2 - x_1} \\
&= C_k^1[x_1] + x_2 C_{k-1}^2[x_1, x_2], \qquad (k \geq 1)
\end{aligned}
$$

and also $C_0^2[x_1, x_2] = 1$. Hence, the recursion (2.30) holds for the case $q = 2$.

The proof can now be completed by induction on q. Under the induction hypothesis, the divided difference $f^q(z_m, z_{m+1}, \ldots, z_{m+q})$, being invariant under any permutation of its arguments, can be written as

$$
f^q(z_m, z_{m+1}, \ldots, z_{m+q}) = \frac{1}{z_{m+q} - z_m}
$$

$$
\sum_{k=0}^{\infty} (C_{k+1}^{q-1}[z_{m+1} - z_j, \ldots, z_{j-1} - z_j, z_{j+1} - z_j, z_{m+q} - z_j]
$$

$$
- C_{k+1}^{q-1}[z_m - z_j, \ldots, z_{j-1} - z_j, z_{j+1} - z_j, \ldots, z_{m+q-1} - z_j])
$$

$$
\times \frac{f^{(k+q)}(z_j)}{(k+q)!},
$$

$$
(m \leq j \leq m+q),
$$

which is of the form (2.29) if one defines :

$$
C_k^q[x_1, x_2, \ldots, x_q] := \frac{1}{x_q - x_1} \left(C_{k+1}^{q-1}[x_2, \ldots, x_q] - C_{k+1}^{q-1}[x_1, \ldots, x_{q-1}] \right).
$$

By the induction hypothesis and taking into account the invariance of C upon permutations of its arguments, one can apply on the right hand side twice the property (2.30) in the following way :

$$
\begin{aligned}
C_k^q[x_1, x_2, \ldots, x_q] &= \frac{1}{x_q - x_1} (C_{k+1}^{q-2}[x_2, \ldots, x_{q-1}] \\
&+ x_q C_k^{q-1}[x_2, \ldots, x_q] - C_{k+1}^{q-2}[x_2, \ldots, x_{q-1}] - x_1 C_k^{q-1}[x_1, \ldots, x_{q-1}]) \\
&= C_k^{q-1}[x_1, \ldots, x_{q-1}] \\
&+ \frac{x_q}{x_q - x_1} (C_k^{q-1}[x_2, \ldots, x_q] - C_k^{q-1}[x_1, \ldots, x_{q-1}]) \\
&= C_k^{q-1}[x_1, \ldots, x_{q-1}] + x_q C_{k-1}^q[x_1, \ldots, x_q], \qquad (k \geq 1).
\end{aligned}
$$

The fact that also the equality $C_0^q[x_1, x_2, \ldots, x_q] = 1$ holds, completes the proof. Q.E.D.

For convenience, let us assume that the points $z_1, z_2, \ldots, z_n$ are in monotonically increasing order, $z_1 \leq z_2 \leq \cdots \leq z_n$. (If this is not the case with the original form of (2.17), an extra permutation of the equations is needed.)

The theorem suggests the following procedure for a safe evaluation of f^q. As before, a threshold value $\Delta > 0$ should be introduced. If $z_{m+q} - z_m > \Delta$ the function $f^q(z_m, \ldots, z_{m+q})$ is computed by its very definition (equation (2.24) or equation (2.25)) while, if $z_{m+q} - z_m \leq \Delta$, the series (2.29) must be used. The value of Δ is dictated by the shape of function $f(z)$, by the required accuracy in the evaluation of f^q and by the number of terms to be retained in the series expansion. If the latter is kept fixed, then Δ may depend on q and m. To generate all elements of $\mathbf{A}_R$ and $\mathbf{b}_R$, the procedure must be repeated for all $n+1$ functions and for all q-s involved there.

The theorem also shows that, if $z_1 = z_2 = \ldots = z_{q+1}$, then

$$f^q = \frac{1}{q!} f^{(q)}(z_1).$$

With the n by n diagonal matrix $\mathbf{M}$ with the elements $M_{ii} := (i-1)!$ one defines $\mathbf{A}_{R'} := \mathbf{M}\mathbf{A}_R$ and $\mathbf{b}_{R'} := \mathbf{M}\mathbf{b}_R$. They read

$$\mathbf{A}_{R'} = \begin{bmatrix} f_1(z_1) & f_2(z_1) & \cdots & f_n(z_1) \\ f_1'(z_1) & f_2'(z_1) & \cdots & f_n'(z_1) \\ f_1^{(2)}(z_1) & f_2^{(2)}(z_1) & \cdots & f_n^{(2)}(z_1) \\ \vdots & \vdots & & \vdots \\ f_1^{(n-1)}(z_1) & f_2^{(n-1)}(z_1) & \cdots & f_n^{(n-1)}(z_1) \end{bmatrix} \tag{2.31}$$

and

$$\mathbf{b}_{R'} = [g(z_1), g'(z_1), g^{(2)}(z_1), \cdots, g^{(n-1)}(z_1)]^T. \tag{2.32}$$

The equations in the system $\mathbf{A}_{R'}\mathbf{x} = \mathbf{b}_{R'}$ do not coincide with each other, thus contrasting with the behaviour of the original system in the quoted limiting case.

2.1 Subroutine REGSOLV

The construction of the fully regularized forms (2.27) and (2.28) of $\mathbf{A}$ and $\mathbf{b}$ is always possible but it is actually unavoidable only when the z arguments are close to each other. If the divided differences are calculated by their definition formulae the computation of the $n-1$ divided differences which appear in each column of $\mathbf{A}_R$ and in $\mathbf{b}_R$ requires the evaluation of the whole set of $n(n-1)/2$ divided differences of Table 2.1 but if the series expansions are used (this is compulsory when the z-s are tightly packed) the same set of data requires the calculation of only $n-1$ truncated series.

The two ways of computing are therefore quite different for efficiency (for big n, at least), and the former should be avoided. The regularization procedure will then gain in efficiency if only two regimes are activated, which means either keeping the original $\mathbf{A}$ and $\mathbf{b}$ unchanged if the z-s are well separated or using the series representation of their divided differences otherwise. The price to be paid consists in taking a large number of terms in the series and in general the calculation of such terms is not so easy; it is sufficient to notice that this would involve the evaluation of the high order derivatives of the involved functions.

Another problem is that such a treatment must be applied simultaneously on all $n + 1$ functions involved in the system. Some of these functions may be slowly varying while others may exhibit a strong variation. One and the same set of z-s may be then seen as sparse for some functions but tightly packed for others and, for such reasons, writing a safe regularization subroutine to cover the general case is not an easy task.

Fortunately, the forms of the functions involved in our problems exhibit two helpful features. On one hand, analytic expressions are available for any of their derivatives and, moreover, these expressions are rather simple and easy to compute. This means that we are allowed to take a big number of terms in the series expansions for the divided differences without a serious decrease in speed. On the other hand, the slopes of various functions are very comparable in size and then, once a selection between the two regimes was made for one of these functions, one may admit that this will be also acceptable for all others.

REMARK 2.1 *In what follows the sample programs are presented in the programming language FORTRAN 95.*

The subroutine REGSOLV exploits the stated features. It consists of the following six steps:

Step 1.
    The n input values of z are placed in monotonically ascending order.

Step 2.
    A piecewise regularization procedure is constructed. The n equations of the system are first grouped into families. Each family k ($k = 1, 2, \ldots$), consists of the equations with $i = i_k + 1, i_k + 2, \ldots, i_{k+1}$ ($i_1 = 0$ is taken by default), where the index i_{k+1} of the last member of the family is selected as the biggest i which meets the condition

$$\frac{z_{i_{k+1}} - z_{i_k+1}}{1 + \sqrt{|z_{i_k+1}|}} \leq \Delta.$$

This form is suggested by the specific behaviour of the f-s involved in our specific applications. We take $\Delta = 0.25$. If K is the total number of such families (this implies that $i_{K+1} = n$) and we denote $m_k := i_{k+1} - i_k - 1$, the regularized form of **A** (let it be denoted $\mathbf{A}_{R^*}$) reads:

$$
\begin{bmatrix}
f_1(z_1) & f_2(z_1) & \cdots & f_n(z_1) \\
f_1^1(z_1,z_2) & f_2^1(z_1,z_2) & \cdots & f_n^1(z_1,z_2) \\
f_1^2(z_1,z_2,z_3) & f_2^2(z_1,z_2,z_3) & \cdots & f_n^2(z_1,z_2,z_3) \\
\vdots & \vdots & & \vdots \\
f_1^{m_1}(z_1,\ldots,z_{i_2}) & f_2^{m_1}(z_1,\ldots,z_{i_2}) & \cdots & f_n^{m_1}(z_1,\ldots,z_{i_2}) \\
& & & \\
f_1(z_{i_2+1}) & f_2(z_{i_2+1}) & \cdots & f_n(z_{i_2+1}) \\
f_1^1(z_{i_2+1},z_{i_2+2}) & f_2^1(z_{i_2+1},z_{i_2+2}) & \cdots & f_n^1(z_{i_2+1},z_{i_2+2}) \\
\vdots & \vdots & & \vdots \\
f_1^{m_2}(z_{i_2+1},\ldots,z_{i_3}) & f_2^{m_2}(z_{i_2+1},\ldots,z_{i_3}) & \cdots & f_n^{m_2}(z_{i_2+1}\ldots,z_{i_3}) \\
& & & \\
\vdots & \vdots & & \vdots \\
& & & \\
f_1(z_{i_K+1}) & f_2(z_{i_K+1}) & \cdots & f_n(z_{i_K+1}) \\
f_1^1(z_{i_K+1},z_{i_K+2}) & f_2^1(z_{i_K+1},z_{i_K+2}) & \cdots & f_n^1(z_{i_K+1},z_{i_K+2}) \\
\vdots & \vdots & & \vdots \\
f_1^{m_K}(z_{i_K+1},\ldots,z_{i_{K+1}}) & f_2^{m_K}(z_{i_K+1),\ldots,z_{i_{K+1}}}) & \cdots & f_n^{m_K}(z_{i_K+1}\ldots,z_{i_{K+1}})
\end{bmatrix}
$$

$$\tag{2.33}$$

and the regularized form of **b**, denoted $\mathbf{b}_{R^*}$, is:

$$
[g(z_1), g^1(z_1,z_2), \cdots, g^{m_1}(z_1,\ldots,z_{i_1}), g(z_{i_1+1}), g^1(z_{i_1+1},z_{i_1+2}),
$$
$$
\cdots, g^{m_2}(z_{i_1+1},\ldots,z_{i_2}), \cdots, g(z_{i_K+1}), g^1(z_{i_K+1},z_{i_K+2}), \cdots,
$$
$$
g^{m_K}(z_{i_K+1},\ldots,z_{i_{K+1}})]^T .
$$

$$\tag{2.34}$$

Step 3.

The elements of $\mathbf{A}_{R^*}$ and $\mathbf{b}_{R^*}$ are computed. Only the truncated series representations are used for the divided differences. In the program we retain thirty terms.

Step 4.

The system $\mathbf{A}_{R^*}\mathbf{x} = \mathbf{b}_{R^*}$ is solved by a standard numerical solver. We opted for a solver based on the LU decomposition by Crout's method with partial pivoting, see [10], and wrote the pair of subroutines LUDEC and LUSBS which cover the LU decomposition of **A** and the forwards or backwards substitution, respectively.

Step 5.

The set of z-s is restored to its original order.

Step 6.

A global estimator is used to measure the error. This is

$$
err := \frac{\|\mathbf{Ax} - \mathbf{b}\|}{\displaystyle\max_{i,j=1,\ldots,n} |A_{ij}|}
$$

where $\| \cdot \|$ is the L_1 norm.

Calling this subroutine is:

$$\text{CALL REGSOLV(N, F, G, Z, X, ERR)}$$

The arguments are:

- N: the number of equations in the system (integer, input); N must be smaller than 20.

- F and G : external modules which specify the functions $f_j(z)$ $(j = 1, 2 \ldots, n)$ and $g(z)$, and their derivatives. F is of the form

$$\text{DOUBLE PRECISION FUNCTION F(J, IDER, Z)}$$

  and, for each input set of three parameters (i.e. the pair of integers $1 \leq J \leq n$ and $\text{IDER} \geq 0$, and double precision variable Z) it must return the value of $f_J^{(\text{IDER})}(Z)$. For $\text{IDER} = 0$ it must return $f_J(Z)$. The form of G is

$$\text{DOUBLE PRECISION FUNCTION G(IDER, Z)}$$

  and, for each input IDER and Z, it must return the value of $g^{(\text{IDER})}(Z)$.

- Z : double precision vector (input) which contains the values of z_1, z_2, ..., z_n.

- X : double precision vector (output). It returns the solution vector. In the calling program the dimensions of Z and of X should be declared as N, at least.

- ERR : double precision parameter (output) which furnishes the error estimate.

A sample program follows in which the system (2.20) is solved for three different values in the set $\{z_1, z_2\}$. The computed solution is compared with the analytical one (equation (2.21)). The resultant err is also listed.

```
implicit double precision(a-h,o-z)
dimension x(2),z(2),refx(2)
data n/2/
external f,g
write(*,*)'  z1    z2        x1        x2        errx1
!',' errx2    L_1 error '
write(*,*)
do i=1,3
  if(i.eq.1)then
    z(1)=0.d0
    z(2)=0.d0
```

```
          else
            z(1)=1.d0
            z(2)=i-2
          endif
          call regsolv(n,f,g,z,x,err)
          factor=2.d0/dcos(.5d0*(z(1)-z(2)))
          refx(1)=factor*dcos(z(1))*dcos(z(2))
     !     *dcos(.5d0*(z(1)+z(2)))
          refx(2)=factor*dsin(z(1))*dsin(z(2))
     !     *dsin(.5d0*(z(1)+z(2)))
          write(*,50)(z(k),k=1,n),(x(k),k=1,n),
     !    ((refx(k)-x(k)),k=1,n),err
          enddo
50        format(2(2x,f3.1),1x,2(2x,f7.5),1x,2(2x,d9.2),
          !2x,d9.2)
          end

          double precision function f(j,ider,z)
          implicit double precision(a-h,o-z)
          iref=ider-(ider/4)*4
          if(j.eq.1)then
            if(iref.eq.0)f=sin(z)
            if(iref.eq.1)f=cos(z)
            if(iref.eq.2)f=-sin(z)
            if(iref.eq.3)f=-cos(z)
            return
          endif
          if(j.eq.2)then
            if(iref.eq.0)f=cos(z)
            if(iref.eq.1)f=-sin(z)
            if(iref.eq.2)f=-cos(z)
            if(iref.eq.3)f=sin(z)
            return
          endif
          end

          double precision function g(ider,z)
          implicit double precision(a-h,o-z)
          iref=ider-(ider/4)*4
          zz=2.d0*z
          if(iref.eq.0)gg=sin(zz)
          if(iref.eq.1)gg=cos(zz)
```

```
if(iref.eq.2)gg=-sin(zz)
if(iref.eq.3)gg=-cos(zz)
g=gg*2.d0**ider
return
end
```

The output is

z1	z2	x1	x2	errx1	errx2	L_1 error
0.0	0.0	2.00000	0.00000	0.00D+00	0.00D+00	0.00D+00
1.0	0.0	1.08060	0.00000	-0.22D-15	0.00D+00	0.11D-15
1.0	1.0	0.31546	1.19165	0.56D-16	0.00D+00	0.00D+00

3.   An outline of the exponential fitting procedure

The things are best seen on a simple example. We consider the approximation of the first derivative of $y(x)$ at x_1 by the three-point formula

$$y'(x_1) \approx \frac{1}{h}[a_0 y(x_0) + a_1 y(x_1) + a_2 y(x_2)], \quad x_i = x_1 + (i-1)h, \quad (2.35)$$

and we want to determine the three coefficients a_0, a_1 and a_2 which ensure the best approximation. A particular formula of this type is the so-called central difference formula. Its coefficients are

$$a_0 = -\frac{1}{2}, \, a_1 = 0, \, a_2 = \frac{1}{2}. \qquad (2.36)$$

What about the features to be satisfied by an approximation formula in order to be called 'the best' ? We may take, for example, $a_0 = a_2 = 0$ and $a_1 = nh/x_1$, $x_1 \neq 0$. The formula will be exact for $y(x) = x^n$, but this is not of the type we wish because it is exact only for this function and also because its coefficients depend on the value of x_1. In general, instead of looking for a set of abscissa dependent coefficients which make a formula exact for one or for a discrete set of individual functions, we like to obtain a set of coefficients which is independent of the position of the mesh point taken for reference, upon the condition that the formula becomes convenient for functions in some predetermined class. For such formulae the typical h dependence of the leading term of the error is of the form h^p, and the condition that p is the biggest attainable value represents a central condition to call such an approximation the best.

The class of polynomials (of low degree, in general) is perhaps the most popular among the classes of functions of potential interest. Approximation formulae suited for this case are presented in almost every standard book, see e.g. [1], [4], [9]. The practical importance of such formulae originates in the fact that low degree polynomials produce good piecewise approximations for smooth functions and therefore a formula with the coefficients derived on this basis will be convenient when $y(x)$ is smooth enough, as it is in some applications, and/or when h is sufficiently small.

Other classes of functions can be also considered. In particular, the exponential fitting approach enables treating linear combinations of functions of the form (2.9), with polynomials as a limiting class.

The quality of the approximation (2.35) is measured by the difference between its left and right hand sides. If, for some $y(x)$, that difference equals zero irrespective of what values we choose for x_1 and for the step size h, then the formula is exact for that function. This suggests introducing an operator $\mathcal{L}$, acting on $y(x)$ and depending parametrically on h and on the set $\mathbf{a} = [a_0, a_1, a_2]$, in the following way:

$$\mathcal{L}[h, \mathbf{a}]y(x) := y'(x) - \frac{1}{h}[a_0 y(x - h) + a_1 y(x) + a_2 y(x + h)], \quad (2.37)$$

and then asking for the determination of the coefficients upon the condition that the function $\mathcal{L}[h, \mathbf{a}]y(x)$ is identically vanishing for some prerequisite forms for $y(x)$. Since (2.37) is linear in y it will be identically vanishing also for any linear combination of such forms.

For the polynomial case it is natural to consider the set of power functions

$$1, \ x, \ x^2, \ x^3, \dots \quad (2.38)$$

Indeed, any polynomial is a linear combination of power functions.

We have:

$$\mathcal{L}[h, \mathbf{a}]1 = -\frac{1}{h}(a_0 + a_1 + a_2),$$

$$\mathcal{L}[h, \mathbf{a}]x = -\frac{x}{h}(a_0 + a_1 + a_2) + (1 + a_0 - a_2),$$

$$\mathcal{L}[h, \mathbf{a}]x^2 = -\frac{x^2}{h}(a_0 + a_1 + a_2) + 2x(1 + a_0 - a_2) - h(a_0 + a_2),$$

$$\mathcal{L}[h, \mathbf{a}]x^3 = -\frac{x^3}{h}(a_0 + a_1 + a_2) + 3x^2(1 + a_0 - a_2) \quad (2.39)$$
$$-3xh(a_0 + a_2) + h^2(a_0 - a_2),$$

$$\mathcal{L}[h, \mathbf{a}]x^4 = -\frac{x^4}{h}(a_0 + a_1 + a_2) + 4x^3(1 + a_0 - a_2) - 6x^2h(a_0 + a_2)$$
$$+4xh^2(a_0 - a_2) - h^3(a_0 + a_2)$$

etc.

We also need the expressions of $\mathcal{L}[h, \mathbf{a}]x^m$ ($m = 0, 1, 2, \ldots$) at $x = 0$. These will be denoted as $L_m(h, \mathbf{a})$ and called moments. We have:

$$L_0(h, \mathbf{a}) = -\frac{1}{h}(a_0 + a_1 + a_2), \ L_1(h, \mathbf{a}) = 1 + a_0 - a_2,$$

$$L_{2k}(h, \mathbf{a}) = -h^{2k-1}(a_0 + a_2), \ L_{2k+1}(h, \mathbf{a}) = h^{2k}(a_0 - a_2), \quad (2.40)$$

$$k = 1, 2, \ldots$$

Since $\mathcal{L}$ is a linear operator it follows that, upon taking $y(x)$ as a linear combination of power functions, $y(x) = y_0 + y_1 x + y_2 x^2 + y_3 x^3 + \ldots$, we have

$$\mathcal{L}[h, \mathbf{a}]y(x) = \sum_{m=0} y_m \mathcal{L}[h, \mathbf{a}]x^m$$

$$= L_0(h, \mathbf{a})(y_0 + y_1 x + y_2 x^2 + y_3 x^3 + \ldots)$$

$$+ L_1(h, \mathbf{a})(y_1 + 2y_2 x + 3y_3 x^2 + \ldots)$$

$$+ L_2(h, \mathbf{a})(y_2 + 3y_3 x + 6y_4 x^2 + \ldots) + \ldots \quad (2.41)$$

$$= \sum_{m=0}^{\infty} \frac{1}{m!} L_m(h, \mathbf{a}) D^m y(x).$$

We now address the problem of determining the values of the coefficients a_0, a_1 and a_2 such that the function $\mathcal{L}[h, \mathbf{a}]y(x)$ is identically vanishing at any x and at any $h \in (0, H]$ for as many successive terms as possible in the classical set (2.38). Condition $\mathcal{L}[h, \mathbf{a}]1 = 0$ is fully equivalent to $L_0(h, \mathbf{a}) = 0$, the pair of conditions $\mathcal{L}[h, \mathbf{a}]1 = \mathcal{L}[h, \mathbf{a}]x = 0$ is equivalent to $L_0(h, \mathbf{a}) = L_1(h, \mathbf{a}) = 0$ and, in general, the set of M conditions

$$\mathcal{L}[h, \mathbf{a}]x^m = 0, \ m = 0, 1, 2, \ldots, M - 1 \quad (2.42)$$

is equivalent to

$$L_m(h, \mathbf{a}) = 0, \ m = 0, 1, 2, \ldots, M - 1, \quad (2.43)$$

which is a set of M linear equations for three unknowns.

The stated problem is then equivalent to that of finding the biggest M such that system (2.43) is compatible. We use the expressions listed in (2.40) and solve the linear system formed by the first three equations to get $a_2 = -a_0 = 1/2$ and $a_1 = 0$. The values thus determined will be denoted $\bar{a}_i$, $i = 1, 2, 3$.

REMARK 2.2 *Throughout this chapter we will use the notation* $\mathbf{a}$ *for the set of coefficients to be determined and* $\bar{\mathbf{a}}$ *for the particular set resulting from the determination.*

Since $L_3(h, \bar{a}) = -h^2 \neq 0$, the biggest admissible M is $M = 3$, that is equal to the number of unknowns. (This is not trivial. We shall see later on that there are cases when the maximal M and the number of unknowns may be different.) We also see that $L_{2k}(h, \bar{a}) = 0$ for $k = 2, 3, 4, \ldots$. Only derivatives of an odd order appear in (2.41), viz.

$$
\begin{aligned}
\mathcal{L}[h, \bar{a}]y(x) &= \frac{1}{3!}L_3(h, \bar{a})y^{(3)} + \frac{1}{5!}L_5(h, \bar{a})y^{(5)} + \cdots \\
&= -\frac{1}{6}h^2 y^{(3)} - \frac{1}{120}h^4 y^{(5)} + \cdots
\end{aligned}
\tag{2.44}
$$

To summarize, the above derivation had the polynomial form for starting and it used the biggest admissible M to lead to the parameters (2.36). This means that the central difference formula represents the best approximation of form (2.35) in the class of polynomials; it is exact for $y(x) \in \mathcal{P}_2$. As a matter of fact, other polynomial based approximations can be derived by taking smaller values of M. If, for example, $M = 2$, one of the three coefficients can be taken as a free parameter and the first two equations have to be solved for the other two coefficients. If $a_0 = p$ is chosen as the free parameter then we get $\bar{a}_0 = -(p+1)/2$ and $\bar{a}_2 = -(p-1)/2$. This formula is weaker than the previous one because it is exact only when $y(x) \in \mathcal{P}_1$.

Quite often it happens that the input information available for $y(x)$ is richer than assumed before, i.e., that $y(x)$ is smooth enough on the interval of interest. For example, it may happen that $y(x)$ is of the form

$$
y(x) = f_1(x)\exp(\mu_1 x) + f_2(x)\exp(\mu_2 x) + f_3(x)\exp(\mu_3 x),
\tag{2.45}
$$

where $f_1(x)$, $f_2(x)$, $f_3(x)$ are known as smooth (although their explicit expressions may not be available) and reasonable approximations of the constant factors μ_1, μ_2, μ_3 of x in the exponential functions are also known. The formula (2.35) with the classical coefficients given by (2.36) will be inadequate for functions of this form, but it makes sense to derive the coefficients on a basis which exploits the mentioned specific information on $y(x)$. This is what we do below.

Since now the function to be differentiated is of the form (2.45) where the weights $f_1(x)$, $f_2(x)$ and $f_3(x)$ are smooth, one may accept that these weight functions are well approximated by low degree polynomials. It follows that the set

$$
x^m \exp(\mu_1 x),\ x^m \exp(\mu_2 x),\ x^m \exp(\mu_3 x),\ m = 0, 1, \ldots
\tag{2.46}
$$

will be more suited to evaluate the coefficients a_i than the classical set (2.38).

From now on, the constant coefficient of x in the argument of an exponential function will be called frequency, thus conforming the usual terminology in the literature on the exponential fitting although, as already mentioned in Chapter 1, this does not overlap with the usual acceptance of the term, which typically refers to the real coefficient ω in a trigonometric function like $\sin(\omega x)$. However, they are correlated. Since $\sin(\omega x)$ is a linear combination of $\exp(i\omega x)$ and $\exp(-i\omega x)$ then, with the new acceptance in mind, this trigonometric function is characterized by a pair of two imaginary frequencies, $\pm i\omega$.

We use the same procedure as before, i.e. we consider the operator $\mathcal{L}$ but apply it on functions of form (2.46). For $m = 0$ and for some generic μ we have

$$\mathcal{L}[h, \mathbf{a}] \exp(\mu x) = \frac{1}{h} \exp(\mu x) E_0^*(z, \mathbf{a}), \tag{2.47}$$

where $z := \mu h$ and

$$E_0^*(z, \mathbf{a}) := z - a_0 \exp(-z) - a_1 - a_2 \exp(z). \tag{2.48}$$

We now impose the condition that (2.47) is identically vanishing for any x and any $h \in (0, H]$, if $\mu = \mu_i$, $i = 1, 2, 3$. This implies solving the linear algebraic system

$$E_0^*(z_i, \mathbf{a}) = 0, \; i = 1, 2, 3 \tag{2.49}$$

for the three unknowns a_0, a_1 and a_2. With $e_i := \exp(z_i)$ and $q_i := e_i z_i$ the solution is

$$\bar{a}_0 = -\frac{e_1 e_2 e_3 [z_1(e_3 - e_2) + z_2(e_1 - e_3) + z_3(e_2 - e_1)]}{(e_2 - e_1)(e_3 - e_2)(e_1 - e_3)},$$

$$\bar{a}_1 = \frac{q_1(e_3^2 - e_2^2) + q_2(e_1^2 - e_3^2) + q_3(e_2^2 - e_1^2)}{(e_2 - e_1)(e_3 - e_2)(e_1 - e_3)}, \tag{2.50}$$

$$\bar{a}_2 = -\frac{q_1(e_3 - e_2) + q_2(e_1 - e_3) + q_3(e_2 - e_1)}{(e_2 - e_1)(e_3 - e_2)(e_1 - e_3)}.$$

We see that each component of $\bar{\mathbf{a}}$ depends on the frequencies μ_i, $i = 1, 2, 3$, to be collected in vector μ, and on the step size h through the set of compressed arguments z_i, $i = 1, 2, 3$, to be collected in vector $\mathbf{z}$. This set will be denoted $\bar{\mathbf{a}}(\mathbf{z})$.

REMARK 2.3 *The set (2.46) with $m = 0$, used to determine $\bar{\mathbf{a}}(\mathbf{z})$, is the set of the three linearly independent solutions of the RDE*

$$(D - \mu_1)(D - \mu_2)(D - \mu_3)y = 0,$$

which in detailed form reads

$$y^{(3)} - (\mu_1 + \mu_2 + \mu_3)y^{(2)} + (\mu_1\mu_2 + \mu_2\mu_3 + \mu_3\mu_1)y' - \mu_1\mu_2\mu_3 y = 0. \tag{2.51}$$

When μ_1, μ_2, μ_3 tend to zero (we write this as $\mu \to 0$), this equation reduces to $y^{(3)} = 0$. The three linearly independent solutions of the latter are 1, x and x^2, i.e. the functions used to generate the classical set of coefficients (2.36). Also, since $\mu \to 0$ implies $z \to 0$, an alternative way of evaluation of the classical coefficients is suggested, that is by taking $\lim_{z \to 0} \bar{a}(z)$. As a matter of fact, this way of evaluation of the classical coefficients can be checked independently by using the expressions in (2.50). (Use the l'Hospital rule to eliminate the indeterminacy.) With this point in mind, the set $\bar{a}$ of the classical coefficients can be denoted as $\bar{a}(z = 0)$.

The fact that the parameters h, z and μ are linked may generate some ambiguity for the meaning of a process like $z \to 0$ because this is consistent with $h \to 0$ and/or with $\mu \to 0$. However, if not mentioned explicitly otherwise, $z \to 0$ will be meant to have solely $\mu \to 0$ behind it.

Quite instructive is to examine a closely related problem. Since the classical coefficients are antisymmetric, $\bar{a}_0 = -\bar{a}_2$ and $\bar{a}_1 = 0$, one may expect that the same results will be obtained also if the following simplified version of $\mathcal{L}$ is introduced:

$$\mathcal{L}[h, a_2]y(x) := y'(x) - \frac{1}{h}a_2[y(x+h) - y(x-h)], \qquad (2.52)$$

instead of the original (2.37). (The set a now consists of only one component, a_2.) The set (2.40) becomes:

$$L_0(h, a_2) = 0, \; L_1(h, a_2) = 1 - 2a_2, \qquad (2.53)$$
$$L_{2k}(h, a_2) = 0, L_{2k+1}(h, a_2) = -2h^{2k}a_2, \; k = 1, \, 2, \, \ldots$$

and it yields $\bar{a}_2 = 1/2$, i.e. the original and the simplified forms of $\mathcal{L}$ are indeed equivalent when treating the classical case.

However, this is not true when exponential functions are involved. In fact, for the new $\mathcal{L}$ we have $E_0^*(z, a_2) := z + a_2[\exp(-z) - \exp(z)]$, and asking that $E_0^*(z_1, a_2) = 0$ gives

$$\bar{a}_2(z_1) = \frac{z_1}{\exp(z_1) - \exp(-z_1)}. \qquad (2.54)$$

It is easy to check that $\lim_{z_1 \to 0} \bar{a}_2(z_1) = 1/2$, i.e. the classical coefficient is re-obtained.

The important point is that now the equality $E_0^*(z, \bar{a}_2(z_1)) = 0$ holds also for $z = 0$ and for $z = -z_1$, and therefore the condition $\mathcal{L}[h, \bar{a}_2]y(x) = 0$ is satisfied for $y(x) = 1$ and $\exp(\pm \mu_1 x)$ (and for their linear combinations, of course). This is inconvenient for our purpose because only one of these functions belongs to the set (2.46) and thus the approximation formula (2.35) with $a_0 = -a_2 = -\bar{a}_2(z_1)$ and $a_1 = 0$ is not suited for functions of the form (2.45). It becomes convenient only when some particular links between the three frequencies happen to exist, for example when $\mu_1 = \mu_2 = \mu_3$, or when $\mu_1 = -\mu_2$ and $\mu_3 = 0$. The above examination also suggests that the general form of $\mathcal{L}$ represents the most appropriate form to start with when approaching the case when exponential functions with uncorrelated frequencies are involved.

Coming back to the original definition of $\mathcal{L}$, equation (2.37), the error in the value of $y'(x_1)$ produced by the formula (2.35) with arbitrary **a** is given by $\mathcal{L}[h, \mathbf{a}]y(x)$ evaluated at $x = x_1$. The equation (2.41) allows expressing it as an infinite sum over the successive derivatives of $y(x)$,

$$error = \mathcal{L}[h, \mathbf{a}]y(x_1) = \sum_{m=0}^{\infty} \frac{1}{m!} L_m(h, \mathbf{a}) D^m y(x_1). \qquad (2.55)$$

If the classical set $\bar{\mathbf{a}}$ is taken for **a**, then the sum runs from $m = 3$ on, see the equation (2.44). The leading term of the error is

$$lte_{class} = -\frac{1}{6} h^2 y^{(3)}(x_1), \qquad (2.56)$$

and thus it behaves as h^2. For $\bar{\mathbf{a}}(\mathbf{z} \neq \mathbf{0})$, the h dependence of the leading term is not visible on (2.55) but the equation (2.41) can be conveniently re-arranged for this purpose. We write

$$\mathcal{L}[h, \bar{\mathbf{a}}(\mathbf{z})]y(x) = h^2 \sum_{k=0}^{\infty} h^k T_k^*(\mathbf{z}, \bar{\mathbf{a}}(\mathbf{z})) D^k \qquad (2.57)$$

$$(y^{(3)} - (\mu_1 + \mu_2 + \mu_3)y^{(2)} + (\mu_1\mu_2 + \mu_2\mu_3 + \mu_3\mu_1)y' - \mu_1\mu_2\mu_3 y),$$

a form which makes obvious the fact that $\mathcal{L}[h, \bar{\mathbf{a}}(\mathbf{z})]y(x)$ is identically vanishing for any x and $h \in (0, H]$ for $y(x) = \exp(\mu_i x)$, $i = 1, 2, 3$. The coefficients T_k^*, $k = 0, 1, \ldots$ result by identification of the like derivatives in this and in the original expansion (2.41). For $k = 0$ we get

$$T_0^*(\mathbf{z}, \bar{\mathbf{a}}(\mathbf{z})) = -\frac{h L_0(h, \bar{\mathbf{a}}(\mathbf{z}))}{z_1 z_2 z_3} = \frac{\bar{a}_1(\mathbf{z}) + \bar{a}_2(\mathbf{z}) + \bar{a}_3(\mathbf{z})}{z_1 z_2 z_3},$$

which shows that T_0^* depends on **z** both directly and indirectly (through $\bar{\mathbf{a}}(\mathbf{z})$). The same remains true for any T_k^*. Note also that, since (2.57) tends to (2.44) when $\mathbf{z} \to \mathbf{0}$, each T_k^* has a finite limit, viz.:

$$\lim_{\mathbf{z} \to 0} T_k^*(\mathbf{z}, \bar{\mathbf{a}}(\mathbf{z})) = \frac{h^{-(k+2)}}{(k+3)!} L_{k+3}(h, \bar{\mathbf{a}}(0)).$$

There is no h dependence in the right-hand side (see the expression of L_m).

The leading term of the error for this choice of $\bar{\mathbf{a}}$ then is

$$lte_{ef} = h^2 T_0^*(\mathbf{z}, \bar{\mathbf{a}}(\mathbf{z}))(y^{(3)}(x_1) - (\mu_1 + \mu_2 + \mu_3)y^{(2)}(x_1)$$
$$+(\mu_1\mu_2 + \mu_2\mu_3 + \mu_3\mu_1)y'(x_1) - \mu_1\mu_2\mu_3 y(x_1)). \qquad (2.58)$$

The expressions of the *lte* of the formula (2.35) for the two sets of coefficients, classical and exponential fitting (ef) (that is equations (2.56) and (2.58), respectively), are sufficient to get a qualitative insight on how the two versions behave for accuracy. Each of these expressions is a product of three factors. The h dependence is the same, h^2. The constant factor $-1/6$ in the former is replaced by $T_0^*(z, \bar{a}(z))$ in the latter. As said, it tends to $-1/6$ when $z \to 0$ and, anyway, it is close to this value when $|z_i| \le 1$, $i = 1, 2, 3$. For fixed μ_1, μ_2 and μ_3 the latter condition is always fulfilled if we care to use a conveniently small h.

The real difference in accuracy is due to the third factor. To see this, let us consider the case when $|\mu_i|$, $i = 1, 2, 3$ are big numbers. The first three derivatives of $y(x)$ of the form (2.45) are:

$$y'(x) = \sum_{i=1}^{3} \exp(\mu_i x)[\mu_i f_i(x) + f_i'(x)],$$

$$y''(x) = \sum_{i=1}^{3} \exp(\mu_i x)[\mu_i^2 f_i(x) + 2\mu_i f_i'(x) + f_i''(x)],$$

$$y^{(3)}(x) = \sum_{i=1}^{3} \exp(\mu_i x)[\mu_i^3 f_i(x) + 3\mu_i^2 f_i'(x) + 3\mu_i f_i''(x) + f_i^{(3)}(x)].$$

The terms with μ_i^3 are then dominating in the last factor of the lte_{class} but these cubic terms cancel each other in the corresponding factor in lte_{ef}. The dominating μ dependence is therefore only quadratic in the latter case.

For a numerical illustration we take the function

$$y(x) := f_1(x) \exp(\lambda_1 x) + f_2(x) \exp(\lambda_2 x) + f_3(x) \exp(\lambda_3 x),$$

where

$$f_1(x) := \exp(-0.3x), \quad f_2(x) := \exp(-0.2x), \text{ and } f_3(x) := \exp(-0.1x),$$

with given λ_2 and with $\lambda_1 = \lambda_2 - 1$ and $\lambda_3 = \lambda_2 + 1$. We take $x_1 = 1$ and compute $y'(x_1)$ by formula (2.35) with three sets of coefficients: (*i*) classical (equation (2.36)), (*ii*) exponential fitting $\bar{a}(z)$ (equation (2.50)) with $\mu_i = \lambda_i$ for input, and (*iii*) the same $\bar{a}(z)$, but with $\mu_i = \lambda_i + 1$ for input. The latter case is intended to simulate the practical situation when the function is known as being of the form (2.45), but only approximate values are available for the frequencies. The values of $y'(x_1)$ produced by the three versions are denoted as y'_{class}, y'_{ef1}, and y'_{ef2}, respectively.

In Table 2.2 we collect these data for $h = 0.1$ and various values for λ_2. The exact y' values are also listed for reference.

The results confirm the expectations. Thus, the errors of the three outputs are more or less the same only if λ_2 is around zero. When $|\lambda_2|$ is increased

Table 2.2. The numerical values of $y'(x_1 = 1)$, calculated by the three point formula with classical coefficients (y'_{class}) and with exponential fitting based coefficients in two versions (y'_{ef1} and y'_{ef2}) at different values of λ_2. Fixed step size $h = 0.1$ was used throughout.

λ_2	$exact\ y'$	y'_{class}	y'_{ef1}	y'_{ef2}
-8.0	$-8.96065(-3)$	$-9.85438(-3)$	$-8.96093(-3)$	$-8.96226(-3)$
-6.0	$-4.86073(-2)$	$-5.12444(-2)$	$-4.86094(-2)$	$-4.86192(-2)$
-4.0	$-2.29089(-1)$	$-2.34461(-1)$	$-2.29104(-1)$	$-2.29177(-1)$
-2.0	$-7.31640(-1)$	$-7.36571(-1)$	$-7.31754(-1)$	$-7.32290(-1)$
0.0	$1.69560(0)$	$1.69759(0)$	$1.69475(0)$	$1.69080(0)$
2.0	$6.50040(1)$	$6.58059(1)$	$6.49977(1)$	$6.49685(1)$
4.0	$8.68060(2)$	$8.99317(2)$	$8.68014(2)$	$8.67797(2)$
6.0	$9.27918(3)$	$9.96402(3)$	$9.27884(3)$	$9.27725(3)$
8.0	$8.97344(4)$	$1.01103(5)$	$8.97319(4)$	$8.97201(4)$

the quality of the results from the classical version quickly deteriorates while the accuracy of the other two versions remains almost unaltered. If $\lambda_2 = 8$, for example, the classical version is unable to reproduce any exact figure while the other two versions give results correct in four and three significant figures, respectively. Also as expected, the accuracy is slightly worse for (*iii*) than for (*ii*).

If h is such that $T_0^*(\mathbf{z}, \bar{\mathbf{a}}(\mathbf{z})) \approx -1/6$ the ratio lte_{class}/lte_{ef} reduces to the ratio of the last factors in the two expressions of the lte and then it is independent of h. Any alteration in the h dependence of the ratio $(y'(x_1) - y'_{class})/(y'(x_1) - y'_{ef})$ (we call this the accuracy enhancement factor of the exponential fitting version of the approximation formula over the classical one) would indicate the influence of the higher order terms in the error. This dependence is shown in Figure 2.1 for $\lambda_2 = 6$ and $h < 0.1$, for the two choices (*ii*) and (*iii*) of the nonvanishing frequencies. No significant variation is detected, i.e. the higher order terms are negligible, indeed. It is also seen that the errors from (*ii*) and form (*iii*) are smaller by a factor of about two thousand and of about four hundred, respectively, than the error from the classical version.

Finally the λ_2 dependence of the accuracy enhancement factor at fixed h is plotted on Figure 2.2 for $h = 0.1$. As expected, the enhancement factor is negligible for λ_2 around zero, to increase substantially with $|\lambda_2|$.

As a technical detail we mention that in the program the coefficients have not been calculated by the analytic expressions but by the subroutine REGSOLV. The system (2.49) is of form (2.17) with $n = 3$ equations for the unknowns $x_1 = a_0$, $x_2 = a_1$ and $x_3 = a_3$, where $f_1(z) = \exp(-z)$, $f_2(z) = 1$, $f_3(z) = \exp(z)$ and $g(z) = z$. Suited values for z_1, z_2 and z_3 were used to

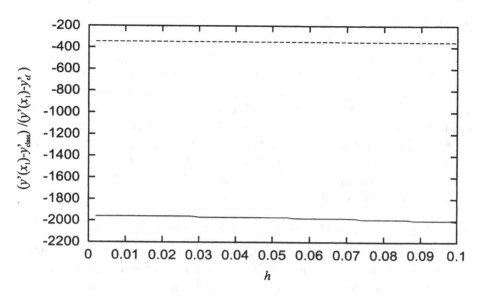

Figure 2.1. The h dependence, $h \in [0.02, 0.1]$, of the accuracy enhancement factor $(y'(x_1) - y'_{class})/(y'(x_1) - y'_{ef})$ for $\lambda_2 = 6$. Solid: version *(ii)*, broken: version *(iii)*.

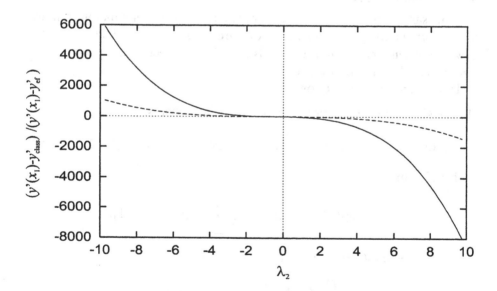

Figure 2.2. The λ_2 dependence, $\lambda_2 \in [-9.8, 9.8]$, of the accuracy enhancement factor $(y'(x_1) - y'_{class})/(y'(x_1) - y'_{ef})$ for $h = 0.1$. Solid: version *(ii)*, broken: version *(iii)*.

compute each set of coefficients. In particular, the values $z_1 = z_2 = z_3 = 0$ were taken to get the classical coefficients.

To conclude, the exponential fitting procedure is a technique devised for the construction of the coefficients such that the approximation formula becomes exact if $y(x)$ is satisfying a RDE. The explicit form of the convenient RDE must be given by the user in terms of the properties of the class of functions of interest. With the approximation formula (2.35) in mind, for functions of the form (2.46) the associated RDE was (2.51) but it can be different for other functions. If $y(x) = f(x)\exp(\mu_1 x)$ with smooth f, a linear combination of $\exp(\mu_1 x)$, $x\exp(\mu_1 x)$ and $x^2\exp(\mu_1 x)$ represents a good approximation and therefore the convenient RDE is $y^{(3)} - \mu_1^3 y = 0$.

4.   The theory of the exponential fitting approach

The main steps of the exponential fitting approach, formulated in the previous section on a test case, are now examined from a more general perspective. We follow in spirit the theoretical investigations of Lyche, [8], on multistep solvers for differential equations and of Ixaru, [5], for other operations (numerical differentiation, quadrature etc.)

4.1   The $\mathcal{L}$ operator

In the same way as we proceeded for the three point differentiation formula, to each formula whose coefficients have to be evaluated we associate an operator $\mathcal{L}$ which, when applied on $y(x)$, yields the difference between the left and the right hand sides of that formula.

Three more examples follow:

1 The simplest formula for the second derivative, that is:

$$y''(x_i) \approx \frac{1}{h^2}[a_0 y(x_0) + a_1 y(x_1) + a_2 y(x_2)], \quad x_i = x_1 + (i-1)h, \quad (2.59)$$

has $\mathcal{L}$ defined by

$$\mathcal{L}[h, \mathbf{a}]y(x) := y''(x) - \frac{1}{h^2}\sum_{i=0}^{2} a_i y(x + (i-1)h).$$

2 To the Newton-Cotes quadrature rule

$$\int_a^b y(x)dx \approx h\sum_{i=1}^{n} a_i y(X + x_i h),$$

where $X = (a+b)/2$ and $h = (b-a)/2$, we associate

$$\mathcal{L}[h, \mathbf{a}]y(x) := \int_{x-h}^{x+h} y(x')dx' - h\sum_{i=1}^{n} a_i y(x + x_i h).$$

3 To the n point Gauss-Legendre quadrature rule

$$\int_{-1}^{1} y(x)dx \approx \sum_{i=1}^{n} w_i y(x_i),$$

we associate

$$\mathcal{L}[h, \mathbf{a}]y(x) := \int_{x-h}^{x+h} y(x')dx' - \sum_{i=1}^{n} w_i y(x + x_i h).$$

Operator $\mathcal{L}$ depends on the step-size $h \in (0, H]$ and on the vector **a** of the coefficients to be determined. The latter has three components, viz: $\mathbf{a} := [a_0, a_1, a_2]$, for the case 1, n components for the case 2, and $2n$ components for the case 3, that is $\mathbf{a} := [w_1, w_2 \ldots, w_n, x_1, x_2, \ldots x_n]$. There is no specific pattern for the **a** dependence of $\mathcal{L}$; it may be either linear (as in 1 and 2) or nonlinear (as in 3). Note also that $\mathcal{L}$ is a linear operator in all these examples.

REMARK 2.4 *In many papers* $\mathcal{L}[h, \mathbf{a}]y(x)$ *is denoted* $\mathcal{L}[y(x); h, \mathbf{a}]$ *and called a functional.*

Operator $\mathcal{L}$ is called linear if for any two functions $y_1(x)$ and $y_2(x)$ on which it can be applied and for any two constants α_1 and α_2, one has

$$\mathcal{L}[h, \mathbf{a}][\alpha_1 y_1(x) + \alpha_2 y_2(x)] = \alpha_1 \mathcal{L}[h, \mathbf{a}]y_1(x) + \alpha_2 \mathcal{L}[h, \mathbf{a}]y_2(x),$$

for any $h \in (0, H]$ and **a**.

The subsequent theory covers the cases when :

1 $\mathcal{L}[h, \mathbf{a}]$ is a linear operator. **a** is assumed to have N components.

2 Any function $y(x)$ on which this is applied is differentiable indefinitely many times.

Nonlinear approximations are therefore not covered by this theory.

$\mathcal{L}[h, \mathbf{a}]y(x)$ will be used to formulate conditions for the determination of the suited components of the vector **a** and, in the final stage, for finding the expression of the error of the considered formula. The latter will be given by $\mathcal{L}[h, \bar{\mathbf{a}}]y(X)$, where $\bar{\mathbf{a}}$ is the set of the coefficients produced by such a determination and X is the particular value of x which secures the direct link between the formula and the associated $\mathcal{L}y(x)$. This is $X = x_1$ for case 1, $X = (a+b)/2$ for case 2, and $X = 0$ for the case 3. For the latter case we also have to take $h = 1$ in the final stage.

In the exponential fitting approach the application of $\mathcal{L}$ on functions of the form $y(x) = x^m \exp(\mu x)$, $m = 0, 1, 2 \ldots$ plays a central role. We define

$$E_m := \mathcal{L}[h, \mathbf{a}]x^m \exp(\mu x)|_{x=0}, \qquad (2.60)$$

and call it the μ moment of order m of $\mathcal{L}$. E_m depends on h, μ and **a**. The classical moment L_m is the particular case $\mu = 0$.

LEMMA 2.1 *The following relation holds:*

$$\mathcal{L}[h, \mathbf{a}] \exp(\mu x) = \exp(\mu x) \cdot E_0. \qquad (2.61)$$

Proof We apply $\mathcal{L}$ on $y(x) = \exp(\mu(x - \bar{x}))$. Since $\mathcal{L}$ is linear, we have

$$\mathcal{L}[h, \mathbf{a}] \exp(\mu(x - \bar{x})) = \exp(-\mu\bar{x}) \mathcal{L}[h, \mathbf{a}] \exp(\mu x). \qquad (2.62)$$

On multiplying both sides by $\exp(\mu\bar{x})$ and then taking $\bar{x} = x$ we obtain (2.61).

LEMMA 2.2 *Higher order μ moments are given by*

$$E_m = \frac{\partial^m E_0}{\partial \mu^m}, \ m = 1, 2, \ldots. \qquad (2.63)$$

Proof Equation (2.61) is differentiated m times with respect to μ,

$$\mathcal{L}[h, \mathbf{a}] \frac{\partial^m \exp(\mu x)}{\partial \mu^m} = \sum_{j=0}^{m} \binom{m}{j} \frac{\partial^j \exp(\mu x)}{\partial \mu^j} \cdot \frac{\partial^{m-j} E_0}{\partial \mu^{m-j}}. \qquad (2.64)$$

On using that

$$\frac{\partial^j \exp(\mu x)}{\partial \mu^j} = x^j \exp(\mu x),$$

and on taking $x = 0$ we get (2.63).

COROLLARY 2.1 *Successive moments are generated by*

$$E_m = \frac{\partial E_{m-1}}{\partial \mu}, \ m = 1, 2, \ldots. \qquad (2.65)$$

COROLLARY 2.2 *The following relation holds:*

$$\mathcal{L}[h, \mathbf{a}] x^m \exp(\mu x) = \sum_{j=0}^{m} \binom{m}{j} x^j \exp(\mu x) \cdot E_{m-j}. \qquad (2.66)$$

4.2 Dimensional consistency

There is a custom among physicists to examine their formulae for dimensional consistency. The rationale behind it is that, since any such formula is meant to link the components of one and the same physical effect, then the physical dimension must be the same in the two sides of the formula. For example, the formula $s(t) = v \cdot (t - t_0) + s_0$ represents the position at the moment t of a mobile in uniform motion with velocity v, with respect to its position at the moment t_0. The dimension of $s(t)$ and of s_0 is the length, that of t and of t_0 is the time, and the velocity is measured in units of length over time. The two sides of the formula then have one and the same dimension, the length.

A few natural rules are used to evaluate the dimension:

- If v is a physical quantity and its physical dimension is d (we write $dim\{v\} = d$) then $dim\{v^n\} = d^n$.

- If $dim\{v_1\} = d_1$ and $dim\{v_2\} = d_2$ then $dim\{v_1 v_2\} = d_1 d_2$.

- Both the argument and the value of an elementary function are dimensionless. For example, a relation like $v = v_1 \exp(x)$ implies that $dim\{v\} = dim\{v_1\}$ and also that x is dimensionless.

The examination of the formulae in terms of dimensional consistency is useful in that a relation like $v = v_1 + v_2$ necessarily implies that $dim\{v\} = dim\{v_1\} = dim\{v_2\}$ and any deviation from this represents an indication that something must be wrong in that formula. Alternatively, if the formula is known as correct but the dimension of one component is unknown, the approach allows its determination directly. For the same example, if it is known that $dim\{v\} = dim\{v_1\} = d$, then $dim\{v_2\} = d$, as well.

We now examine some expressions which appear in our context from this perspective. Let us assume that the function $y(x)$ has its own dimension Y and that the argument x is measured in units of the step size h. We write $dim\{y(x)\} = Y$ and $dim\{x\} = h$. As for the dimensions of the derivatives of $y(x)$ and of its integral, it is easy to see that $dim\{y^{(m)}\} = h^{-m}Y$ and that $dim\{\int y(x)dx\} = hY$.

The dimension of the left-hand side of equation (2.59) is $dim\{y''\} = h^{-2}Y$ while the right-hand side consists of a sum of three terms and for each of them we have $dim\{a_i y(x_i)/h^2\} = dim\{a_i\}h^{-2}Y$. The dimensional consistency then implies that each a_i is dimensionless. The set $\mathbf{a}$ is also dimensionless for the cases 2 and 3, and the common dimension of the two sides is hY in both cases. It follows that the relation

$$dim\{\mathcal{L}[h, \mathbf{a}]y(x)\} = h^l Y, \tag{2.67}$$

is valid for all the three cases, with $l = -2$ for the case 1 and $l = 1$ for the cases 2 and 3. Also, the set $\mathbf{a}$ is dimensionless in all these cases.

Note the importance of conveniently choosing the power of h when formulating the expression of the approximation sum. Any other power of h would have led to h dependent **a**. As a matter of fact, such a care on choosing the right power of h will be preserved throughout the whole book. Relation (2.67) with suited l and dimensionless **a** will be then satisfied by all considered $\mathcal{L}$-s and from now on we shall refer to h^l as to the dimension of the operator $\mathcal{L}$.

We now examine the μ moment E_m. Since $\exp(\mu x)$ appears in its definition, equation (2.60), and since $dim\{x\} = h$, it follows that $dim\{\mu\} = h^{-1}$ and then $z := \mu h$ is dimensionless. The dependence of E_m, originally formulated in terms of h, μ and **a**, can be alternatively formulated in terms of h, z and **a**, with the advantage that now the last two arguments are dimensionless. The latter formulation allows deriving the structure of E_m. In fact, the equation (2.60) yields

$$dim\{E_m(h, \mu, \mathbf{a})\} = dim\{\mathcal{L}[h, \mathbf{a}]\} \cdot dim\{x^m \exp(\mu x)\} = h^{l+m} \quad (2.68)$$

and, since one of the new arguments is just h while the others are dimensionless, the h dependence factorizes out, that is

$$E_m(h, \mu, \mathbf{a}) = h^{l+m} E_m^*(z, \mathbf{a}), \quad (2.69)$$

with dimensionless E_m^*. As for a direct evaluation of E_m^* when the expresion of E_0^* is known, we can use

$$E_m^*(z, \mathbf{a}) = \frac{\partial^m E_0^*(z, \mathbf{a})}{\partial z^m}, \quad m = 1, 2, \ldots, \quad (2.70)$$

which results from (2.63).

4.3   Evaluation of the coefficients

The classical moment of $\mathcal{L}[h, \mathbf{a}]$ is defined by

$$L_m(h, \mathbf{a}) := \mathcal{L}[h, \mathbf{a}]x^m \mid_{x=0} . \quad (2.71)$$

We have $L_m(h, \mathbf{a}) = E_m(h, \mu = 0, \mathbf{a}) = h^{l+m} E_m^*(z = 0, \mathbf{a})$, and therefore L_m is of the form

$$L_m(h, \mathbf{a}) = h^{l+m} L_m^*(\mathbf{a}), \quad (2.72)$$

where $L^*(\mathbf{a}) = E^*(z = 0, \mathbf{a})$ is dimensionless.

The following theorem represents the basis for the evaluation of the coefficients for the classical formulae:

THEOREM 2.2 *If $y(x)$ is differentiable indefinitely many times then*

$$\mathcal{L}[h, \mathbf{a}]y(x) = \sum_{m=0}^{\infty} \frac{1}{m!} L_m(h, \mathbf{a}) y^{(m)}(x). \quad (2.73)$$

Proof We use the Taylor series expansion of $y(x)$ around $\bar{x}$

$$y(x) = \sum_{m=0}^{\infty} \frac{1}{m!}(x - \bar{x})^m y^{(m)}(\bar{x}), \tag{2.74}$$

and apply $\mathcal{L}$ on both sides:

$$\mathcal{L}[h, \mathbf{a}]y(x) = \sum_{m=0}^{\infty} \frac{1}{m!} y^{(m)}(\bar{x})\mathcal{L}[h, \mathbf{a}](x - \bar{x})^m. \tag{2.75}$$

This must hold for any $\bar{x}$, in particular for $\bar{x} = x$, Q.E.D.

By virtue of (2.72) the equation (2.73) reads

$$\mathcal{L}[h, \mathbf{a}]y(x) = h^l \sum_{m=0}^{\infty} \frac{h^m}{m!} L_m^*(\mathbf{a})y^{(m)}(x), \tag{2.76}$$

and this suggests that the requirement of vanishing successive $L_m^*(\mathbf{a})$-s represents the natural way of determining $\mathbf{a}$. Let us take some positive integer M and assume that a particular set of coefficients, denoted as $\bar{\mathbf{a}} = [\bar{a}_1, \bar{a}_2, \ldots, \bar{a}_N]$, exists such that the conditions

$$L_m^*(\bar{\mathbf{a}}) = 0 \text{ for } m = 0, 1, \ldots, M - 1 \text{ but } L_M^*(\bar{\mathbf{a}}) \neq 0, \tag{2.77}$$

are satisfied. For this particular $\bar{\mathbf{a}}$ the sum in the expansion (2.76) begins with $m = M$,

$$\mathcal{L}[h, \bar{\mathbf{a}}]y(x) = h^{l+M} \sum_{k=0}^{\infty} \frac{h^k}{(M + k)!} L_{M+k}^*(\bar{\mathbf{a}})y^{(M+k)}(x). \tag{2.78}$$

Its leading term then behaves as h^{l+M}.

In applications it is important to construct the set $\bar{\mathbf{a}}$ such that M is as big as possible in (2.77). Then it makes sense to ask how big may be the first M, call it $\underline{M}$, such that the set $\bar{\mathbf{a}}$ can be constructed uniquely. In essence, $\bar{\mathbf{a}}$ is the solution of the system of $\underline{M}$ algebraic equations

$$L_m^*(\mathbf{a}) = 0, \quad m = 0, 1, \ldots, \underline{M} - 1 \tag{2.79}$$

and, since the number of unknowns is N, we may expect that $\underline{M} = N$.

This is indeed the case when the system is linear (as we assume for the moment) and compatible. However, there are situations when not all these equations are distinct (when some $L_m^*(\mathbf{a})$ and $L_{m'}^*(\mathbf{a})$, $m < m' \leq N - 1$, are proportional, for example). If so, only the distinct equations must be retained in the system and $\underline{M}$ will be the first integer, bigger than N, such that $L_{\underline{M}-1}^*(\mathbf{a}) = 0$ is just the N-th distinct equation. Again, the solution will be unique.

Cases when two equations $L_m^*(\mathbf{a}) = 0$ and $L_{m'}^*(\mathbf{a}) = 0$, $m < m' \leq N - 1$ are mutually conflicting may also appear. [Equations $a_0 + a_1 - 1 = 0$ and $a_0 + a_1 + 1 = 0$ are mutually conflicting, for example.] The system then consists of only $\underline{M} = m' < N$ equations, and this will lead to a family of parameter dependent solutions. In fact, $N - \underline{M}$ unknowns have to be considered as free parameters while the remained $\underline{M}$ result by solving the system in terms of these parameters.

There exist also applications when the system to be solved is nonlinear. In these cases there is no general guarantee that the solution does always exist. However, in all applications considered in this book the solution exists and it is unique if $\underline{M}$ is the first M such that the number of distinct equations equals N.

Altogether, (a) there is no sharp correlation between $\underline{M}$ and N, and (b) the biggest M for which a set $\bar{\mathbf{a}}$ can be constructed to satisfy (2.77) equals $\underline{M}$ if it is checked that $L_{\underline{M}}^*(\bar{\mathbf{a}}) \neq 0$ and is greater than $\underline{M}$ if $L_{\underline{M}}^*(\bar{\mathbf{a}}) = 0$. For illustration, if $\mathcal{L} y$ is as in (2.52) there is only one unknown, a_2, and then $N = 1$. Equation (2.53) shows that $L_0^*(a_2) = 0$, $L_1^*(a_2) = 1 - 2a_2$, $L_2^*(a_2) = 0$, $L_3^*(a_2) = -2a_2$ etc. The first equation which is effective in the determination of a_2 is $L_1^*(a_2) = 0$, with the solution $\bar{a}_2 = 1/2$. It follows that $\underline{M} = 2$ but since $L_2^*(\bar{a}_2) = 0$ and $L_3^*(\bar{a}_2) \neq 0$ one has $M = 3 > \underline{M} > N$.

The following theorem holds :

THEOREM 2.3 *If M and $\bar{\mathbf{a}}$ are chosen to satisfy (2.77), then $\mathcal{L}[h, \bar{\mathbf{a}}]y(x)$ identically vanishes in x and in $h \in (0, H]$ if and only if $y(x) \in P_{M-1}$.*

Proof The statement is true if $y(x) \in P_{M-1}$ because the differential equation $y^{(M)} = 0$ is automatically satisfied. As for the 'only if' part of the theorem, let us assume that some non identically vanishing $y(x) \notin P_{M-1}$ exists for which the statement would be still true. The structure of the equation (2.78) indicates that $\mathcal{L}[h, \bar{\mathbf{a}}]y(x)$ is identically vanishing in x and in h only when each term in the sum is also identically vanishing. Since, in particular, $L_M^*(\bar{\mathbf{a}}) \neq 0$, it follows that $y^{(M)} = 0$, that is a contradiction. Q. E. D.

A direct consequence of this theorem is that the approximation formula to which the considered $\mathcal{L}$ was associated, and with the coefficients obtained in this way, is exact irrespective of the values assigned to the reference abscissa X and to the step size h if and only if $y(x) \in P_{M-1}$. However, this does not exclude a better behaviour for some particular values of X and/or h. For example, the central difference formula for the first derivative, equations (2.35) and (2.36), has $M = 3$. For arbitrary X and h it is exact for and only for second degree polynomials, but if $X = 0$ it is also exact for any symmetric function, irrespective of h.

We now consider the problem of constructing the coefficients such that equation $\mathcal{L}[h, \mathbf{a}]y(x) = 0$ holds identically in x and in h if $y(x)$ satisfies the full RDE. This means requiring that $\mathcal{L}[h, \mathbf{a}]y(x) = 0$ for the following set of M

functions:

$$x^k \exp(\mu_i x), \; k = 0, 1, 2, \ldots, m_i - 1, \tag{2.80}$$
$$i = 1, 2, \ldots, I \quad \text{and} \quad m_1 + m_2 + \cdots + m_I = M,$$

see (2.9).

The following result is of help:

LEMMA 2.3 *The set of conditions*

$$\mathcal{L}[h, \mathbf{a}]x^k \exp(\mu x) = 0 \text{ for } k = 0, 1, \ldots, m - 1 \tag{2.81}$$

identically in x and in h implies that

$$E_k^*(z, \mathbf{a}) = 0 \text{ for } k = 0, 1, \ldots, m - 1 \tag{2.82}$$

where $z := \mu h$, *and viceversa.*

Proof Use corollary 2.2 and equation (2.69). Q.E.D.

On this basis, the suited coefficients are determined by solving the algebraic system of M equations

$$E_k^*(z_i, \mathbf{a}) = 0, \; k = 0, 1, \ldots, m_i - 1, \tag{2.83}$$
$$i = 1, 2, \ldots, I, \; m_1 + m_2 + \cdots + m_I = M,$$

where $z_i := \mu_i h$.

For brevity reasons we shall use the following compressed notations:

$$\mu := [\mu_1, \mu_2, \ldots, \mu_I], \; \mathbf{m} := [m_1, m_2, \ldots, m_I] \text{ and } \mathbf{z} := [z_1, z_2, \ldots, z_I] .$$

The solution $\bar{\mathbf{a}}$ of this system will depend on $\mathbf{z}$ and on $\mathbf{m}$. In the subsequent considerations we assume that $\mathbf{m}$ is fixed and then $\bar{\mathbf{a}}$ depends only on $\mathbf{z}$; we denote it by $\bar{\mathbf{a}}(\mathbf{z})$. We also assume that the solution exists and is unique for arbitrary $\mathbf{z}$.

It is convenient to rewrite equation (2.73) in a form in which the left hand side of the RDE appears explicitly. We write:

$$\mathcal{L}[h, \bar{\mathbf{a}}(\mathbf{z})]y(x) = h^{l+M} \sum_{k=0}^{\infty} h^k T_k^*(\mathbf{z}, \bar{\mathbf{a}}(\mathbf{z})) D^k (y^{(M)} + c_1 y^{(M-1)} + \cdots + c_M y).$$

$$\tag{2.84}$$

To determine the dimensionless coefficients T_k^* we proceed in the following way. First, we replace the original coefficients in the RDE by the dimensionless coefficients $c_j^* := c_j h^j$, $j = 1, 2, \ldots, M$. They depend on $\mathbf{z}$; for example, $c_1^* = -(z_1 + z_2 + \ldots + z_M)$, see (2.13). The equation (2.84) then reads

$$\mathcal{L}[h, \bar{\mathbf{a}}(\mathbf{z})]y(x) = h^{l+M} \sum_{k=0}^{\infty} h^k T_k^*(\mathbf{z}, \bar{\mathbf{a}}(\mathbf{z})) \tag{2.85}$$

$$\times D^k (y^{(M)} + h^{-1} c_1^* y^{(M-1)} + \cdots + h^{-M} c_M^* y).$$

Next, we identify the coefficients of the like derivatives in this and in the original form (2.76). On doing so for the coefficients of y and of y' we get

$$L_0^*(\bar{a}(z)) = T_0^*(z, \bar{a}(z))c_M^*(z),$$

and

$$L_1^*(\bar{a}(z)) = T_1^*(z, \bar{a}(z))c_M^*(z) + T_0^*(z, \bar{a}(z))c_{M-1}^*(z),$$

respectively. In general we have:

$$\frac{1}{m!}L_m^*(\bar{a}(z)) = T_m^*(z, \bar{a}(z))c_M^*(z) + T_{m-1}^*(z, \bar{a}(z))c_{M-1}^*(z) \qquad (2.86)$$
$$+ \cdots + T_1^*(z, \bar{a}(z))c_{M-m+1}^*(z) + T_0^*(z, \bar{a}(z))c_{M-m}^*(z)$$

for $m < M$,

$$\frac{1}{M!}L_M^*(\bar{a}(z)) = T_M^*(z, \bar{a}(z))c_M^*(z) + T_{M-1}^*(z, \bar{a}(z))c_{M-1}^*(z) \qquad (2.87)$$
$$+ \cdots + T_1^*(z, \bar{a}(z))c_1^*(z) + T_0^*(z, \bar{a}(z)),$$

and

$$\frac{1}{m!}L_m^*(\bar{a}(z)) = T_m^*(z, \bar{a}(z))c_M^*(z) + T_{m-1}^*(z, \bar{a}(z))c_{M-1}^*(z) \qquad (2.88)$$
$$+ \cdots + T_{m-M+1}^*(z, \bar{a}(z))c_1^*(z) + T_{m-M}^*(z, \bar{a}(z))$$

for $m > M$.

The last three relations allow determining the coefficients T_k^* in terms of the classical L^*s. Let μ_1 be the smallest in absolute value from the components of μ. The same is true for z_1 in the set z. Two cases should be considered separately:

(i) $z_1 \neq 0$. $c_M^*(z)$ is different from zero and then T_0^* results directly from L_0^*,

$$T_0^*(z, \bar{a}(z)) = \frac{L_0^*(\bar{a}(z))}{c_M^*(z)} = (-1)^M \frac{L_0^*(\bar{a}(z))}{z_1^{m_1} z_2^{m_2} \cdots z_I^{m_I}}, \qquad (2.89)$$

and the subsequent T_k^* are determined by recurrence.

(ii) $z_1 = 0$. In this case $c_m^*(z) \neq 0$ if $m \leq M - m_1$ but $c_m^*(z) = 0$ if $m > M - m_1$. T_0^* is now determined in terms of $L_{m_1}^*$,

$$T_0^*(z, \bar{a}(z)) = \frac{L_{m_1}^*(\bar{a}(z))}{m_1! c_{M-m_1}^*(z)} = \frac{(-1)^{M-m_1} L_{m_1}^*(\bar{a}(z))}{m_1! z_2^{m_2} z_3^{m_3} \cdots z_I^{m_I}}, \qquad (2.90)$$

and, again, the subsequent T_k^* are determined by recurrence.

These relations show that each T_k^* depends on z both directly and through $\bar{a}(z)$, in agreement with the notation used for its arguments.

An important property is that the classical (polynomial based) case, treated separately before, becomes a particular case of the latter. The link is given by the following theorem:

THEOREM 2.4 *Let M and* m *be given and let* $\bar{a}(z)$ *be the unique solution of the algebraic system (2.83), and* $\bar{a}_{class}$ *be the unique solution of*

$$L_m^*(a) = 0 \quad \text{for} \quad m = 0, 1, \ldots, M - 1. \tag{2.91}$$

Then we have

$$\lim_{z \to 0} \bar{a}(z) = \bar{a}_{class}, \tag{2.92}$$

Proof By hypotesis $\bar{a}(z)$ is the only set which ensures that $\mathcal{L}[h, \bar{a}(z)]y(x)$ is identically vanishing in x and in h if $y(x)$ satisfies the RDE uniquely determined by $\mu = z/h$ and m. In particular, this holds also when $z \to 0$. In the latter case, the RDE reaches the simple form $y^{(M)} = 0$ and then $\lim_{z \to 0} \bar{a}(z)$ is satisfying (2.91). Q.E.D.

A direct consequence is that the T^* coefficients have finite limits:

$$\lim_{z \to 0} T_k^*(z, \bar{a}(z)) = \frac{1}{(M+k)!} L_{M+k}^*(\bar{a}_{class}), \; k = 0, 1, \ldots. \tag{2.93}$$

but this does not necessarily imply that $L_M^*(\bar{a}_{class}) \neq 0$. However, if it is checked that this condition is fulfilled then an analogue of Theorem 2.3 can be formulated:

THEOREM 2.5 *Let the conditions of Theorem 2.4 be satisfied. If the condition* $L_M^*(\bar{a}_{class}) \neq 0$ *is also satisfied, then* $\mathcal{L}[h, \bar{a}(z)]y(x)$ *identically vanishes in x and in $h \in (0, H]$ if and only if $y(x)$ is a solution of the RDE consistent with the input* m *and with $\mu = z/h$.*

Proof The 'if' part of the statement is true due to the very way in which $\bar{a}(z)$ is constructed. To prove the 'only if' part we apply the reductio ad absurdum. Let us then assume that some $y(x)$ exists, which does not satisfy the RDE but for which the statement is still true. This means that a non identically vanishing function $g(x)$ exists such that $(D^M + c_1 D^{M-1} + \ldots + c_M)y(x) = g(x)$. We use the expansion of $T_k^*(z, \bar{a}(z))$ in powers of z_i, $i = 1, 2, \ldots, I$. The first term of the expansion is a constant. Its value is given by the right-hand side of equation (2.93). Since $z_i = \mu_i h$, this expansion can be re-organized in powers of h,

$$T_k^*(z, \bar{a}(z)) = \frac{1}{(M+k)!} L_{M+k}^*(\bar{a}_{class}) + w_1 h + w_2 h^2 + \ldots$$

where w_1, w_2, ... depend on the μ-s. Upon introducing this in the equation (2.84) we get a series expansion in powers of h for $\mathcal{L}[h, \bar{\mathbf{a}}(\mathbf{z})]y(x)$. The key point is that the first term of this expansion is

$$\frac{h^{l+M}}{M!} L_M^*(\bar{\mathbf{a}}_{class}) \cdot g(x),$$

and it must vanish identically in x and in h. However, since $L_M^*(\bar{\mathbf{a}}_{class}) \neq 0$, this condition can be fulfilled only if $g(x)$ is identically vanishing for any x, that is a contradiction, Q. E. D.

The above results suggest that a natural scheme to implement the exponential fitting procedure consists in first considering the classical case. This allows determining M by the condition that $L_M^*(\bar{\mathbf{a}}_{class})$ is the first nonvanishing classical moment and it is admitted that this M is also the order of the RDE in the exponential fitting extension. For the latter, the values of the input μ and m are fixed in terms of the particular structure of $y(x)$ to be processed by the approximation formula, and the coefficients are determined by solving the algebraic system (2.83).

The scheme works very well in many cases of practical interest but it is important to be aware that it has some limitations. Let the step size h be fixed. Theorem 2.5 assumes that the system (2.83) has a solution but, with M fixed as mentioned, there is no general guarantee that this system is compatible for an arbitrary choice of the set of frequencies μ and of m. The compatibility depends on the forms of $\mathcal{L}$ and of the RDE and particularly important in this context is to keep these consistent for symmetry. Thus, an RDE whose solution space is not closed with respect to the reflection $x \rightarrow -x$ will be not compatible with a form of $\mathcal{L}$ in which symmetry conditions are imposed.

The test case discussed in the previous section provides a good illustration for this. The form (2.37) is free of any symmetry restriction. It is then consistent with the general RDE of order three $y^{(3)} + c_1 y^{(2)} + c_2 y' + c_3 y = 0$ and, for this reason, arbitrary values of μ_1, μ_2 and μ_3 can be used. In contrast, a symmetry condition was imposed in the form (2.52). This $\mathcal{L}$ will be then consistent with an RDE of the same order, but of a form such that the solution space remains closed with respect to the reflection. As explained in Section 1, this form is $y^{(3)} + c_2 y' = 0$, where c_2 is an arbitrary constant. This explains why the form (2.52) of $\mathcal{L}$ is suitable for the classical case (indeed, this is because for $c_2 = 0$ we have $\mu_1 = \mu_2 = \mu_3 = 0$), and it also helps explaining why the same form of $\mathcal{L}$ accepts only correlated frequencies, viz.: $\mu_1 = -\mu_2$, $\mu_3 = 0$. This discussion also suggests that the safest way to approach the things is the one which starts simply from the most general form of $\mathcal{L}$.

Also, some expressions of $\mathcal{L}$ may lead to situations when some equations in the system (2.83) are in conflict for any $\mu \neq 0$ and m. In such situations the order of the RDE will be necessarily smaller than M and then the whole procedure for the construction of the coefficients has to be correspondingly adapted. An example of this type will be presented in Section 2.5 of Chapter 4.

The z dependence of the solution of (2.83) also deserves some comment. To fix the ideas we assume that the values of μ and of $\mathbf{m}$, used for input, are such that the forms of $\mathcal{L}$ and of the RDE are mutually consistent, and then the compatibility is beyond doubt. The coefficients of the system depend on the product $\mathbf{z} = \mu h$, and the same holds for the solution.

For reasons of clarity let us also assume that (2.83) is a linear system. Its determinant is a function of $\mathbf{z}$. The mutual consistency between $\mathcal{L}$ and RDE implies that this function is not identically vanishing but it does not exclude the possibility that it may have some roots. For such special values of $\mathbf{z}$, called critical values, the system is singular and therefore there is no solution. Also, around such values the shape of the solution components exhibits some characterictic pole-like behaviour. The interesting point is that critical values are present for some forms of $\mathcal{L}$ but not for other forms and therefore there is a direct link between the form of $\mathcal{L}$ and the presence or absence of critical values of $\mathbf{z}$. For example, no critical value appears in the usual tuned forms of the Gauss quadrature formula and in one of the interpolation formulae. The mathematical reason of such behaviours is open for further investigation.

Finally, it is important to insist that the discussed scheme is heavily based on the assumption that the algebraic systems (2.91) and (2.83) have unique solutions. As said before, for the applications considered in this book this condition is fulfilled for the system (2.91) no matter whether this system is linear of nonlinear. The situation is comparatively more complicated for the system (2.83) corresponding to the ef extension. A solution also exists in all considered applications but when this system is nonlinear it happens that this solution is not unique; for such an example see Section 3.2.2 in Chapter 6 and Figure 6.8. When $\mathbf{z} \to 0$ one of these solutions tends to $\bar{\mathbf{a}}_{class}$ but the other solutions do not have a classical counterpart. The investigation of such extra solutions is relevant from a theoretical point of view but only the solution which has a classical counterpart is important in practice.

4.4   Error formulae

The error of the approximation formula to which the operator $\mathcal{L}$ was associated is given directly by the expression of $\mathcal{L}[h, \bar{\mathbf{a}}]y(x)$ evaluated at $x = X$, where X is the particular abscissa point in the formula which is replaced by the variable x in the definition of $\mathcal{L}$. The error for the exponential fitting version results from the equation (2.84):

$$err_{ef} = h^{l+M} \sum_{k=0}^{\infty} h^k T_k^*(\mathbf{z}, \bar{\mathbf{a}}(\mathbf{z})) D^k (D - \mu_1)^{m_1} \cdots (D - \mu_I)^{m_I} y(X),$$

$$(2.94)$$

while the error for the associated classical version results by taking the limit $\mu, z \to 0$. It reads

$$err_{class} = h^{l+M} \sum_{k=0}^{\infty} \frac{h^k}{(M+k)!} L_{M+k}^*(\bar{a}_{class}) D^k y^{(M)}(X). \qquad (2.95)$$

(In all these considerations it is tacitly assumed that the conditions of Theorems 2.4 and 2.5 are satisfied.)

The leading terms of these are:

$$lte_{ef} := h^{l+M} T_0^*(\mathbf{z}, \bar{\mathbf{a}}(\mathbf{z}))(D - \mu_1)^{m_1}(D - \mu_2)^{m_2} \cdots (D - \mu_I)^{m_I} y(X), \qquad (2.96)$$

where T_0^* is given by (2.89) or (2.90), and

$$lte_{class} := \frac{h^{l+M}}{M!} L_M^*(\bar{a}_{class}) y^{(M)}(X), \qquad (2.97)$$

respectively.

The expressions of the two leading terms suggest formulating some qualitative expectations.

On one hand, one and the same front factor h^{l+M} appears in each of them (l is associated with the dimension of the operator $\mathcal{L}$ while M is the order of the RDE used as the basis for the construction of the coefficients). Since $l + M > 0$, the error is expected to decrease with h at one and the same rate.

On the other hand, the exponential fitting version is clearly more flexible than the classical one. The frequencies and their multiplicities, used as input when calculating the coefficients, are fixed by a preliminary analysis of the peculiarities of the function $y(x)$ to be processed by the approximation formula. For small h the two errors will dominantly behave as $K_{class}h^{l+M}$ and as $K_{ef}h^{l+M}$, respectively, where K_{class} and K_{ef} are some constants, but for that $y(x)$ it is expected that $|K_{ef}| < |K_{class}|$.

Such expectations are nicely confirmed in practice although, strictly speaking, the relevance of the lte as the dominating term in the whole error must be taken with some caution. For example, this is not the case for the lte_{class} if it happens that $y^{(M)}(X) = 0$. However, by analogy with the Lagrange form of the reminder of a Taylor series, it is tempted to conjecture that there is some η in the vicinity of X such that the whole err_{class} is given by the expression of lte_{class} if argument X is replaced by η, viz:

$$err_{class} = \frac{h^{l+M}}{M!} L_M^*(\bar{a}_{class}) y^{(M)}(\eta). \qquad (2.98)$$

Indeed, such a form would provide a better justification for the expectations mentioned above for the classical formulae.

There is a vast literature around this conjecture. It has been proved that it holds for some approximation formulae but not for the others. For an examination of the linear multistep algorithms for ordinary differential equations see Section 3.6 of [7].

A similar problem can be addressed also for the exponential fitting version. Again, it is conjectured that some η exists such that

$$err_{ef} = h^{l+M} T_0^*(\mathbf{z}, \bar{\mathbf{a}}(\mathbf{z}))(D - \mu_1)^{m_1}(D - \mu_2)^{m_2} \cdots (D - \mu_I)^{m_I} y(\eta).$$
$$(2.99)$$

The first investigation along this line has been undertaken by De Meyer, Vanthournout, Vanden Berghe and Vanderbauwhede , [3], who considered the approximation $y_n(x)$ of $y(x)$ by a mixed type of interpolation. Specifically, given the $n + 1$ abscissa points $x_0 < x_1 < \ldots < x_n$ they constructed the function $y_n(x)$ by linear interpolation on the conditions that $y_n(x_j) = y(x_j)$, $j = 0, 1, \ldots, n$ and that $y_n(x)$ satisfies the following RDE of order $n + 1$:

$$D^{n-1}(D - \mu_1)(D - \mu_2)y = 0, \qquad (2.100)$$

where $\mu_1 = -\mu_2 = i\omega$, with real ω. The error err_{ef} obviously depends on x. These authors established a theorem which says that, if $\omega \in (0, \pi/(x_n - x_0))$, then for any $x \in (x_0, x_n)$ there exists a smooth function $\psi(x)$ which depends parametrically on the positions of the partition points and on ω such that, if $y(x) \in C^{n+1}([x_0, x_n])$, one can find $\eta \in (x_0, x_n)$ such that

$$err_{ef}(x) = \psi(x)D^{n-1}(D - \mu_1)(D - \mu_2)y(\eta). \qquad (2.101)$$

A recent investigation in the same area is due to Coleman [2]. By conveniently adapting a technique originally developed for the treatment of the quadrature rules, he examined a series of exponential fitting based formulae for numerical differentiation, for quadrature and for the numerical solution of ordinary differential equations. Coleman's technique is flexible enough to be applied to other formulae, as well.

He has proved that the conjecture holds for all the investigated cases if the components of $\mathbf{z}$ belong to some sharply delimited, case dependent intervals, and also suggested some cases where the conjecture may be false.

References

[1] Burden, R. L. and Faires J. D. (1981). *Numerical Analysis*. Prindle, Weber and Schmidt, Boston.

[2] Coleman, J. P. (2003). Private communication.

[3] De Meyer, H., Vanthournout, J., Vanden Berghe, G. and Vanderbauwhede, A. (1990). On the error estimation for a mixed type of interpolation. *J. Comput. Appl. Math.*, 32: 407–415.

[4] Gerald, G. F. and Wheatley, P. O. (1994). *Applied Numerical Analysis*. Addison-Wesley, Reading.

[5] Ixaru, L. Gr. (1997). Operations on oscillatory functions. *Comput. Phys. Commun.*, 105: 1–19.

[6] Ixaru, L. Gr. , De Meyer, H. , Vanden Berghe, G. and Van Daele, M. (1996). A regularization procedure for $\sum_{i=1}^{n} f_i(z_j)x_i = g(z_j)(j = 1, 2, \cdots, n)$. *Numer. Linear Algebra Appl.*, 3: 81–90.

[7] Lambert, J. D. (1991). *Numerical Methods for Ordinary Differential Equations*. Wiley, New York.

[8] Lyche, T. (1974). Chebyshevian multistep methods for ordinary differential equations. *Numer. Math.*, 19: 65–75.

[9] Phillips, J. M. and Taylor, P. J. (1973). *Theory and Applications of Numerical Analysis*. Academic Press. London and New York.

[10] Press, W. H., Teukolsky, S. A., Vetterling, W. T. and Flannery, B. P. (1986). *Numerical Recipes, The Art of Scientific Computing*. Cambridge University Press, Cambridge.

[11] Wolfram, S. (1992). *MATHEMATICA   A system for doing mathematics by computer*, Addison–Wesley, Reading.

Chapter 3

CONSTRUCTION OF EF FORMULAE
FOR FUNCTIONS WITH
OSCILLATORY OR HYPERBOLIC VARIATION

The considerations of the previous chapter referred to the general features of the exponential fitting technique. When the functions to be approximated are oscillatory or with a variation well described by hyperbolic functions the technique exhibits some helpful features. This chapter aims at presenting these features and at formulating a simple algorithm-like flow chart to be followed in the current practice.

In Section 3.1 we consider a series of numerical problems associated to the solution of the Schrödinger equation and which may benefit from such a treatment. As a matter of fact, this equation represents the starting point in an important number of papers on the exponential fitting technique. Section 3.2 introduces a set of ad-hoc functions which are of help for a unified treatment of the two types of functions to be approximated (oscillatory, and with hyperbolic variation). Finally, in Section 3.3 we present the adapted, six-step algorithm-like flow chart and a subroutine for the calculation of the solution of the linear algebraic systems generated in this frame.

1. The Schrödinger equation

Many physical phenomena exhibit a pronounced oscillatory character. The behaviour of the pendulum-like systems and vibrations in classical mechanics, planet orbiting in celestial mechanics, wave propagation and resonances in electromagnetism, and the behaviour of the quantum particles are all phenomena of this type. Transient phenomena, well approximated by functions with hyperbolic variation, are also often met in current practice. The investigation of such phenomena then necessarily implies numerical operations such as differentiation, quadrature or interpolation of functions with the quoted behaviours.

53

The Schrödinger equation provides a good illustration for this. It is the fundamental equation of the nonrelavistic quantum mechanics, with direct applications in the description of various effects in nuclear, atomic and molecular physics; see, e. g., [6].

To fix the ideas we consider a simple form of this equation. This is the time independent equation for the behaviour of a spinless particle of mass μ in a potential $v(\mathbf{x})$, where $\mathbf{x}$ is a three-dimensional vector ($\mathbf{x} = [x_1, x_2, x_3]$ in cartesian coordinates). The equation reads

$$-\frac{\hbar^2}{2\mu}\Delta\Psi + (v(\mathbf{x}) - e)\Psi = 0, \quad \mathbf{x} \in \Re^3 \tag{3.1}$$

where $\hbar$ is the reduced Planck's constant, e is the particle energy and Δ is the Laplace operator

$$\Delta := \frac{\partial^2}{\partial x_1^2} + \frac{\partial^2}{\partial x_2^2} + \frac{\partial^2}{\partial x_3^2}.$$

This is a partial differential equation and it must be solved upon the physical condition that $\Psi(\mathbf{x})$ must be finite for any $\mathbf{x}$ in $\Re^3$. Solution $\Psi(\mathbf{x})$ is called the particle wavefunction and it characterizes the localization of the particle in the sense that $|\Psi(\mathbf{x})|^2 d\mathbf{x}$ describes the probability of the presence of the particle in the infinitesimal volume $d\mathbf{x}$ centered at $\mathbf{x}$. Of course, some more specifications should be added here as, for example, that a norm of $\Psi(\mathbf{x})$ has to be defined; see again [6].

For simplicity we shall assume that the potential function $v(\mathbf{x})$ is spherically symmetric, that is it depends only on $x = (x_1^2 + x_2^2 + x_3^2)^{1/2}$. It is then convenient to use spherical coordinates $x \in [0, \infty)$, $\theta \in [0, \pi]$ and $\phi \in [0, 2\pi]$. The cartesian and spherical coordinates are linked as

$$x_1 = x\sin(\theta)\cos(\phi), \quad x_2 = x\sin(\theta)\sin(\phi), \quad x_3 = x\cos(\theta),$$

$$dx = dx_1 dx_2 dx_3 = x^2 dx \sin(\theta)d\theta d\phi.$$

$\Psi(\mathbf{x})$ is expanded over the complete set of the spherical functions $Y_{lm}(\theta, \phi)$,

$$\Psi(\mathbf{x}) = \frac{1}{x}\sum_{l=0}^{\infty}\sum_{m=-l}^{l}\psi_l(x)Y_{lm}(\theta, \phi). \tag{3.2}$$

It is recalled that the spherical functions satisfy the equation

$$\left[\frac{1}{\sin(\theta)}\frac{\partial}{\partial\theta}(\sin(\theta)\frac{\partial}{\partial\theta}) + \frac{1}{\sin^2(\theta)}\frac{\partial^2}{\partial\phi^2} + l(l+1)\right]Y_{lm}(\theta, \phi) = 0, \tag{3.3}$$

$l = 0, 1, \ldots,$ $m = -l, -l+1, \ldots, l+1, l$, with the normalization condition

$$\int_0^{2\pi} d\phi \int_0^{\pi} Y_{lm}^*(\theta, \phi)Y_{l'm'}(\theta, \phi)\sin(\theta)d\theta = \delta_{ll'}\delta_{mm'},$$

and that the Laplace operator in spherical coordinates reads

$$\Delta = \frac{1}{x^2}\left[\frac{\partial}{\partial x}(x^2\frac{\partial}{\partial x}) + \frac{1}{\sin(\theta)}\frac{\partial}{\partial\theta}(\sin(\theta)\frac{\partial}{\partial\theta}) + \frac{1}{\sin^2(\theta)}\frac{\partial^2}{\partial\phi^2}\right].$$

On introducing the expansion (3.2) into (3.1), multiplying by $Y^*_{l'm'}(\theta, \phi)$ and interating over θ and ϕ, it is found that $\psi_l(x)$ satisfy

$$\psi_l'' + (E - V_l(x))\psi_l = 0, \; l = 0, 1, \ldots \tag{3.4}$$

where

$$E = \frac{2\mu}{\hbar^2}e, \quad V_l(x) = \frac{l(l+1)}{x^2} + \frac{2\mu}{\hbar^2}v(x).$$

The solution of the original partial differential equation (3.1) is then reduced to that of a set of separate ordinary differential equations of form (3.4). Strictly speaking, the number of equations of this set is infinite (see (3.2)) but in practice it is cut in terms of the characteristics of the considered physical problem.

Each equation of the form (3.4) is called a radial equation. The important peculiarities of the radial Schrödinger equation are that $V_l(x)$ is singular at the origin when $l \neq 0$ and that, due to the presence of x^{-1} in the expansion (3.2) and to the physical requirement that $\Psi(x)$ must be finite everywhere (in particular at the origin), the solution of (3.4) must vanish at this point, $\psi_l(0) = 0, \; l = 0, 1 \ldots$.

To fix the ideas, let us concentrate on one radial equation. We select some arbitrary value for l, denote $\psi_l(x)$ by $y(x)$ and omit the subscript in $V_l(x)$. We also assume that $v(x)$ is a smooth function which tends to zero sufficiently fast when x tends to infinity. An example of such a function is the Woods-Saxon potential frequently used in nuclear physics. This reads

$$v(x) = -\frac{v_0}{1 + \exp[(x - x_0)/a]}, \tag{3.5}$$

where v_0, x_0 and a are given parameters (see, e. g., [5]). Some typical values are $v_0 = 50$, $x_0 = 7$ and $a = 0.6$. For other potentials see the list of problems in the book [7].

The problem to be solved then is

$$y'' + (E - V(x))y = 0, \; x > 0, \tag{3.6}$$

where $\lim_{x\to\infty} V(x) = 0$, with the conditions

$$y(0) = 0, \tag{3.7}$$

$$y(x) \text{ is finite for any } x > 0. \tag{3.8}$$

The last condition is automatically satisfied except possibly when $x \to \infty$. We distinguish two regimes for E.

$E < 0$. One can always choose some x_{max} such that for any $x \geq x_{max}$ we have $|E| >> |V(x)|$. The approximation $E - V(x) \approx E$ is justified and therefore the equation becomes $y'' + Ey = 0$, with the general solution

$$y(x) = A \exp((-E)^{1/2}x) + B \exp(-(-E)^{1/2}x), \qquad (3.9)$$

where A and B are arbitrary constants. The first component increases with x, and thus the condition (3.8) can be satisfied only if $A = 0$. Equivalently, that condition can be written as

$$y(x) \sim \exp(-(-E)^{1/2}x) \text{ for } x \geq x_{max}. \qquad (3.10)$$

or as

$$y'(x_{max}) + (-E)^{1/2}y(x_{max}) = 0. \qquad (3.11)$$

Equation (3.6) and the two conditions (3.7) and (3.10) (or (3.11)) define an eigenvalue problem. (For a discussion of the link between this problem and the standard Sturm - Liouville problem see Section 4.5 in [1].) Solutions, if any, exist only for a discrete set of eigenvalues E_0, E_1, ... and let y_0, y_1, ... be the associated eigenfunctions.

$E > 0$. The general solution of the asymptotic equation $y'' + Ey = 0$ now reads

$$y(x) = A \exp(ikx) + B \exp(-ikx), \qquad (3.12)$$

where $k = E^{1/2}$, and this solution is bounded. The physical condition (3.8) is then satisfied for any positive E.

For a numerical approach the $x > 0$ half-axis is divided into three intervals, $I_1 := (0, x_{min}]$, $I_2 := (x_{min}, x_{max}]$ and $I_3 := (x_{max}, \infty)$, see [5]. x_{min} is placed close enough to the origin that on I_1 the singular term $l(l + 1)/x^2$ dominates in $V(x)$. The solution which satisfies (3.7) then behaves as x^{l+1}. On I_3 the asymptotic form of the equation is valid and, as seen, this admits an analytic solution.

Numerical algorithms are unavoidable for the solution on I_2 and here the ef procedure may play an important role in making the computation more efficient. To see this, let us introduce the following grid on I_2: $x_0 = x_{min} < x_1 < x_2 < ... < x_n = x_{max}$. On each elementary interval of this partition the function $V(x)$ exhibits a smooth variation. We take some arbitrary subinterval $[x_{k-1}, x_k]$, $(1 \leq k \leq n)$, approximate $V(x)$ by its average value $\bar{V}$ on this subinterval and replace $V(x)$ by this $\bar{V}$ in equation (3.6). We get a new equation,

$$\bar{y}'' + (E - \bar{V})\bar{y} = 0, \ x \in [x_{k-1}, x_k], \qquad (3.13)$$

which mimics the original one but it has the advantage of admitting analytic solution. If $E > \bar{V}$ the solution is

$$\bar{y}(x) = f_1 \sin(\omega x) + f_2 \cos(\omega x), \qquad (3.14)$$

where $\omega := (E - \bar{V})^{1/2}$, while if $E < \bar{V}$ the solution is

$$\bar{y}(x) = f_1 \sinh(\lambda x) + f_2 \cosh(\lambda x), \qquad (3.15)$$

where $\lambda := (\bar{V} - E)^{1/2}$. In both cases f_1 and f_2 are constants fixed in terms of the initial conditions at x_{k-1} or at x_k, according to the direction of propagation of the solution (forwards or backwards). Now, in so much as $\bar{V}$ is a good approximation of $V(x)$ on the quoted elementary interval (this is always the case if we care to take x_{k-1} and x_k conveniently close to each other), the same holds with $y(x)$ with respect to $\bar{y}(x)$. The structure of $y(x)$ is then the same as that of $\bar{y}(x)$ except that the constants must be replaced by some functions of x, whose variation, which reflects the small difference between $V(x)$ and $\bar{V}(x)$, will be smooth enough. The solution of equation (3.6) on that subinterval is then of the form

$$y(x) = f_1(x) \sin(\omega x) + f_2(x) \cos(\omega x) \text{ with } \omega := (E - \bar{V})^{1/2}, \qquad (3.16)$$

if $E > \bar{V}$, and of the form

$$y(x) = f_1(x) \sinh(\lambda x) + f_2(x) \cosh(\lambda x) \text{ with } \lambda := (\bar{V} - E)^{1/2}, \qquad (3.17)$$

if $E < \bar{V}$, where $f_i(x)$, $(i = 1, 2)$ are slowly varying functions.

This result, due to Ixaru and Rizea, [4], was used as the basis for a number of exponential fitting extensions of the algorithms for the numerical solution of the radial Schrödinger equation. Indeed, the quoted structure suggests that it is more advantageous to construct and use algorithms which are exact for as many as possible functions in the set

$$x^i \exp(\pm \mu x), \; i = 0, 1, \ldots \qquad (3.18)$$

with $\mu = i\omega$ or $\mu = \lambda$, than using the classical algorithms, constructed on the condition of being exact for functions in the power functions set

$$1, \; x, \; x^2, \; \ldots . \qquad (3.19)$$

Various applications of this type will be presented in Chapters 5 and 6.

The mentioned structure is of help for other operations as well. Some examples follow.

Quite often there is a need for the interpolation of the wavefunction between the partition points. The interpolation procedures which effectively use the mentioned piecewise structure will certainly be better suited than the classical, polynomial based ones.

There are also cases when the wavefunction $y(x)$ is used to evaluate physical effects associated with the so-called momentum dependent perturbations. This implies the numerical calculation of the first derivative of $y(x)$. The classical formulae, such as the central difference formula, are easily extended for such a purpose.

The numerical evaluation of integrals of the form

$$M_{ij} = \int_0^\infty y_i^*(x)\Delta V(x)y_j(x)dx \qquad (3.20)$$

is also needed in practice, where $y_i(x)$ and $y_j(x)$ are two eigenfunctions of (3.6) and $\Delta V(x)$ is some given function. In particular, if $i = j$ and $\Delta V(x) = 1$ this yields the normalization constant of $y_i(x)$. Again the mentioned piecewise structure of the wavefunctions provides a good basis for the construction of ad hoc, efficient quadrature formulae.

2. Functions $\eta_s(Z)$, $s = -1, 0, 1, 2, \ldots$

The derivation in terms of the standard functions sin and cos or sinh and cosh of the approximation formulae adapted for functions of the forms (3.16) or (3.17) is rather cumbersome. Also the final expressions may be long and difficult to manipulate. Much more convenient is to use some other functions, particularly adapted for this purpose. We have opted for the set $\eta_{-1}(Z)$, $\eta_0(Z)$, $\eta_1(Z)$, $\ldots$ which was originally introduced in Section 3.4 of [1]. In that book, these functions were denoted as $\overline{\xi}(Z)$, $\overline{\eta}_0(Z)$, $\overline{\eta}_1(Z)$, $\ldots$ and used for the construction of CP methods for the numerical solution of the Schrödinger equation. (CP is the short for constant-based perturbation.)

As originally defined, the η functions are real functions of the real variable Z. The functions $\eta_{-1}(Z)$ and $\eta_0(Z)$ are introduced by the formulae:

$$\eta_{-1}(Z) := \frac{1}{2}[\exp(Z^{1/2}) + \exp(-Z^{1/2})] = \begin{cases} \cos(|Z|^{1/2}) & \text{if } Z < 0 \\ \cosh(Z^{1/2}) & \text{if } Z \geq 0 \end{cases}$$
$$(3.21)$$

and

$$\eta_0(Z) := \begin{cases} \dfrac{1}{2Z^{1/2}}[\exp(Z^{1/2}) - \exp(-Z^{1/2})] & \text{if } Z \neq 0 \\ 1 & \text{if } Z = 0 \end{cases}$$

$$= \begin{cases} \dfrac{\sin(|Z|^{1/2})}{|Z|^{1/2}} & \text{if } Z < 0 \\ 1 & \text{if } Z = 0 \\ \dfrac{\sinh(Z^{1/2})}{Z^{1/2}} & \text{if } Z > 0 \end{cases} \qquad (3.22)$$

while $\eta_s(Z)$ with $s > 0$ are subsequently generated by recurrence

$$\eta_s(Z) := \frac{1}{Z}[\eta_{s-2}(Z) - (2s-1)\eta_{s-1}(Z)], \quad s = 1, 2, 3, \ldots \qquad (3.23)$$

if $Z \neq 0$, and by the following values at $Z = 0$:

$$\eta_s(0) := \frac{1}{1 \cdot 3 \cdot 5 \cdots (2s + 1)} = \frac{2^s s!}{(2s + 1)!}, \quad s = 1, 2, 3, \ldots \quad (3.24)$$

(The recurrence cannot be used for $Z = 0$ because of the presence of Z in the denominator.)

These functions satisfy the following properties:

(i) Power series:

$$\eta_s(Z) = 2^s \sum_{q=0}^{\infty} \frac{g_{sq}}{(2q + 2s + 1)!} Z^q \quad (3.25)$$

with

$$g_{sq} = \begin{cases} 1 & \text{if } s = 0 \\ (q + 1)(q + 2)\ldots(q + s) & \text{if } s > 0 \end{cases} \quad (3.26)$$

(ii) Behaviour at large $|Z|$:

$$\eta_s(Z) \simeq \begin{cases} \eta_{-1}(Z)/Z^{(s+1)/2} & \text{for odd } s \\ \eta_0(Z)/Z^{s/2} & \text{for even } s \end{cases} \quad (3.27)$$

(iii) Differentiation:

$$\eta_s'(Z) = \frac{1}{2}\eta_{s+1}(Z), \quad s = -1, 0, 1, 2, \ldots \quad (3.28)$$

(iv) Generating differential equation: $\eta_s(Z)$ ($s = 0, 1, \ldots$) is the suitably normalized regular solution of

$$Zw'' + \frac{1}{2}(2s + 3)w' - \frac{1}{4}w = 0. \quad (3.29)$$

(v) Relation with the spherical Bessel functions:

$$\eta_s(-x^2) = x^{-s}j_s(x), \quad s = 0, 1, 2, \ldots \quad (3.30)$$

The proof of any of the properties (i)–(iii) can be done by induction. Property (iii), for example, can be derived as follows. First, it is easy to check directly that this property holds for $s = -1$ and for $s = 0$. We take some arbitrary integer $m > 0$ and assume that (iii) also holds for all $s < m$. It remains to show that (iii) also holds for m.

First we consider the case $Z \neq 0$. The η functions are defined by recurrence and we have in order:

$$
\begin{aligned}
\eta'_m(Z) &= \frac{1}{Z}[\eta'_{m-2}(Z) - (2m-1)\eta'_{m-1}(Z)] \\
&\quad - \frac{1}{Z^2}[\eta_{m-2}(Z) - (2m-1)\eta_{m-1}(Z)] \\
&= \frac{1}{2Z}(\eta_{m-1}(Z) - (2m-1)\eta_m(Z)) - \frac{1}{Z}\eta_m(Z) \\
&= \frac{1}{2Z}(\eta_{m-1}(Z) - (2m+1)\eta_m(Z)) = \frac{1}{2}\eta_{m+1}(Z) .
\end{aligned}
$$

The first and the last members in this chain show that indeed property (iii) holds for m. To see that (iii) remains valid also for $Z = 0$ we assume that the series representation of $\eta_s(Z)$ is valid for all $s = 0, 1, \ldots, m$. Then

$$
2\eta'_m(Z) = 2^{m+1} \sum_{q=0}^{\infty} \frac{(q+1)g_{m\,q+1}}{(2q+2m+3)!} Z^q = 2^{m+1} \sum_{q=0}^{\infty} \frac{g_{m+1\,q}}{(2q+2m+3)!} Z^q .
$$

Since the derivative of a function (if it exists) is unique and we have just established that $\eta'_m(Z) = \eta_{m+1}(Z)/2$ for $Z \neq 0$ it results that the last member represents the series expansion of $\eta_{m+1}(Z)$ defined by recurrence for $Z \neq 0$. Moreover, the first term of this series coincides with the value imposed in the definition of $\eta_{m+1}(0)$. Altogether, property (iii) holds for all real Z. To check for properties (iv) and (v) use the series expansion.

REMARK 3.1 *In a series of papers the function $\eta_{-1}(Z)$ was denoted $\xi(Z)$ to underline its somewhat special position in the set; for example, series expansion (3.25) does not hold for this function. However, we give preferrence to the first notation although the second will be also used on some occasions.*

In this book the η functions will be used predominantly for negative values of Z because this is the range corresponding to oscillatory functions but the formulae expressed in terms of these functions can be used as well for positive Z, to describe functions with exponential behaviour. As a matter of fact, if the spherical Bessel functions would be used in the latter case, then the argument x would be imaginary.

Functions $\eta_s(Z)$ exhibit a definite hierarchy in terms of s. $\eta_{-1}(Z)$ is the biggest in amplitude and it is followed in order by $\eta_0(Z)$, $\eta_1(Z)$, $\cdots$. This behaviour may be of help when analysing sums over η-s with coefficients of comparable size: only the terms with the smallest indices are important from numerical point of view.

The things are seen on Figure 3.1, where $\eta_{-1}(Z)$, $\eta_0(Z)$ and the conveniently scaled $\eta_s(Z)$, $s = 1, 2$ (that is $3\eta_1(Z)$, and $15\eta_2(Z)$) are presented. On the upper graph the range is $-300 \leq Z \leq 50$ and, for $Z > 0$,

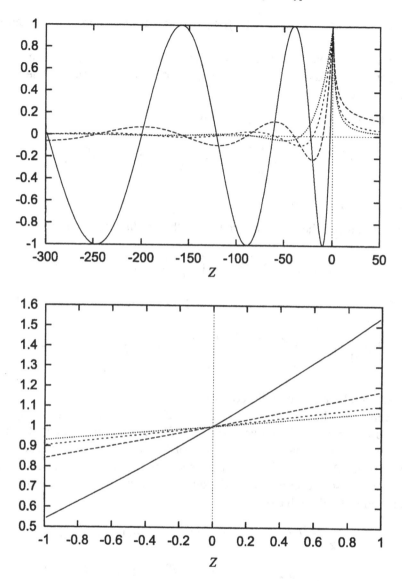

Figure 3.1. The Z dependence of the first four η functions for $-300 \leq Z \leq 50$ (up) and for $-1 \leq Z \leq 1$ (down). Up: for $Z \leq 0$: $\eta_{-1}(Z)$ (solid), $\eta_0(Z)$ (long dashed), $3\eta_1(Z)$ (short-dashed) and $15\eta_2(Z)$ (dots) ; for $Z > 0$: $\eta_0(Z)/\eta_{-1}(Z)$ (long-dashed), $3\eta_1(Z)/\eta_{-1}(Z)$ (short-dashed) and $15\eta_2(Z)/\eta_{-1}(Z)$ (dots). Down: $\eta_{-1}(Z)$ (solid), $\eta_0(Z)$ (long dashed), $3\eta_1(Z)$ (short-dashed) and $15\eta_2(Z)$ (dots) are represented for all Z.

where the functions increase exponentially, a different scaling is used. We plot $\eta_0(Z)/\eta_{-1}(Z)$, $3\eta_1(Z)/\eta_{-1}(Z)$ and $15\eta_0(Z)/\eta_{-1}(Z)$. The lower graph helps seeing better the behaviour of these functions around the origin. The range is

$-1 \leq Z \leq 1$ and $\eta_{-1}(Z)$, $\eta_0(Z)$, $3\eta_1(Z)$ and $15\eta_2(Z)$ are represented for all Z.

The expressions of the derivatives with respect to Z of products like $Z^n \eta_s(aZ)$ (integer $n > 0$ and real a) are required quite often in applications. The following result is of use:

LEMMA 3.1 *For $s = -1, 0, 1, \ldots$ we have*

$$[Z^n \eta_s(aZ)]^{(m)} = \frac{1}{2^m} \sum_{j=0}^{J} 2^j \binom{m}{j} \frac{n!}{(n-j)!} Z^{n-j} a^{m-j} \eta_{m-j+s}(aZ),$$

$$(3.31)$$

where $J = \min\{n, m\}$.

Proof On using the Leibniz formula for the product $f(Z) \cdot g(Z)$,

$$[f(Z) \cdot g(Z)]^{(m)} = \sum_{j=0}^{m} \binom{m}{j} f^{(j)}(Z) \cdot g^{(m-j)}(Z),$$

for $f(Z) = Z^n$ and $g(Z) = \eta_s(aZ)$ and the relations

$$[Z^n]^{(j)} = \begin{cases} \dfrac{n!}{(n-j)!} Z^{n-j} & \text{for } j \leq n \\ 0 & \text{for } j > n \end{cases} \quad \text{and} \quad [\eta_s(aZ)]^{(i)} = \frac{1}{2^i} a^i \eta_{i+s}(aZ),$$

$$(3.32)$$

the stated relation results directly. Q. E. D.

Expressions like $(\eta_{-1}(Z) - 1)/Z$, $(\eta_0(Z) - 1)/Z$ or $(\eta_0(4Z) - \eta_0(Z))/Z$ also appear in some formulae. For small Z these can be evaluated accurately by series but the separate construction of the series expansion requires some extra work. An alternative way consists in re-expressing them in terms of η functions of scaled arguments. Specifically, on using the definition relations (3.21) and (3.22) we have

$$\frac{1}{2} Z\eta_0^2(Z/4) = \frac{1}{2}[\exp(Z^{1/2}/2) - \exp(-Z^{1/2}/2)]^2 \qquad (3.33)$$

$$= \frac{1}{2}[\exp(Z^{1/2}) + \exp(-Z^{1/2})] - 1 = \eta_{-1}(Z) - 1,$$

while by using the definition of $\eta_1(Z)$ we have

$$\eta_0(Z) - 1 = \eta_{-1}(Z) - Z\eta_1(Z) - 1. \qquad (3.34)$$

On this basis it results that

$$\frac{\eta_{-1}(Z) - 1}{Z} = \frac{1}{2}\eta_0^2(Z/4), \quad \frac{\eta_0(Z) - 1}{Z} = \frac{1}{2}[\eta_0^2(Z/4) - 2\eta_1(Z)], \quad (3.35)$$

$$\frac{\eta_0(4Z) - \eta_0(Z)}{Z} = \frac{1}{2}[4(\eta_0^2(Z) - 2\eta_1(4Z)) - \eta_0^2(Z/4) + 2\eta_1(Z)].$$

2.1   **Subroutines** GEBASE **and** GEBASEV

These subroutines compute blocks of η functions for given argument Z.

Subroutine GEBASE computes the set $\eta_s(Z)$ with $-1 \leq s \leq 6$. Forwards or backwards recurrence is used in terms of $|Z|$. If $|Z| < 50$ the subroutine first computes $\eta_6(Z)$ and $\eta_5(Z)$. The truncated series representation is used. Seven terms are taken if $|Z| < 0.9$, nine if $0.9 \leq |Z| < 3.5$, fourteen if $3.5 \leq |Z| < 22$, and ninteen if $22 \leq |Z| < 50$. The values of the other η functions are computed by backwards recurrence, viz.

$$\eta_s(Z) = Z\eta_{s+2}(Z) + (2s+3)\eta_{s+1}(Z), \ s = \ 4, \ 3, ..., \ 0, \ -1$$

If $|Z| \geq 50$ $\eta_{-1}(Z)$ and $\eta_0(Z)$ are computed first and the forwards recurrence (3.23) is used for the other five functions.

Calling it is:

CALL GEBASE(Z, CSI, ETA0, ETA1, ETA2, ETA3, ETA4, ETA5, ETA6)

where Z is the input value for Z and CSI, ETA0, ETA1, ETA2, ETA3, ETA4, ETA5 and ETA6 are the output values for $\eta_s(Z), s = -1, 0, 1, \ldots, 6$. All parameters are in double precision.

Subroutine GEBASEV computes the set $\eta_s(Z)$ for $-1 \leq s \leq n_{max}$ where $n_{max} \geq 2$ is given by the user. Integer $i_{cut} := \min\{\lfloor(|Z| - 6)/16\rfloor, n_{max}\}$ is used to separate two ways of computation ($\lfloor u \rfloor$ is the biggest integer smaller than or equal to u). If $i_{cut} = n_{max}$ the forwards recurrence is used for the whole set, while if $i_{cut} < n_{max}$ the backwards recurrence is used. In the latter case the integer $i_{max} := \lfloor 3n_{max}/2 \rfloor$ is introduced to compute the functions $\eta_{i_{max}}(Z)$ and $\eta_{i_{max}-1}(Z)$ by power series with thirty terms. These are subsequently used to compute the functions $\eta_s(Z)$ with $s = i_{max} - 2, i_{max} - 3, \ldots, \ldots, 0, -1$ by the backwards recurrence, but only the values with $s \leq n_{max}$ are actually stored for the output.

Calling this subroutine is

CALL   GEBASEV(Z,ETA,NMAX)

Input parameters: Z (double precision) and integer NMAX for n_{max} (remember that $n_{max} \geq 2$).

Output parameter: vector ETA (double precision, to be declared as DIMENSION ETA(-1:NN) where NN $\geq$ NMAX). For any integer s between -1 and n_{max} element ETA(s) contains the value of $\eta_s(Z)$.

The accuracy is of fourteen significant figures in each computed $\eta_s(Z)$ except around the zeros; zeros appear only for negative Z. The tested range was

$-300 \leq Z \leq 50$ in both subroutines and $n_{max} \leq 50$ in GEBASEV. Extended precision versions, in which the number of terms in the power series was also increased, have been used for comparison.

A short sample program follows in which GEBASE and GEBASEV with NMAX= 50 are activated for Z= -10, and $\eta_4(Z)$ is printed:

```
implicit double precision (a-h,o-z)
dimension eta(-1:50)
z=-10.d0
call gebase(z,csi,eta0,eta1,eta2,eta3,eta4,eta5,eta6)
nmax=50
call gebasev(z,eta,nmax)
write(*,*)' gebase  : ',eta4
write(*,*)' gebasev : ',eta(4)
stop
end
```

The output is

```
gebase   :   6.601403571786596E-04
gebasev :   6.601403571786592E-04
```

2.2   Subroutines CGEBASE and CGEBASEV

The set $\eta_s(Z)$, $s = -1, 0, 1, \ldots$, can be also defined for complex Z, see [3]. To this aim the functions $\eta_{-1}(Z)$ and $\eta_0(Z)$ are defined via functions sin and cos of a complex argument,

$$\eta_{-1}(Z) := \cos(iZ^{1/2}), \tag{3.36}$$

and

$$\eta_0(Z) := \begin{cases} \dfrac{\sin(iZ^{1/2})}{iZ^{1/2}} & \text{if } Z \neq 0 \\ 1 & \text{if } Z = 0 \end{cases} \tag{3.37}$$

while the equations (3.23) and (3.24) are used to define $\eta_s(Z)$ with $s \geq 1$. The properties (i)–(iv) remain unchanged.

Subroutines CGEBASE and CGEBASEV compute these functions. They are double complex extensions of GEBASE and GEBASEV. Calling CGEBASE is

```
CALL    CGEBASE(Z,CSI,ETA0,ETA1,ETA2,ETA3,ETA4,ETA5,ETA6)
```

where the input Z and the output CSI, ETA0, ..., ETA6 are double complex arguments.
Calling CGEBASEV is

```
CALL    CGEBASEV(Z,ETA,NMAX)
```

where Z and the vector ETA are double complex arguments, as well.
A potential application of these subroutines is in the numerical evaluation of
the coefficients in ef formulae for integrals like (3.20) in which $y_i(x)$ is of form
(3.16) while $y_j(x)$ is of form (3.17).

The following driving program computes the first seven η-s (CGEBASE) and
$\eta_s(s = -1, 0, 1, \ldots, 50)$ (CGEBASEV) for four Z-s: $Z = 1 \pm i, -1 \pm i$ and
prints the results for $\eta_4(Z)$.

```
implicit double complex(a-h,o-q,s-z)
implicit double precision(r)
dimension eta(-1:50),z(4)
data z/ (1.d0,1.d0),(1.d0,-1.d0),(-1.d0,1.d0),(-1.d0,-1.d0)/
nmax=50
do i=1,4
  write(*,*)' z = ',z(i)
  call cgebase(z(i),csi,eta0,eta1,eta2,eta3,eta4,eta5,eta6)
  call cgebasev(z(i),eta,nmax)
  write(*,*)' cgebase :',eta4
  write(*,*)' cgebasev:',eta(4)
enddo
stop
end
```

The output is:

```
z =   (1.000000000000000,1.000000000000000)
cgebase : (1.106280246846162E-03,4.997060392250612E-05)
cgebasev: (1.106280246846160E-03,4.997060392250606E-05)
z =   (1.000000000000000,-1.000000000000000)
cgebase : (1.106280246846162E-03,-4.997060392250612E-05)
cgebasev: (1.106280246846160E-03,-4.997060392250606E-05)
z =   (-1.000000000000000,1.000000000000000)
cgebase : (1.010121264980187E-03,4.627060024775621E-05)
cgebasev: (1.010121264980187E-03,4.627060024775619E-05)
z =   (-1.000000000000000,-1.000000000000000)
cgebase : (1.010121264980187E-03,-4.627060024775621E-05)
cgebasev: (1.010121264980187E-03,-4.627060024775619E-05)
```

3.   A six-step flow chart

An important special feature when deriving approximation formulae suited for operations on functions of the forms

$$y(x) = f_1(x)\sin(\omega x) + f_2(x)\cos(\omega x) \tag{3.38}$$

or

$$y(x) = f_1(x)\sinh(\lambda x) + f_2(x)\cosh(\lambda x), \tag{3.39}$$

(with smooth $f_1(x)$ and $f_2(x)$ and given ω or λ) is that a pair of correlated exponential fitting frequencies is involved. In fact, since $\sin(\omega x)$ and $\cos(\omega x)$ are linear combinations of $\exp(\pm i\omega x)$, while $\sinh(\lambda x)$ and $\cosh(\lambda x)$ are linear combinations of $\exp(\pm\lambda x)$, the natural set to be considered in an ef scheme consists of pairs of functions, viz.:

$$x^p \exp(\pm\mu x), \; p = 0, 1, \ldots, \tag{3.40}$$

where $\mu = i\omega$ or $\mu = \lambda$.

The application of the ef approach in the way presented in Chapter 2 remains possible, of course, but it will force us to work with complex functions if $\mu = i\omega$. Also, some extra effort will be needed to reconvert the expressions of the coefficients in terms of the standard trigonometric or hyperbolic functions, with different final formulae in the two cases.

The following adaptation of the ef procedure has the advantage of treating the two cases on equal footing.

Let $\mathcal{L}[h, \mathbf{a}]$ be the operator associated to the approximation formula whose coefficients are searched for, and let us assume that $\mathcal{L}[h, \mathbf{a}]y(x)$ is vanishing identically in x and in h for the pairs in (3.40) with $p = 0, 1, \ldots, P$. The corresponding RDE is of order $M = 2(P + 1)$. It reads

$$(D^2 - \mu^2)^{P+1}y = 0. \tag{3.41}$$

We see that μ^2 appears here, and this is real irrespective of whether μ alone is real or imaginary. It is suggested that the ef predure will become simpler and more compact if the $\mathbf{z}$ dependence ($\mathbf{z} := [z_1 = z, z_2 = -z]$ where $z := \mu h$) of quantities like $\mathbf{a}(\mathbf{z})$ or $T(\mathbf{z})$ is rewritten in terms of a single real variable, that is $Z := z^2 = \mu^2 h^2$. The expressions for $\mathbf{a}(Z)$ and $T(Z)$ obtained in this way are real functions of a real variable and they cover both cases, that is the oscillatory form (3.38) if $Z < 0$, and the form (3.39) if $Z > 0$. The classical case will be re-obtained for $Z = 0$.

Following Ixaru, [2], we introduce the functions

$$G^+(Z, \mathbf{a}) := \frac{1}{2}[E_0^*(z, \mathbf{a}) + E_0^*(-z, \mathbf{a})], \tag{3.42}$$

and

$$G^-(Z, \mathbf{a}) := \frac{1}{2z}[E_0^*(z, \mathbf{a}) - E_0^*(-z, \mathbf{a})], \qquad (3.43)$$

and let $G^{\pm(p)}(Z, \mathbf{a})$ be the p-th derivative of $G^{\pm}(Z, \mathbf{a})$ with respect to Z. (Of course, $G^{\pm(0)}(Z, \mathbf{a})$ means simply $G^{\pm}(Z, \mathbf{a})$.)

In the frame of the general ef procedure, the coefficients to be determined for the discussed problem have to satisfy the system

$$E_p^*(\pm z, \mathbf{a}) = 0, \quad p = 0, 1, \ldots, P, \qquad (3.44)$$

see (2.82).

In the new version this is replaced by

$$G^{\pm(p)}(Z, \mathbf{a}) = 0, \quad p = 0, 1, \ldots, P. \qquad (3.45)$$

Indeed,

LEMMA 3.2 *The systems (3.44) and (3.45) are equivalent.*

Proof This is obvious for $P = 0$, see the expressions of $G^{\pm}(Z, \mathbf{a})$, while to check it for $P > 0$ it is sufficient to show that the expressions of $G^{\pm(p)}(Z, \mathbf{a})$, $(p = 0, 1, \ldots, P)$ are linear in $E_m^*(\pm z, \mathbf{a})$, $m = 0, 1, \ldots, p$. This is true because $\dfrac{\partial}{\partial Z} = \dfrac{1}{2z}\dfrac{\partial}{\partial z}$ and also $\dfrac{\partial}{\partial z}E_m^*(\pm z, \mathbf{a}) = \pm E_{m+1}^*(\pm z, \mathbf{a})$. Q.E.D.

Just for illustration,

$$G^{+(1)}(Z, \mathbf{a}) = \frac{1}{4z}\frac{\partial}{\partial z}[E_0^*(z, \mathbf{a}) + E_0^*(-z, \mathbf{a})] = \frac{1}{4z}[E_1^*(z, \mathbf{a}) - E_1^*(-z, \mathbf{a})]$$

and

$$G^{-(1)}(Z, \mathbf{a}) = \frac{1}{4z^3}[-E_0^*(z, \mathbf{a}) + E_0^*(-z, \mathbf{a}) + z(E_1^*(z, \mathbf{a}) + E_1^*(-z, \mathbf{a}))].$$

The following result helps to establish the behaviour of $G^{\pm(p)}(Z, \mathbf{a})$ when $Z \to 0$:

LEMMA 3.3 *Functions $G^{\pm}(Z, \mathbf{a})$ have the following expansion in powers of Z:*

$$G^+(Z, \mathbf{a}) = \sum_{k=0}^{\infty} \frac{1}{(2k)!}L_{2k}^*(\mathbf{a})Z^k, \quad G^-(Z, \mathbf{a}) = \sum_{k=0}^{\infty} \frac{1}{(2k+1)!}L_{2k+1}^*(\mathbf{a})Z^k,$$

$$(3.46)$$

Proof We evaluate $\mathcal{L}[h, \mathbf{a}]\exp(\pm\mu x)$. Equation (2.73) gives directly

$$\mathcal{L}[h, \mathbf{a}]\exp(\pm\mu x) = h^l \exp(\pm\mu x)\sum_{m=0}^{\infty} \frac{1}{m!}L_m^*(\mathbf{a})(\pm z)^m.$$

On using (2.61) and (2.69) we have in order:

$$E_0^*(\pm z, \mathbf{a}) = h^{-l} \exp(\mp \mu x) \mathcal{L}[h, \mathbf{a}] \exp(\pm \mu x) = \sum_{m=0}^{\infty} \frac{1}{m!} L_m^*(\mathbf{a})(\pm z)^m.$$

This is introduced in $G^+(Z, \mathbf{a})$ and in $G^-(Z, \mathbf{a})$ to obtain

$$G^+(Z, \mathbf{a}) = \frac{1}{2} \sum_{m=0}^{\infty} \frac{1}{m!} L_m^*(\mathbf{a})[z^m + (-z)^m] = \sum_{k=0}^{\infty} \frac{1}{(2k)!} L_{2k}^*(\mathbf{a}) Z^k \quad (3.47)$$

and

$$G^-(Z, \mathbf{a}) = \frac{1}{2z} \sum_{m=0}^{\infty} \frac{1}{m!} L_m^*(\mathbf{a})[z^m - (-z)^m] = \sum_{k=0}^{\infty} \frac{1}{(2k+1)!} L_{2k+1}^*(\mathbf{a}) Z^k.$$

$$(3.48)$$

Q. E. D.

Equation (3.46) shows that $G^{\pm}(Z, \mathbf{a})$ and their derivatives have finite limits when $Z \to 0$, viz.:

$$G^{+(p)}(Z = 0, \mathbf{a}) = \frac{p!}{(2p)!} L_{2p}^*(\mathbf{a}), \ G^{-(p)}(Z = 0, \mathbf{a}) = \frac{p!}{(2p+1)!} L_{2p+1}^*(\mathbf{a}).$$

$$(3.49)$$

$G^{+(p)}(0, \mathbf{a})$ and $G^{-(p)}(0, \mathbf{a})$ are then proportional to the starred classical moments with even and odd indices, respectively. A direct practical consequence of this property is that $G^{\pm}(0, \mathbf{a})$ and their derivatives can be used directly as an alternative formulation of the conditions to be satisfied by the classical coefficients. With the classical power function set in mind,

$$1, \ x, \ x^2, \ \dots \quad (3.50)$$

the set of M conditions (2.79) based on L^* moments can be replaced by the G-based conditions

$$G^{\pm(p)}(0, \mathbf{a}) = 0, \ p = 0, 1, \dots, P = M/2 - 1, \quad (3.51)$$

if M is even, and by

$$G^{\pm(p)}(0, \mathbf{a}) = 0, \ p = 0, 1, \dots, P = (M - 1)/2 - 1, \ G^{+(P+1)}(0, \mathbf{a}) = 0,$$

$$(3.52)$$

if M is odd.

The following six-step procedure, taken from [2], has to be used to get tuned formulae when the functions are of the forms (3.38) or (3.39), or of such forms plus a smoothly varying function:

Step *i*.

Choose the appropriate form of $\mathcal{L}[h, \mathbf{a}]$ and find the expressions of its classical moments

$$L_m(h, \mathbf{a}), \ m = 0, 1, 2, \dots.$$

Each $L_m(h, \mathbf{a})$ is of the form (2.72).

Step *ii*.

Examine the algebraic system

$$L_m(h, \mathbf{a}) = 0 \quad (\text{or } L_m^*(\mathbf{a}) = 0), \ m = 0, 1, 2, \dots, M - 1 \quad (3.53)$$

to find out the maximal M for which it is compatible.

Step *iii*.

Denote $z := \mu h$, construct the formal expression of $E_0^*(z, \mathbf{a})$ and, on this basis, write the expressions of $G^{\pm}(Z, \mathbf{a})$ where $Z := z^2$. Also write the expressions of their derivatives $G^{\pm(p)}(Z, \mathbf{a})$, $p = 1, 2, \dots$ with respect to Z.

Hint: express $G^{\pm}(Z, \mathbf{a})$ in terms of the functions $\eta_s(Z)$, $s = -1, 0, 1, \dots$. One of the advantages will be a simpler evaluation of the derivatives.

Step *iv*.

Choose the reference set of M functions which is appropriate for the given form of $y(x)$. This is in general a hybrid set:

$$y = 1, x, x^2, \dots, x^K,$$
$$\exp(\pm \mu x), x \exp(\pm \mu x), x^2 \exp(\pm \mu x), \dots, x^P \exp(\pm \mu x), \quad (3.54)$$

with

$$K + 2P = M - 3. \quad (3.55)$$

The reference set is thus characterized by two integer parameters, K and P. The set in which there is no classical component is identified by $K = -1$ while the set in which there is no exponential fitting component is identified by $P = -1$. Parameter P will be called the level of tuning and the formula which is the best tuned on functions of the form (3.38) or (3.39) will be the one with $P = [M/2] - 1$, because this is the biggest value consistent with the positive M resulted from step *ii*.

Note 1: the reference set should contain *all* successive functions inbetween. Lacunary sets are not allowed. For example, the set $y = 1, x, x^2, x^4$, $x \exp(\pm \mu x)$ is illegitimate because x^3 is absent from the classical part and also because $\exp(\pm \mu x)$ is absent from the exponential fitting part.

Note 2: The relation $K + 2P = M - 3$, called selfconsistency condition, implies that M and K are of different parities: if M is even/odd then K is odd/even. Now, M is already fixed, while K must be chosen in terms of the form of the function. When the function is purely of the form (3.38), one should normally take $K = -1$ but if M is even we are forced to take $K = 0$. Expressed in other words, K can be always used also as an adjustment parameter for preventing the violation of the selfconsistency condition.

Note 3: This flow chart closely follows in spirit the general scheme introduced in Section 4.3 of Chapter 2 and therefore it will preserve the limitations mentioned for the latter. In essence this means that accidental situations do exist when the selfconsistency condition (3.55) is violated. A situation of this type and a procedure adapted for the treatment of such a special case will be presented in Section 2.5 of Chapter 4.

Step v.

Solve the algebraic system

$$L_k^*(\mathbf{a}) = 0, \ 0 \le k \le K, \ G^{\pm(p)}(Z, \mathbf{a}) = 0, \ 0 \le p \le P \qquad (3.56)$$

for the Z dependent coefficients and let $\mathbf{a}(Z) = [a_0(Z), a_1(Z), \ldots]$ be its solution. (The conditions which involve L_k^* may be well replaced by the $G^{\pm}$-based conditions at $Z = 0$.)

Step vi.

The leading term of the error of the formula obtained in this way reads

$$lte_{ef} = (-1)^{P+1} h^{l+M} \frac{L_{K+1}^*(\mathbf{a}(Z))}{(K+1)! Z^{P+1}} D^{K+1}(D^2 - \mu^2)^{P+1} y(X),$$

$$(3.57)$$

where $\mu = i\,\omega$ or $\mu = \lambda$, according to the case. For the derivation of this formula use (2.90) and (2.96).

The scheme can be easily extended to cover the case when different μ-s are present. The set (3.54) now becomes

$$
\begin{aligned}
y \ = \ & 1, \, x, \, x^2, \ldots, \, x^K, \\
& \exp(\pm\mu_i x), \, x\exp(\pm\mu_i x), \, \ldots, \, x^{P_i}\exp(\pm\mu_i x) \quad (3.58) \\
& (i = 1, \, 2, \, \ldots, \, I),
\end{aligned}
$$

where the selfconsistency condition

$$K + 2(P_1 + P_2 + \cdots + P_I) = M - 2I - 1 \qquad (3.59)$$

has to be satisfied. The algebraic system to be solved then is

$$L_k^*(\mathbf{a}) = 0, \ 0 \le k \le K, \ G^{\pm(p)}(Z_i, \mathbf{a}) = 0, \ 0 \le p \le P_i, \, i = 1, 2, \ldots, I,$$

$$(3.60)$$

where $Z_i := \mu_i^2 h^2$, and the leading term of the error reads

$$lte_{ef} = (-1)^{P^*+I} h^{l+M} \frac{L_{K+1}^*(\bar{\mathbf{a}}(\mathbf{Z}))}{(K+1)! Z_1^{P_1+1} \cdots Z_I^{P_I+1}} D^{K+1} O_1 O_2 \cdots O_I y(X),$$

$$(3.61)$$

where

$$P^* := P_1 + \cdots + P_I, \ O_i := (D^2 - \mu_i^2)^{P_i+1}, \ \mathbf{Z} := [Z_1, Z_2, \ldots, Z_I].$$

3.1 Subroutine REGSOLV2

The fact that the starred classical moments can be expressed by the values at $Z = 0$ of $G^{\pm(p)}(Z, \mathbf{a})$ and of their derivatives suggests that the system (3.60) consists of two blocks of equations, one with G^{+} and its derivatives, and another with G^{-} and its derivatives. In general the $\mathbf{a}$ dependence may be either linear or nonlinear but in many cases to be investigated later on the two-block system is linear.

It then makes sense to adapt the regularization procedure presented in Section 2.2 for a single block system. The problem to be solved is

$$\mathbf{A}\,\mathbf{x} = \mathbf{b}, \qquad (3.62)$$

in which $\mathbf{A}$ and $\mathbf{b}$ are an n by n matrix and a column vector with n components, respectively, whose elements are generated in terms of two sets of $n + 1$ given functions, viz. $f_m^{ib}(u)$, $m = 1, 2, \ldots, n$ and $g^{ib}(u)$, $(i_b = 1, 2)$ in the following way: $A_{km} := f_m^1(u_k)$ and $b_k := g^1(u_k)$, if $1 \le k \le k_{line}$, and $A_{km} := f_m^2(u_k)$ and $b_k := g^2(u_k)$, if $k_{line} + 1 \le k \le n$. The system then is

$$f_1^1(u_k)x_1 + f_2^1(u_k)x_2 + \cdots + f_n^1(u_k)x_n = g^1(u_k) \text{ for } 1 \le k \le k_{line} \quad (3.63)$$
$$f_1^2(u_k)x_1 + f_2^2(u_k)x_2 + \cdots + f_n^2(u_k)x_n = g^2(u_k) \text{ for } k_{line} + 1 \le k \le n.$$

The regularization procedure is applied on each block separately and in this way the potential difficulties which may appear when the u_k-s are packed together in either of the two blocks are eliminated. For example, if $u_1 = u_2 = \ldots = u_{\bar{k}}$ and $u_{\bar{k}+1} = \ldots = u_{k_{line}}$ the first block takes the regularized form

$$f_1^{1(k-1)}(u_1)x_1 + f_2^{1(k-1)}(u_1)x_2 + \cdots + f_n^{1(k-1)}(u_1)x_n = g^{1(k-1)}(u_1)$$
$$\text{for } 1 \le k \le \bar{k}, \qquad (3.64)$$
$$f_1^{1(k-\bar{k}-1)}(u_{\bar{k}+1})x_1 + f_2^{1(k-\bar{k}-1)}(u_{\bar{k}+1})x_2 + \cdots + f_n^{1(k-\bar{k}-1)}(u_{\bar{k}+1})x_n$$
$$= g^{1(k-\bar{k}-1)}(u_{\bar{k}+1}) \text{ for } \bar{k} + 1 \le k \le k_{line}.$$

The subroutine REGSOLV2 performs the two-block regularization of the system (3.63) and then solves it numerically. Calling this subroutine is

$$\text{CALL REGSOLV2 (N, KLINE, INFF, INFG, U, X)}$$

The arguments are:

- N: the number of equations in the system (integer, input).

- KLINE: the number of equations in the first block (integer, input).

- U : double precision vector (input) which contains the values of $u_1, u_2, \ldots, u_n$.

- X : double precision vector (output). It returns the solution vector. In the calling program the dimensions of U and of X should be declared as N, at least.

- INFF and INFG : external subroutines to be written by the user. They must furnish the values of $f_m^1(u)$, $f_m^2(u)$ $(1 \leq m \leq n)$ and of $g^1(u)$, $g^2(u)$, respectively, and also of the first fifty derivatives with respect to u of these functions.

  INFF is

  SUBROUTINE INFF (IBLOCK, M, U, VF)

  For each input pair of integers IBLOCK (this is for i_b and has the values 1 or 2), M (this is the column index m), and the double precision U, the subroutine produces the double precision vector VF dimensioned as VF(0:50), where VF(I) contains the value of the i-th order derivative of $f_m^{i_b}(u)$.

  INFG is

  SUBROUTINE INFG (IBLOCK, U, VG)

  The input IBLOCK and U are as described before. The output is the double precision vector VG with the same dimension as VF; VG(I) contains the i-th order derivative of $g^{i_b}(u)$.

For test we consider the following system of three equations

$$\eta_1(Z_i)x_1 + x_2 + \eta_1(Z_i)x_3 = Z_i, \quad i = 1, 2$$
$$\eta_0(Z_3)x_1 - \eta_0(Z_3)x_3 = 0, \tag{3.65}$$

which is of the form (3.63) with $k_{line} = 2$, $f_1^1(Z) = f_3^1(Z) = \eta_{-1}(Z)$, $f_2^1(Z) = 1$, $g^1(Z) = Z$, and $f_1^2(Z) = -f_3^2(Z) = \eta_0(Z)$, $f_2^2(Z) = g^2(Z) = 0$. For the expressions of their derivatives use (3.28).
The solution is

$$x_1 = x_3 = \frac{Z_2 - Z_1}{2(\eta_{-1}(Z_2) - \eta_{-1}(Z_1))}, \quad x_2 = \frac{Z_1\eta_{-1}(Z_2) - Z_2\eta_{-1}(Z_1)}{\eta_{-1}(Z_2) - \eta_{-1}(Z_1)}. \tag{3.66}$$

It is independent of Z_3. We take three particular cases:

1. $Z_1 = 0$, $Z_2 = Z$: $x_1 = x_3 = \dfrac{Z}{2(\eta_{-1}(Z) - 1)} = \dfrac{1}{\eta_0(Z/4)}$, $x_2 = -2x_1$,

where equation (3.35) was also used.

2. $Z_1 = Z_2 = Z$: $x_1 = x_3 = \dfrac{1}{\eta_0(Z)}$, $x_2 = \dfrac{Z\eta_0(Z) - 2\eta_{-1}(Z)}{\eta_0(Z)}$,

(to derive this use the l'Hospital rule in (3.66)).

3. $Z_1 = Z_2 = 0$: $x_1 = x_3 = 1$, $x_2 = -2$.

The following driving program solves the system by this subroutine and compares the computed solution with the exact one, for the three mentioned cases. We take $Z = -1$ and the default value $Z_3 = Z$ in all cases.

```
implicit double precision (a-h,o-z)
dimension z(3),x(3),xref(3)
external inff,infg
kline=2
n=3
zref=-1.d0
z(1)=zref
z(2)=zref
z(3)=zref
write(*,*)'   k ',' 	   x_1 ','   x_2 ','   x_3 ',
!'      err_1 ','      err_2 ','      err_3 '
write(*,*)
do k=1,3
 if(k.eq.1) then
call gebase(zref,csi,eta0,eta1,eta2,eta3,eta4,eta5,eta6)
  xref(1)=1.d0/eta0
  xref(2)=(zref*eta0-2.d0*csi)/eta0
 endif
 if(k.eq.2)then
  z(1)=0.d0
  call gebase(zref/4.d0,cs4,e04,e14,e24,e34,e44,e54,e64)
  xref(1)=1.d0/(e04*e04)
  xref(2)=-2.d0*xref(1)
 endif
 if(k.eq.3) then
  z(2)=0.d0
  xref(1)=1.d0
  xref(2)=-2.d0
 endif
 xref(3)=xref(1)
 call regsolv2(n,kline,inff,infg,z,x)
 write(*,50)k,(x(i),i=1,3),((xref(i)-x(i)),i=1,3)
 enddo
50  format(2x,i3,2x,3(1x,f7.4),3(2x,e10.3))
 stop
 end
```

The associated subroutines INFF and INFG are:

```
subroutine inff(iblock,m,z,vf)
```

```
      implicit double precision(a-h,o-z)
      dimension vf(0:50),eta(-1:50)
      if(m.eq.1.or.m.eq.3) then
       call gebasev(z,eta,50)
       do i=0,50
        if(iblock.eq.1)then
         vf(i)=eta(i-1)/2.d0**i
        else
         vf(i)=eta(i)/2.d0**i
         if(m.eq.1)vf(i)=-vf(i)
        endif
       enddo
      else
       do i=0,50
        vf(i)=0.d0
       enddo
       if(iblock.eq.1)vf(0)=1.d0
      endif
      return
      end

      subroutine infg(iblock,z,vg)
      implicit double precision(a-h,o-z)
      dimension vg(0:50)
      do i=0,50
       vg(i)=0.d0
      enddo
      if(iblock.eq.1)then
       vg(0)=z
       vg(1)=1.d0
      endif
      return
      end
```

The output is:

k	x_1	x_2	x_3	err_1	err_2	err_3
1	1.1884	-2.2842	1.1884	-0.178E-14	-0.444E-15	-0.178E-14
2	1.0877	-2.1753	1.0877	-0.666E-15	-0.888E-15	-0.666E-15
3	1.0000	-2.0000	1.0000	-0.111E-14	0.000E+00	-0.111E-14

References

[1] Ixaru, L. Gr. (1984). *Numerical Methods for Differential Equations and Applications*. Reidel, Dordrecht - Boston - Lancaster.

[2] Ixaru, L. Gr. (1997). Operations on oscillatory functions. *Comput. Phys. Commun.*, 105: 1–19.

[3] Ixaru, L. Gr. (2002). LILIX – A program for the solution of the coupled channel Schrödinger equation. *Comput. Phys. Commun.*, 147: 834–852.

[4] Ixaru, L. Gr. and Rizea, M. (1980). A Numerov - like scheme for the numerical solution of the Schroedinger equation in the deep continuum spectrum of energies. *Comput. Phys. Commun.*, 19: 23 - 27.

[5] Ixaru, L. Gr., Rizea, M. and Vertse, T. (1995). Piecewise perturbation methods for calculating eigensolutions of a complex optical potential. *Comput. Phys. Commun.*, 85: 217 - 230.

[6] Messiah, A. (1999). *Quantum Mechanics*. Dover Publications, New York.

[7] Pryce, J. D. (1993). *Numerical Solution of Sturm-Liouville Problems*. Oxford University Press, Oxford.

Chapter 4

NUMERICAL DIFFERENTIATION, QUADRATURE AND INTERPOLATION

A series of ef formulae tuned on functions of the form (3.38) or (3.39) are derived here by the procedure described in the previous chapter. We construct the ef coefficients for approximations of the first and the second derivative of $y(x)$, for a set of quadrature rules, and for some simple interpolation formulae.

1. Numerical differentiation

Techniques for the derivation of classical formulae for the numerical differentiation can be found in several books of numerical analysis, for example in Fröberg [17]. The exponential fitting procedure allows extending such formulae to become efficient for a larger class of functions. Some examples follow.

1.1 Three-point formulae for the first derivative

We consider the approximation of the first derivative of $y(x)$ at x_1 by two versions of the three-point formula.

Central difference formula

This is

$$y'(x_1) \approx \frac{1}{h}[a_0 y(x_0) + a_1 y(x_1) + a_2 y(x_2)], \quad x_i = x_1 + (i-1)h, \quad (4.1)$$

and we want to determine the values of the three coefficients a_0, a_1 and a_2 which ensure that this approximation is best tuned on functions of forms (3.38) or (3.39). The same approximation was considered in Section 3 of Chapter 2 but with functions of the form (2.45) in mind.

77

We follow the six-step flow chart presented in Section 3 of Chapter 3.

Step *i*. We introduce the operator $\mathcal{L}$ by

$$\mathcal{L}[h, \mathbf{a}]y(x) := y'(x) - \frac{1}{h}[a_0 y(x - h) + a_1 y(x) + a_2 y(x + h)], \qquad (4.2)$$

where $\mathbf{a} := [a_0, a_1, a_2]$. We see that $dim\ \{\mathcal{L}[h, \mathbf{a}]\} = h^{-1}$ and that, as desired, the coefficients are dimensionless.

The classical moments of this $\mathcal{L}[h, \mathbf{a}]$ are known,

$$L_0(h, \mathbf{a}) = -\frac{1}{h}(a_0 + a_1 + a_2), \ L_1(h, \mathbf{a}) = 1 + a_0 - a_2,$$
$$L_{2k}(h, \mathbf{a}) = -h^{2k-1}(a_0 + a_2), \ L_{2k+1}(h, \mathbf{a}) = h^{2k}(a_0 - a_2), \qquad (4.3)$$
$$k = 1, 2, \ldots,$$

see (2.40), and therefore

$$L_{2k}^*(\mathbf{a}) = -(a_0 + \delta_{0\,k} a_1 + a_2), \ L_{2k+1}^*(\mathbf{a}) = \delta_{0\,k} + a_0 - a_2, \ k = 0, 1, 2, \ldots \qquad (4.4)$$

Step *ii*. It is easy to see that the maximal M for which the system

$$L_m^*(\mathbf{a}) = 0, \ m = 0, 1, \ldots, M - 1$$

is compatible is $M = 3$. The solution of this system is the set of the classical coefficients $a_2 = -a_0 = 1/2, \ a_1 = 0$.

Step *iii*. The expression of $E_0^*(z, \mathbf{a})$ is already known,

$$E_0^*(z, \mathbf{a}) := z - a_0 \exp(-z) - a_1 - a_2 \exp(z), \qquad (4.5)$$

see (2.48).

The construction of $G^{\pm}(Z, \mathbf{a})$ follows. We have

$$G^+(Z, \mathbf{a}) = \frac{1}{2}[E_0^*(z, \mathbf{a}) + E_0^*(-z, \mathbf{a})] = -a_1$$
$$-\frac{1}{2}[\exp(z) + \exp(-z)](a_0 + a_2) = -a_1 - (a_0 + a_2)\eta_{-1}(Z), \qquad (4.6)$$

and

$$G^-(Z, \mathbf{a}) = \frac{1}{2z}[E_0^*(z, \mathbf{a}) - E_0^*(-z, \mathbf{a})]$$
$$= 1 + \frac{1}{2z}[\exp(z) - \exp(-z)](a_0 - a_2) = 1 + (a_0 - a_2)\eta_0(Z), \qquad (4.7)$$

where the definition relations (3.21) and (3.22) of $\eta_{-1}(Z)$ and of $\eta_0(Z)$, respectively, were used. As for the derivatives of $G^{\pm}(Z, \mathbf{a})$, the differentiation relations (3.28) give directly

$$G^{+(p)}(Z, \mathbf{a}) = -2^{-p}(a_0 + a_2)\eta_{p-1}(Z), \ G^{-(p)}(Z, \mathbf{a}) = 2^{-p}(a_0 - a_2)\eta_p(Z). \qquad (4.8)$$

Step *iv*. According to the selfconsistency condition (3.55), two pairs P, K are consistent with $M = 3$, that is $P = -1$, $K = 2$ (the classical option), and $P = K = 0$.

Step *v*. For the latter option the algebraic system for the coefficients is

$$L_0^*(\mathbf{a}) = 0, \; G^\pm(Z, \mathbf{a}) = 0,$$

with the solution (Ixaru, [19])

$$a_2(Z) = -a_0(Z) = \frac{1}{2\eta_0(Z)}, \; a_1(Z) = 0. \tag{4.9}$$

This differs from the classical set of coefficients by a factor $1/\eta_0(Z)$ which is $\omega h / \sin(\omega h)$ if $\mu = i\omega$ (with the notation $\theta = \omega h$ this becomes $\theta / \sin(\theta)$, as in equation (1.3) in Chapter 1), and $\lambda h / \sinh(\lambda h)$ if $\mu = \lambda$. Note also that $1/\eta_0(0) = 1$ i.e. the classical coefficients are re-obtained for $Z = 0$.

Step *vi*. Since for this set of coefficients we have $L_1^*(\mathbf{a}(Z)) = 1 - 2a_2(Z) = (\eta_0(Z) - 1)/\eta_0(Z)$, the leading term of the error of the formula (4.1) reads

$$lte_{ef} = -h^2 \frac{\eta_0(Z) - 1}{Z\eta_0(Z)} D(D^2 - \mu^2) y(x_1). \tag{4.10}$$

In the limit μ, $Z \to 0$ this becomes (2.56). To see this, l'Hospital rule can be used for the factor in the middle. We have $(\eta_0(Z) - 1)' = \eta_1(Z)/2$ and $(Z\eta_0(Z))' = \eta_0(Z) + Z\eta_1(Z)/2$, and since $\eta_1(0) = 1/3$, the factor in discussion becomes $1/6$. Alternatively, use (3.35) and take $Z = 0$ directly.

In both versions (classical and tuned) the resultant coefficients are correlated, $a_2 = -a_0$ and $a_1 = 0$. This behaviour may have been expected because the mesh points are symmetrically distributed around the central one and also because the reference functions are such that if $y(x)$ satisfies the RDE, then $y(-x)$ also does. As a matter of fact, the latter condition did not hold for functions of the form (2.46) and this was the reason for the coefficients (2.50) being uncorrelated.

For the problem (4.1) the derivation of the coefficients by assuming from the very beginning that $a_2 = -a_0$ and $a_1 = 0$ is fully justified. Since in this way only one coefficient remains to be determined, such an approach has the advantage of being simpler and more direct than the original one. Specifically, the approximation reads

$$y'(x_1) \approx \frac{1}{h} a_2 [y(x_2) - y(x_0)], \; x_i = x_1 + (i - 1)h, \tag{4.11}$$

and the operator $\mathcal{L}[h, a_2]$ associated to it is introduced by

$$\mathcal{L}[h, a_2] y(x) := y'(x) - \frac{1}{h} a_2 [y(x + h) - y(x - h)]. \tag{4.12}$$

The relevant starred moments are

$$L_{2k}^*(a_2) = 0, \ L_{2k+1}^*(a_2) = \delta_{k0} - 2a_2, \ k = 0, 1, 2, \ldots \quad (4.13)$$

and

$$E_0^*(\pm z, a_2) := \pm z \mp a_2[\exp(z) - \exp(-z)], \quad (4.14)$$

whence

$$G^{+(p)}(Z, a_2) = 0, \ G^{-(p)}(Z, a_2) = \delta_{p0} - 2^{-p+1}a_2\eta_p(Z), \ p = 0, 1, \ldots \, . \quad (4.15)$$

Again the maximal M resulting from step *ii* is $M = 3$ although only one equation has to be solved actually. This is $L_1^*(a_2) = 0$ with the solution $a_2 = 1/2$, as expected. The two pairs in step *iv* above remain the same, while the system of three equations in step *v* reduces to only one. This is $G^-(Z, a_2) = 0$, with the expected solution (4.9).

An ad-hoc version of the three-point formula

On some occasions the available information on $y(x)$ is more detailed than is assumed in the central difference formula. For example, if $y(x)$ is the solution of the radial Schrödinger equation (3.6) computed by a multistep method (such as the method of Numerov), then not only the values of the solution at the mesh points are known, but also the values of its second order derivative; they are given by the equation directly, viz.: $y''(x_k) = (V(x_k) - E)y(x_k)$. It makes sense to exploit such an additional piece of information and then to consider an approximation of the form

$$y'(x_1) \approx \frac{1}{h}a_2[y(x_2) - y(x_0)] + hb_2[y''(x_2) - y''(x_0)], \quad (4.16)$$

$(x_i = x_1 + (i-1)h)$, instead of (4.11).
The operator $\mathcal{L}[h, \mathbf{a}]$ is introduced by

$$\mathcal{L}[h, \mathbf{a}]y(x) := y'(x) - \frac{1}{h}a_2[y(x+h) - y(x-h)] - hb_2[y''(x+h) - y''(x-h)], \quad (4.17)$$

where $\mathbf{a} := [a_2, b_2]$; notice that a_2 and b_2 are dimensionless.
To construct the moments we apply $\mathcal{L}[h, \mathbf{a}]$ on $\exp(\pm\mu x)$,

$$\mathcal{L}[h, \mathbf{a}]\exp(\pm\mu x) = \pm\frac{1}{h}\exp(\pm\mu x)[z - (a_2 + z^2 b_2)(\exp(z) - \exp(-z))] \quad (4.18)$$

and therefore

$$E_0^*(\pm z, \mathbf{a}) = \pm[z - (a_2 + z^2 b_2)(\exp(z) - \exp(-z))]. \quad (4.19)$$

On this basis we get

$$G^+(Z, \mathbf{a}) = 0, \quad G^-(Z, \mathbf{a}) = 1 - 2(a_2 + Zb_2)\eta_0(Z), \qquad (4.20)$$

and

$$G^{+(p)}(Z, \mathbf{a}) = 0, \quad p = 1, 2, \ldots$$
$$G^{-(p)}(Z, \mathbf{a}) = -2^{-p+1}[a_2\eta_p(Z) + b_2(\eta_{p-1}(Z) + \eta_{p-2}(Z))], \quad (4.21)$$

(for the first derivative of G^- use $(Z\eta_0(Z))' = \eta_0(Z) + Z\eta_1(Z)/2 = [\eta_{-1}(Z) + \eta_0(Z)]/2$). Their values at $Z = 0$ allow obtaining the expressions of the starred classical moments (use equations (3.46) and (3.49)),

$$L^*_{2k}(\mathbf{a}) = 0, \quad L^*_{2k+1}(\mathbf{a}) = \delta_{k0} - 2[a_2 + 2k(2k+1)b_2], \quad k = 0, 1, 2, \ldots (4.22)$$

and this completes the operations involved in steps *i* and *iii*.
Step *ii*. Equations $L^*_0(\mathbf{a}) = L^*_2(\mathbf{a}) = L^*_4(\mathbf{a}) = 0$ are identically satisfied while $L^*_1(\mathbf{a}) = L^*_3(\mathbf{a}) = 0$ give

$$a_2 = \frac{1}{2}, \quad b_2 = -\frac{1}{12}, \qquad (4.23)$$

which represent the classical coefficients. Since with these values we have $L^*_5(\mathbf{a}) \neq 0$ it results that $M = 5$.
Step *iv*. As the selfconsistency condition (3.55) shows, three pairs are consistent with this M. They are $P = -1$, $K = 4$, $P = 0$, $K = 2$, and $P = 1$, $K = 0$. The results of steps *v* and *vi* are listed below:
  $P = -1$, $K = 4$. This is the classical case. The leading term of the error is:

$$lte_{class} = \frac{7}{360}h^4 y^{(5)}(x_1), \qquad (4.24)$$

This case has been investigated by Blatt, [1].
  $P = 0$, $K = 2$. The coefficients are

$$a_2(Z) = \frac{1}{2}, \quad b_2(Z) = \frac{1 - \eta_0(Z)}{2Z\eta_0(Z)} = -\frac{\eta_0^2(Z/4) - 2\eta_1(Z)}{4\eta_0(Z)}. \qquad (4.25)$$

The last expression of $b_2(Z)$ was obtained by (3.35). It has the advantage that no separate use of the l'Hospital rule is needed for $Z \to 0$. The *lte* is

$$lte_{ef} = h^4 \frac{Z\eta_0(Z) + 6(1 - \eta_0(Z))}{6Z^2\eta_0(Z)} D^3(D^2 - \mu^2)y(x_1). \qquad (4.26)$$

  $P = 1$, $K = 0$. The coefficients and the *lte* are:

$$a_2(Z) = \frac{\eta_{-1}(Z) + \eta_0(Z)}{4\eta_0^2(Z)}, \quad b_2(Z) = -\frac{\eta_1(Z)}{4\eta_0^2(Z)}, \qquad (4.27)$$

and

$$lte_{ef} = h^4 \frac{2\eta_0^2(Z) - \eta_{-1}(Z) - \eta_0(Z)}{2Z^2\eta_0^2(Z)} D(D^2 - \mu^2)^2 y(x_1). \qquad (4.28)$$

It can be checked easily that the Z dependent factors in the last two lte-s tend to the classical value $7/360$ when $Z \to 0$. As for applications, the version with the classical coefficients (4.23) is appropriate when $y(x)$ has a smooth behaviour, while the versions with the coefficients (4.25) and (4.27) are particularly suitable when $y(x)$ is a sum of one of the forms (3.38) or (3.39) and a smooth function, and of form (3.38) or (3.39) plus a constant, respectively.

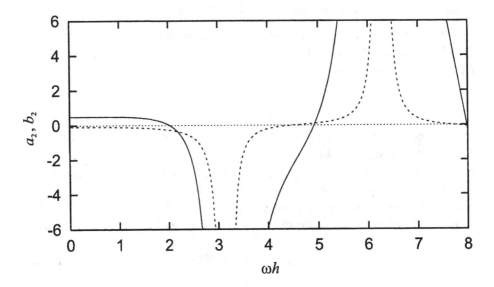

Figure 4.1. The ωh dependence of coefficients $a_2(Z = -\omega^2 h^2)$ (solid line) and $b_2(Z = -\omega^2 h^2)$ (broken line) for the version $P = 1$ of the ad-hoc formula (4.16).

Critical values

A common feature of the tuned versions of both discussed formulae is the appearance of $\eta_0(Z)$ in the denominator of the coefficient expressions. When $Z < 0$ this has the same zeros as $\sin(\omega h)$ for $\omega h > 0$, that is $\omega h = n\pi$, $n = 1, 2, \ldots$. The coefficients of these versions will then exhibit a pole-like behaviour around these critical values, see Figure 4.1, and therefore some care should be taken in practice. Specifically, h must be chosen such that the product ωh is not too close to the critical values.

A numerical illustration

We take the function

$$y(x) = f(x)\cos(\omega x), \quad f(x) = 1/(1+x). \tag{4.29}$$

Its first three derivatives are:

$$y'(x) = -f(x)[f(x)\cos(\omega x) + \omega\sin(\omega x)],$$
$$y''(x) = f(x)[(2f^2(x) - \omega^2)\cos(\omega x) + 2f(x)\omega\sin(\omega x)], \tag{4.30}$$
$$y^{(3)}(x) = f(x)[3f(x)(-2f^2(x)+\omega^2)\cos(\omega x)+\omega(-6f^2(x)+\omega^2)\sin(\omega x)].$$

We compute $y'(x)$ at $x = 1$ and with $h = 0.1$ for various values of ω.

To compare the accuracies of the two versions of the central difference formula, let $y'_{cl}(1)$ and $y'_{ef}(1)$ be the results produced by the classical version and by the tuned version, respectively. The coefficients of the latter are in (4.9). The scaled errors $\Delta_{cl} = (y'(1) - y'_{cl}(1))/s$ and $\Delta_{ef} = (y'(1) - y'_{ef}(1))/s$, where $s = 1$ if $\omega \leq 1$ and $s = \omega$ if $\omega > 1$, are plotted on Figure 4.2. We see that at small ω's the two formulae exhibit comparable accuracies but, when ω is subsequently advanced, the error from the classical formula increases dramatically in amplitude while the error from the new formula stays in reasonable limits except for ω's around 31.4... and 62.8..., which correspond to the critical values $\omega h = \pi$ and $\omega h = 2\pi$, respectively.

On Figure 4.3 we illustrate the accuracy gain of the ad-hoc formula (4.16) on the same test function (4.29) when the level of tuning P is increased. Scaled errors are presented for $P = -1$ (the classical version), 0, and 1. For comparison we add the scaled error from the tuned version of the central difference formula. A general tendency of the error to increase with ω is visible in all cases (this is partly due to the gradual advance of ωh towards the first critical value) but, as expected, the magnitude of its slope diminishes when P is increased. Also as expected, the ad-hoc formula behaves better than the central difference one.

1.2 The five-point formula for the first derivative

The general form of this approximation is

$$y'(x_2) \approx \frac{1}{h}[a_0 y(x_0) + a_1 y(x_1) + a_2 y(x_2) + a_3 y(x_3) + a_4 y(x_4)], \tag{4.31}$$

where $x_i = x_2 + (i - 2)h$. It has been treated as such in [19] but again it is sufficient to assume from the very beginning that $a_4 = -a_0$, $a_3 = -a_1$ and $a_2 = 0$, i.e. to take

$$y'(x_2) \approx \frac{1}{h}[a_3(y(x_3) - y(x_1)) + a_4(y(x_4) - y(x_0))], \quad x_i = x_2 + (i - 2)h, \tag{4.32}$$

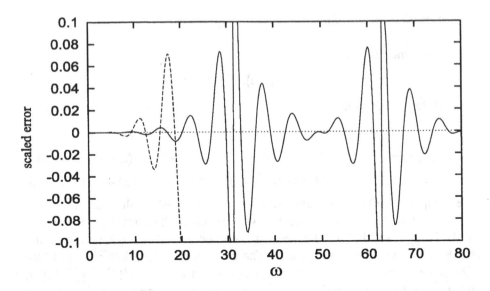

Figure 4.2. The central difference formula (4.1) on function (4.29): the scaled error for the classical (dashed) and ef version (solid).

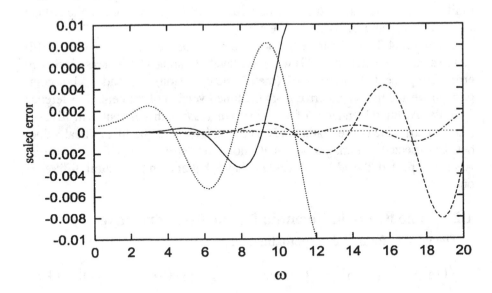

Figure 4.3. Scaled errors for the ef version of the central difference formula (solid), and for the ad-hoc formula (4.16) applied on the test function (4.29): $P = -1$ (dots), $P = 0$ (dashed) and $P = 1$ (dash and dots).

whose classical coefficients are well known,

$$a_3 = \frac{2}{3}, \; a_4 = -\frac{1}{12}. \tag{4.33}$$

In the same way as we did for the ad-hoc formula we perform the steps *i* and *iii* simultaneously. We denote $\mathbf{a} := [a_3, \, a_4]$ and introduce the operator $\mathcal{L}[h, \mathbf{a}]$ by

$$\mathcal{L}[h, \mathbf{a}]y(x) := y'(x) - \frac{1}{h}[a_3(y(x+h)-y(x-h))+a_4(y(x+2h)-y(x-2h))]. \tag{4.34}$$

and, to construct the moments, we apply $\mathcal{L}[h, \mathbf{a}]$ on $\exp(\pm\mu x)$,

$$\begin{aligned}\mathcal{L}[h, \mathbf{a}]\exp(\pm\mu x) \; = \; & \pm\frac{1}{h}\exp(\pm\mu x)[z - a_4(\exp(2z) - \exp(-2z)) \\ & - \; a_3(\exp(z) - \exp(-z))],\end{aligned} \tag{4.35}$$

whence

$$E_0^*(\pm z, \mathbf{a}) = \pm[z - a_4(\exp(2z) - \exp(-2z)) - a_3(\exp(z) - \exp(-z))], \tag{4.36}$$

$$G^{+(p)}(Z, \mathbf{a}) = 0, \; p = 0, \, 1, \, 2, \, \ldots \tag{4.37}$$
$$G^{-(p)}(Z, \mathbf{a}) = \delta_{p0} - 2^{p+2}a_4\eta_p(4Z) - 2^{1-p}a_3\eta_p(Z),$$

and

$$L_{2k}^*(\mathbf{a}) = 0, \; L_{2k+1}^*(\mathbf{a}) = \delta_{k0} - 2[2^{2k+1}a_4 + a_3], \; k = 0, \, 1, \, 2, \, \ldots \tag{4.38}$$

Step *ii*. Equations $L_0^*(\mathbf{a}) = L_2^*(\mathbf{a}) = L_4^*(\mathbf{a}) = 0$ are identically satisfied while the algebraic system $L_1^*(\mathbf{a}) = L_3^*(\mathbf{a}) = 0$ gives the classical coefficients (4.33). Since for these coefficients $L_5^*(\mathbf{a})$ does not vanish (its value is actually $L_5^*(\mathbf{a}) = 4$) it results that $M = 5$.

Step *iv*. Three pairs are consistent with this M, that is $P = -1, \; K = 4$, $P = 0, \; K = 2$, and $P = 1, \; K = 0$.

The results of steps *v* and *vi* are as follows:

$P = -1, \; K = 4$. This is the classical case. As said, the coefficients are (4.33). The leading term of the error is

$$lte_{class} = \frac{1}{30}h^4 y^{(5)}(x_2). \tag{4.39}$$

$P = 0, \; K = 2$. The coefficients are

$$a_3(Z) = \frac{\eta_0(4Z) - 1}{2D_0(Z)}, \; a_4(Z) = -\frac{\eta_0(Z) - 1}{4D_0(Z)}, \tag{4.40}$$

where $D_0(Z) = \eta_0(4Z) - \eta_0(Z)$ (for a safe numerical evaluation of these coefficients use (3.35)), and the *lte* is

$$lte_{ef} = h^4 \frac{\eta_0(4Z) - 4\eta_0(Z) + 3}{6Z D_0(Z)} D^3(D^2 - \mu^2)y(x_2). \qquad (4.41)$$

The Z dependent factor in the middle of this expression tends to $1/30$ when μ, $Z \to 0$, as expected.

$P = 1$, $K = 0$. With the notation $D_1(Z) = 4[\eta_1(Z)\eta_0(4Z) - 4\eta_0(Z)\eta_1(4Z)]$, the coefficients and the *lte* are:

$$a_3(Z) = -\frac{8\eta_1(4Z)}{D_1(Z)}, \quad a_4(Z) = \frac{\eta_1(Z)}{D_1(Z)}, \qquad (4.42)$$

and

$$lte_{ef} = h^4 \frac{D_1(Z) - 4\eta_1(Z) + 16\eta_1(4Z)}{Z^2 D_1(Z)} D(D^2 - \mu^2)^2 y(x_2). \qquad (4.43)$$

respectively, see also Ixaru, [19]. Again the Z dependent factor in the latter tends to $1/30$ when $Z \to 0$.

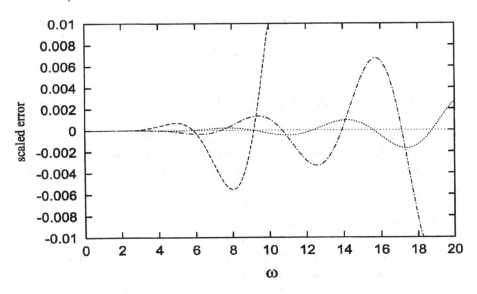

Figure 4.4. Scaled errors for the five-point formula (4.32) on the test function (4.29): $P = -1$ (dashed), $P = 0$ (dash and dots) and $P = 1$ (dots).

The five-point and the ad-hoc formulae for the first derivative are quite similar in accuracy: the same h^4 factor appears in the *lte*, the error constants of their classical versions are comparable in magnitude ($1/30$ in the former while $7/360$

in the latter). This is nicely confirmed on the considered test function (4.29); compare the scaled errors presented in Figures 4.3 and 4.4. However, the tuned versions of the ad-hoc formula should be preferred (if the second derivatives are available, of course) because the length of the reference interval of such formulae is $2h$, that is shorter than the length $4h$ for the five-point formula. Indeed, the shorter the interval the smaller the variation of $f_1(x)$ and of $f_2(x)$ is, and then the original hypothesis on their variation is better justified.

1.3 The three-point formula for the second derivative

We consider the approximation

$$y''(x) \approx \frac{1}{h^2}[a_0 y(x_0) + a_1 y(x_1) + a_2 y(x_2)], \quad x_i = x_1 + (i - 1)h, \quad (4.44)$$

to derive the coefficients such that it becomes tuned on functions of form (3.38) or (3.39).

We denote $\mathbf{a} := [a_0, \ a_1, \ a_2]$ and introduce the operator $\mathcal{L}[h, \mathbf{a}]$ by

$$\mathcal{L}[h, \mathbf{a}]y(x) := y''(x) - \frac{1}{h^2}[a_0 y(x + h) + a_1 y(x) + a_2 y(x - h)]. \quad (4.45)$$

We have $dim\{\mathcal{L}[h, \mathbf{a}]\} = h^{-2}$. Also the coefficients are dimensionless. We first construct the pair $E_0^*(\pm z, \mathbf{a})$. Since

$$\mathcal{L}[h, \mathbf{a}] \exp(\pm \mu x) = \frac{1}{h^2} \exp(\pm \mu x)[z^2 - (a_0 \exp(\pm z) + a_1 + a_2 \exp(\mp z))] \quad (4.46)$$

where $z = \mu h$, as usual, the pair searched for is

$$E_0^*(\pm z, \mathbf{a}) = z^2 - (a_0 \exp(\pm z) + a_1 + a_2 \exp(\mp z)). \quad (4.47)$$

We now form $G^{\pm}(Z, \mathbf{a})$:

$$G^+(Z, \mathbf{a}) = \frac{1}{2}[E_0^*(z, \mathbf{a}) + E_0^*(-z, \mathbf{a})] = Z - (a_0 + a_2)\eta_{-1}(Z) - a_1, \quad (4.48)$$

and

$$G^-(Z, \mathbf{a}) = \frac{1}{2z}[E_0^*(z, \mathbf{a}) - E_0^*(-z, \mathbf{a})] = (a_2 - a_0)\eta_0(Z). \quad (4.49)$$

Their derivatives with respect to Z are

$$G^{+(p)}(Z, \mathbf{a}) = \delta_{1p} - 2^{-p}(a_0 + a_2)\eta_{p-1}(Z),$$

$$G^{-(p)}(Z, \mathbf{a}) = 2^{-p}(a_2 - a_0)\eta_p(Z), \quad p = 1, \ 2, \ \ldots \quad (4.50)$$

The relation (3.49) permits the derivation of the expressions of the starred classical moments, with the result

$$L_{2k}^*(\mathbf{a}) = -a_1\delta_{0k} + 2\delta_{1k} - (a_0 + a_2),$$
$$L_{2k+1}^*(\mathbf{a}) = a_2 - a_0, \quad k = 0, 1, 2, \ldots, \qquad (4.51)$$

and this covers the operations involved in the steps *i* and *iii*.

Step *ii*. The linear system $L_m^*(\mathbf{a}) = 0$, $m = 0, 1, 2$ gives the classical coefficients

$$a_0 = a_2 = 1, \ a_1 = -2. \qquad (4.52)$$

Since with these one still has $L_3^*(\mathbf{a}) = 0$ but $L_4^*(\mathbf{a}) = -2 \neq 0$ it follows that $M = 4$.

Step *iv*. Three pairs P, K are consistent with this M: $P = -1$, $K = 3$ (the classical case), $P = 0$, $K = 1$, and $P = 1$, $K = -1$.

The results of the steps *v* and *vi* are:

$P = -1$, $K = 3$. The coefficients are given in (4.52) and the *lte* is

$$lte_{class} = -\frac{1}{12}h^2 y^{(4)}(x_1). \qquad (4.53)$$

Both are well known.

$P = 0$, $K = 1$. The system (3.56) becomes $L_0^*(\mathbf{a}) = L_1^*(\mathbf{a}) = G^{\pm}(Z, \mathbf{a}) = 0$, that is a system of four equations for three unknowns. However two equations are equivalent, viz.: $L_1^*(\mathbf{a}) = 0$ and $G^-(Z, \mathbf{a}) = 0$. One of them is then eliminated and the remaining system of three equations for the three unknowns has the solution (Ixaru [19])

$$a_0(Z) = a_2(Z) = \frac{Z}{2(\eta_{-1}(Z) - 1)}, \ a_1(Z) = -\frac{Z}{\eta_{-1}(Z) - 1}. \qquad (4.54)$$

To remove the indeterminacy at $Z = 0$ use (3.35).

The *lte* for this version is

$$lte_{ef} = -h^2 \frac{2\eta_{-1}(Z) - 2 - Z}{2Z(\eta_{-1}(Z) - 1)} D^2(D^2 - \mu^2)y(x_1). \qquad (4.55)$$

$P = 1$, $K = -1$. The system $G^{\pm}(z, \mathbf{a}) = G^{\pm(1)}(Z, \mathbf{a}) = 0$ is solved to give

$$a_0(Z) = a_2(Z) = \frac{1}{\eta_0(Z)}, \ a_1(Z) = \frac{Z\eta_0(Z) - 2\eta_{-1}(Z)}{\eta_0(Z)}, \qquad (4.56)$$

and the leading term of the error is

$$lte_{ef} = -h^2 \frac{2 + Z\eta_0(Z) - 2\eta_{-1}(Z)}{Z^2\eta_0(Z)} (D^2 - \mu^2)^2 y(x_1), \qquad (4.57)$$

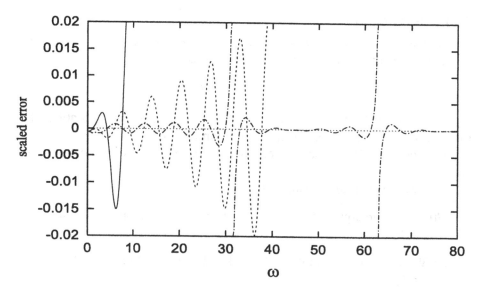

Figure 4.5. Scaled errors for three versions of the formula (4.44) for the second derivative: $P = -1$ (solid), $P = 0$ (dashed) and $P = 1$ (dash and dots).

as in [19]. It is easy to see that each of the two tuned versions tends to the classical one when $Z \rightarrow 0$. Critical values appear in each of them but, in contrast with the formulae presented for the first order derivative, these are not the same in the two. For the version with $P = 0$ the critical values are $wh = 2n\pi$, $n = 1, 2, \ldots$ while for the version with $P = 1$ they are $wh = n\pi$, $n = 1, 2, \ldots$.

For a numerical illustration we take again the test function (4.29) and compute $y''(x = 1)$ by the three versions, for $h = 0.1$. The scaled errors are defined by $\Delta_{comput} = (y''(1) - y''_{comput}(1))/s$, where $s = 1$ if $w \leq 1$, and $s = w^2$ if $w > 1$. They are plotted on Figure 4.5 for $0 \leq w \leq 80$.

The error produced by the classical version is seen to increase dramatically with w. The slope of the increase is significantly smaller for the version $P = 0$, and it is almost negligible for the best tuned version $P = 1$ except around $31.4\ldots$ and $62.8\ldots$ which correspond to the critical values $wh = \pi$, 2π.

We mention finally that the same expressions for the coefficients will be obtained also if it is assumed from the very beginning that $a_0 = a_2$ but we leave this as an exercise.

1.3.1   Subroutine EFDER

This subroutine computes the numerical values of the coefficients of all formulae investigated in this section. Calling it is

$$\text{CALL EFDER(Z, IFORM, IP, W)}$$

Input arguments: Z (double precision), IFORM (an integer which identifies the formula) and IP (an integer for the level of tuning P).
Output argument: W, a double precision vector with two components for the coefficients.
The following values are permitted for IFORM and IP:

IFORM=1 - three-point formula (4.1); IP=-1, 0. Output: $W(1) = a_2 = -a_0, W(2) = a_1 = 0$.

IFORM=2 - ad-hoc formula (4.16); IP=-1, 0, 1. Output: $W(1) = a_2$, $W(2) = b_2$.

IFORM=3 - five-point formula (4.32); IP=-1, 0, 1. Output: $W(1) = a_3$, $W(2) = a_4$.

IFORM=4 - three-point formula (4.44) for the second derivative; IP=-1, 0, 1. Output: $W(1) = a_1, W(2) = a_0 = a_2$.

The following driving program yields all these coefficients for $Z = -1$ and 1:

```
implicit double precision (a-h,o-z)
dimension ww(-1:1,1:2),w(2)
write(*,*)'      z      ','             ip=-1  ',
!'      ip=0  ','      ip=1  '
do iform=1,4
  write(*,*)' iform = ',iform
  do iz=-1,1,2
    z=iz
    ipmax=0
    if(iform.ne.1)ipmax=1
    do ip=-1,ipmax
      call efder(z,iform,ip,w)
      ww(ip,1)=w(1)
      ww(ip,2)=w(2)
    enddo
    write(*,50)z,(ww(ip,1),ip=-1,ipmax)
    write(*,51)z,(ww(ip,2),ip=-1,ipmax)
  enddo
enddo
```

```
50      format(2x,f8.4,'  w(1): ',3(2x,e10.3))
51      format(2x,f8.4,'  w(2): ',3(2x,e10.3))
        stop
        end
```

The output is

z		ip=-1	ip=0	ip=1
iform =	1			
-1.0000	w(1):	0.500E+00	0.594E+00	
-1.0000	w(2):	0.000E+00	0.000E+00	
1.0000	w(1):	0.500E+00	0.425E+00	
1.0000	w(2):	0.000E+00	0.000E+00	
iform =	2			
-1.0000	w(1):	0.500E+00	0.500E+00	0.488E+00
-1.0000	w(2):	-0.833E-01	-0.942E-01	-0.106E+00
1.0000	w(1):	0.500E+00	0.500E+00	0.492E+00
1.0000	w(2):	-0.833E-01	-0.745E-01	-0.666E-01
iform =	3			
-1.0000	w(1):	0.667E+00	0.705E+00	0.731E+00
-1.0000	w(2):	-0.833E-01	-0.102E+00	-0.126E+00
1.0000	w(1):	0.667E+00	0.637E+00	0.600E+00
1.0000	w(2):	-0.833E-01	-0.686E-01	-0.567E-01
iform =	4			
-1.0000	w(1):	0.100E+01	0.109E+01	0.119E+01
-1.0000	w(2):	-0.200E+01	-0.218E+01	-0.228E+01
1.0000	w(1):	0.100E+01	0.921E+00	0.851E+00
1.0000	w(2):	-0.200E+01	-0.184E+01	-0.163E+01

2. Quadrature

Along the time the numerical quadrature received substantially more attention than the numerical differentiation. Chapters devoted to it occupy an important place in any book of numerical methods, see e. g. Phillips and Taylor [31], Burden and Faires [3], and Gerald and Wheatley [15], while a number of reference books are just concentrated on the numerical quadrature, to mention only the books by Davis and Rabinowitz [6], Krylov [29], Ghizzetti and Ossicini [16], Stroud and Secrest [35], and by Evans [12].

The integration of functions with special forms is one of the central concerns in the field. The integration of oscillatory functions enjoyed a particular attention, to mention only Sections 2.10 and 3.8 of [6] and more or less recent contributions by Duris [8], Ehrenmark [9],[10],[11], Vanden Berghe, De Meyer and Vanthournout [39], De Meyer, Vanthournout, Vanden Berghe and Vanderbauwhede [7], Bocher, De Meyer and Vanden Berghe [2], Evans and Webster

[13], [14], Köhler [28], who used various techniques.

The exponential fitting technique was used by Ixaru [19], [23], Ixaru and Paternoster [24], Kim, Cools and Ixaru [26], [27], and by Kim [25].

2.1   Simpson formula

The Simpson formula is for the calculation of the integral of a function $y(x)$ on $[a, b]$ in terms of the values of the function at three predetermined distinct points. With $X = (a + b)/2$ and $h = (b - a)/2$ a three-point formula has the form

$$\int_a^b y(z)dz = \int_{X-h}^{X+h} y(z)dz \approx h \sum_{k=1}^{3} a_k y(X + x_k h), \quad (4.58)$$

where the dimensionless abscissa weights x_k ($k = 1, 2, 3$) can be fixed at will by the user. The Simpson formula is the particular case when these are $x_1 = -1$, $x_2 = 0$ and $x_3 = 1$, which means that the nodes are equidistant and that the two extreme nodes are chosen to coincide with the integration limits. The Simpson formula then reads

$$\int_{X-h}^{X+h} y(z)dz \approx h\left[a_1 y(X - h) + a_2 y(X) + a_3 y(X + h)\right], \quad (4.59)$$

whose coefficients

$$a_1 = a_3 = \frac{1}{3} \text{ and } a_2 = \frac{4}{3}, \quad (4.60)$$

are well-known.

This set of coefficients is convenient only when the integrand is a smooth enough function but we now want to determine the coefficients such that this approximation becomes tuned on functions of the form (3.38) or (3.39). To this aim we apply the six-step flow chart. The same way was used in [19].

Thus we denote $\mathbf{a} := [a_1, a_2, a_3]$ and introduce the operator $\mathcal{L}[h, \mathbf{a}]$ by

$$\mathcal{L}[h, \mathbf{a}]y(x) := \int_{x-h}^{x+h} y(z)dz - h\left[a_1 y(x-h) + a_2 y(x) + a_3 y(x+h)\right]. \quad (4.61)$$

It is easily seen that $dim\{\mathcal{L}[h, \mathbf{a}]\} = h$ and then $l = 1$, and also that the coefficients are dimensionless. We apply $\mathcal{L}[h, \mathbf{a}]$ on $\exp(\pm\mu x)$ to get

$$\mathcal{L}[h, \mathbf{a}]\exp(\pm\mu x) = h\exp(\pm\mu x)E_0^*(\pm z), \quad (4.62)$$

where

$$E_0^*(\pm z) = \frac{1}{z}[\exp(z) - \exp(-z)] - a_1\exp(\mp z) - a_2 - a_3\exp(\pm z), \quad (4.63)$$

whence

$$G^{+(p)}(Z, \mathbf{a}) = 2^{-p}[2\eta_p(Z) - (a_1 + a_3)\eta_{p-1}(Z)] - \delta_{p0}a_2,$$
$$G^{-(p)}(Z, \mathbf{a}) = 2^{-p}(a_1 - a_3)\eta_p(Z), \quad p = 0, 1, 2, \ldots. \quad (4.64)$$

By using (3.49) we get the following expressions for the starred classical moments:

$$L_{2k}^*(\mathbf{a}) = \frac{2}{2k+1} - a_1 - \delta_{k0}a_2 - a_3, \quad L_{2k+1}^*(\mathbf{a}) = a_1 - a_3, \quad k = 0, 1, \ldots$$

(4.65)

and this completes the evaluations required in steps *i* and *iii*.

Step *ii*. The system $L_m^*(\mathbf{a}) = 0$, $m = 0$, 1, 2 has the solution (4.60), whence $L_3^*(\mathbf{a}) = 0$ but $L_4^*(\mathbf{a}) = -4/15 \neq 0$ and therefore $M = 4$.

Step *iv*. The pairs P, K consistent with $M = 4$ are $P = -1$, $K = 3$, $P = 0$, $K = 1$ and $P = 1$, $K = -1$, from which the latter is the best tuned on functions of the form (3.38) or (3.39). The steps *v* and *vi* are now considered for each case:

$P = -1$, $K = 3$ (classical case). The *lte* is

$$lte_{class} = -\frac{1}{90}h^5 y^{(4)}(X),$$

(4.66)

a well known result.

$P = 0$, $K = 1$. The system to be solved is $L_0^*(\mathbf{a}) = L_1^*(\mathbf{a}) = G^{\pm}(Z, \mathbf{a}) = 0$. It consists in four equations for three unknowns but $L_1^*(\mathbf{a})$ and $G^-(Z, \mathbf{a})$ are proportional. Equation $G^-(Z, \mathbf{a}) = 0$ is then removed and the remaining system of three equations has the solution

$$a_1(Z) = a_3(Z) = \frac{1 - \eta_0(Z)}{1 - \eta_{-1}(Z)} = 1 - \frac{2\eta_1(Z)}{\eta_0^2(Z/4)},$$

$$a_2(Z) = \frac{2(\eta_0(Z) - \eta_{-1}(Z))}{1 - \eta_{-1}(Z)} = \frac{4\eta_1(Z)}{\eta_0^2(Z/4)}.$$

(4.67)

The leading term of the error is

$$lte_{ef} = -h^5 \frac{3\eta_0(Z) - \eta_{-1}(Z) - 2}{3Z(1 - \eta_{-1}(Z))} D^2(D^2 - \mu^2)y(X).$$

(4.68)

This version has been considered also by Ehrenmark in [9] and by Vanden Berghe *et al.* [39] on the basis of a different approach. These authors investigated only the oscillatory case $\mu = i\omega$ to obtain the coefficients

$$a_1 = a_3 = \frac{\omega h - \sin(\omega h)}{\omega h(1 - \cos(\omega h))}, \quad a_2 = \frac{2(\sin(\omega h) - \omega h \cos(\omega h))}{\omega h(1 - \cos(\omega h))},$$

(4.69)

which coincide with those written in (4.67) if $Z = -\omega^2 h^2$. They have also shown that, if $\omega h < \pi$, there is some $\eta \in [a, b]$ such that the whole error reads

$$err = \frac{h^3}{6\omega^2}(\frac{6}{\omega h}\cot(\frac{1}{2}\omega h) - 3\cot^2(\frac{1}{2}\omega h) - 1) D^2(D^2 + \omega^2)y(\eta).$$

(4.70)

Again for $Z = -\omega^2 h^2$ the factors in front of D^2 in (4.68) and (4.70) coincide.

$P = 1$, $K = -1$. The system to be solved is $G^{\pm}(Z, \mathbf{a}) = G^{\pm(1)}(Z, \mathbf{a}) = 0$, and this is a system of four equations for the three unknowns. However, as before, the system is compatible and its solution is

$$a_1(Z) = a_3(Z) = \frac{\eta_1(Z)}{\eta_0(Z)}, \quad a_2(Z) = \frac{2[\eta_0^2(Z) - \eta_1(Z)\eta_{-1}(Z)]}{\eta_0(Z)}. \quad (4.71)$$

The *lte* of this version has the expression

$$lte_{ef} = 2h^5 \frac{\eta_0(Z) - \eta_1(Z) - \eta_0^2(Z) + \eta_1(Z)\eta_{-1}(Z)}{Z^2 \eta_0(Z)} (D^2 - \mu^2)^2 y(X).$$

$$(4.72)$$

The versions with $P = 0$ and 1 tend to the classical version when $Z \to 0$. Also, in the range $Z < 0$ they exhibit critical values at $\omega h = 2n\pi$ and at $\omega h = n\pi$, respectively, with $n = 1, 2, \ldots$.

The whole derivation can be repeated by taking $a_1 = a_3$ from the very beginning.

A numerical illustration

The test function is

$$y(x) = -f(x)[f(x)\cos(\omega x) + \omega \sin(\omega x)], \quad f(x) = 1/(1 + x), \quad (4.73)$$

and the analytic expression of its integral is

$$\int_{X-h}^{X+h} y(z)dz = f(X+h)\cos[\omega(X+h)] - f(X-h)\cos[\omega(X-h)]. \quad (4.74)$$

Let I and I_P ($P = -1$, 0 and 1) be the exact integral and its approximate value produced by each of the three versions, respectively. On Figure 4.6 we plot the errors $\Delta_P = I - I_P$ for $X = 1$ and $h = 0.1$ at various values of ω up to 80. All versions produce similar results at small ω's but when ω is increased the variation of the error strongly depends on P and the bigger P the better the accuracy. Also, as expected, the results for $P = 1$ are spurious around the critical values $\omega = 10\pi$ and $\omega = 20\pi$ (corresponding to $\omega h = n\pi$, $n = 1, 2$).

2.2   Quadrature rules with predetermined abscissa points

To compute the integral of $y(x)$ over $[a, b]$ we denote $X = (a + b)/2$ and $h = (b - a)/2$ and assume that the values of $y(x)$ at the points $X + x_n h$, $n = 1, 2, \ldots, N$ are known. The N-point rule is

$$\int_a^b y(z)dz = \int_{X-h}^{X+h} y(z)dz \approx h \sum_{n=1}^{N} a_n y(X + x_n h). \quad (4.75)$$

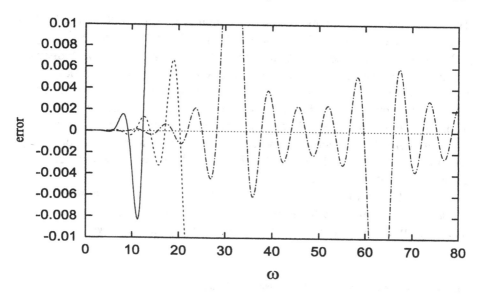

Figure 4.6. Errors for the Simpson quadrature formula : $P = -1$ (solid), $P = 0$ (dashed) and $P = 1$ (dash and dots).

2.2.1 Symmetric abscissas

In principle the position of these abscissa points can be fixed at will except for that they are normally assumed to be placed inside $[a, b]$, that is, $|x_n| \leq 1$ for all n. However, the most familiar case is when they are symmetrically distributed around X and equidistant, according to the formula

$$x_n = 2(n-1)/(N-1) - 1 \quad (n = 1, 2, \ldots, N). \tag{4.76}$$

The rule (4.75) on such a partition is called an N-point Newton–Cotes rule. When $N = 3$ this rule reduces to the Simpson formula.

Rules whose abscissas are distinct, symmetrically distributed but not necessarily equidistant also exist. Examples include:

$$x_1 = 1, \; x_N = -1, \; x_n = \cos[(2n-3)\pi/(2N-2)], n = 2, 3, \ldots, N-1,$$

corresponding to the Basu rule,

$$x_n = \cos[(n-1)\pi/(N-1)], n = 1, 2, \ldots, N,$$

for the Clenshaw-Curtis rule, and

$$x_n = \cos[(2n-1)\pi/(2N)] \text{ or } x_n = \cos[n\pi/(N+1)], n = 1, 2, \ldots, N,$$

for Fejer's first or second rule, respectively. For all these rules the classical, polynomial-based values of the coefficients $a_1, a_2, \ldots, a_N$ are symmetric.

They can be found in many books, for example in Section 2.5 of [6]. The coefficients of the Newton–Cotes rule are also listed in Table 4.1 for $2 \leq N \leq 6$.

We want to determine the coefficients such that the rule (4.75) with distinct, symmetrically distributed but not necessarily equidistant abscissa points becomes tuned on functions with oscillatory or hyperbolic behaviour. The fact that we can now consider bigger values for N than in the Simpson formula leads to a larger area of possible applications than before. Instead of considering the pair P, K which is relevant for functions of form (3.38) or (3.39), we can now consider the larger set P_1, P_2, ..., P_I, K, which will cover linear combinations of such forms.

As said, the coefficients of the polynomial-based rules with symmetric abscissas exhibit symmetry, see also Table 4.1. The following lemma shows that this property holds for the ef-based extension of such rules, as well.

Table 4.1. Classical coefficients for the Newton–Cotes rule.

coefficient	$N = 2$	$N = 3$	$N = 4$	$N = 5$	$N = 6$
a_1	1	$\frac{1}{3}$	$\frac{1}{4}$	$\frac{7}{45}$	$\frac{19}{144}$
a_2		$\frac{4}{3}$	$\frac{3}{4}$	$\frac{32}{45}$	$\frac{75}{144}$
a_3				$\frac{12}{45}$	$\frac{50}{144}$

LEMMA 4.1 *If the nodes of the quadrature rule (4.75) are symmetrically distributed around* X, *that is* $x_n = -x_{N+1-n}$, *and if the coefficients* a_n *are the only ones which ensure that the rule is exact for the functions (3.58), then these coefficients satisfy the symmetry property* $a_n = a_{N+1-n}$.

Proof Without restricting the generality we can take $X = 0$. Since the functions (3.58) form a complete set of linearly independent solutions of the RDE

$$D^{K+1}(D^2 - \mu_1^2)^{P_1+1} \cdots (D^2 - \mu_I^2)^{P_I+1}y = 0, \qquad (4.77)$$

the stated condition can be reformulated by saying that the rule is the only one which is exact for any $y(x)$ which satisfies this RDE. The point is that, due to the particular form of this equation, if $y(x)$ is a solution then $y(-x)$ also is. The rule is then exact for both functions, that is:

$$\int_{-h}^{h} y(x)dx = h \sum_{n=1}^{N} a_n y(x_n h),$$

and

$$\int_{-h}^{h} y(-x)dx = h \sum_{n=1}^{N} a_n y(-x_n h) = h \sum_{n=1}^{N} a_{N+1-n} y(x_n h).$$

The two integrals are equal and, since the quadrature rule is assumed unique, it results that $a_n = a_{N+1-n}$. Q. E. D.

For the classical, polynomial-fitting case the existence and uniqueness of the quadrature rule is guaranteed by Hermite interpolation theory, see Chapter 3 in [29], and for the moment we accept that this holds also in the ef-based case. It will be seen that this is true except for the critical values.

The practical consequence of the stated lemma is that we can reduce the number of the coefficients to be determined. We distinguish two situations and consider the first three steps of the flow chart for each of them.

Even N. With $N^* := N/2$ equation (4.75) becomes

$$\int_{X-h}^{X+h} y(z)dz \approx h \sum_{n=1}^{N^*} a_n[(y(X+x_nh) + y(X-x_nh)] \qquad (4.78)$$

and thus only N^* coefficients have to be determined. With the operations required by steps *i* and *iii* in mind, we define the operator $\mathcal{L}[h, \mathbf{a}]$ by

$$\mathcal{L}[h, \mathbf{a}]y(x) := \int_{x-h}^{x+h} y(z)dz - h \sum_{n=1}^{N^*} a_n[(y(x+x_nh) + y(x-x_nh)], \quad (4.79)$$

and apply it on $\exp(\pm\mu x)$. With $z := \mu h$, as usual, we get

$$\mathcal{L}[h, \mathbf{a}] \exp(\pm\mu x) = h \exp(\pm\mu x) E_0^*(\pm z), \qquad (4.80)$$

where

$$E_0^*(\pm z, \mathbf{a}) = \frac{1}{z}[\exp(z) - \exp(-z)] - \sum_{n=1}^{N^*} a_n[\exp(x_n z) + \exp(-x_n z)]. \qquad (4.81)$$

On this basis, the expressions of $G^{\pm}(Z, \mathbf{a})$ are

$$G^+(Z, \mathbf{a}) = 2[\eta_0(Z) - \sum_{n=1}^{N^*} a_n \eta_{-1}(x_n^2 Z)], \quad G^-(Z, \mathbf{a}) = 0, \qquad (4.82)$$

and then only $G^+(Z, \mathbf{a})$ is of concern. Its derivatives are

$$G^{+(p)}(Z, \mathbf{a}) = 2^{1-p}[\eta_p(Z) - \sum_{n=1}^{N^*} a_n x_n^{2p} \eta_{p-1}(x_n^2 Z)], \quad p = 1, 2, \ldots \qquad (4.83)$$

whence

$$L_{2k}^*(\mathbf{a}) = 2[\frac{1}{2k+1} - \sum_{n=1}^{N^*} a_n x_n^{2k}], \quad L_{2k+1}^*(\mathbf{a}) = 0, \ k = 0, 1, \cdots. \quad (4.84)$$

The fact that $E_0^*(z, \mathbf{a}) = E_0^*(-z, \mathbf{a})$, which in turn leads to vanishing G^- and starred classical moments with odd indices, is a consequence of the symmetric distribution of the nodes and then this property will hold for all versions (that is polynomial and ef-based) with symmetric nodes.

Step ii. The system of N^* equations $L_{2k}^*(\mathbf{a}) = 0, \ k = 0, 1, \ldots, N^* - 1$ is linear and since its matrix is of the Vandermonde type it has a unique solution. As a matter of fact, this way of determining the classical coefficients should be seen as an alternative formulation of the standard one, see e.g. [6]. It is not so elegant as the latter but has the advantages of being in the spirit of the flow chart and of enabling an efficient approach to the genuine ef cases. Note also that the Vandermonde matrix is notorious for its ill conditioning which increases with N^*, see e.g. [33], and for this reason only moderate values of N^* should be considered, $N^* \leq 10$, say. However, this is perhaps sufficient for the current practice.

Once the set $\mathbf{a}$ has been computed it remains to determine the value of M. This means to examine the starred moments $L_{2k}^*(\mathbf{a}), \ k = N^*, N^* + 1, \ldots,$ with the aim of finding the first k such that $L_{2k}^*(\mathbf{a}) \neq 0$. The typical result is that $k = N^*$ and then $M = 2N^* = N$; in particular, this holds for the Newton–Cotes rule. However, some accidental choices of the abscissas may lead to bigger values for k. This is the case, for example, when the abscissa weights are the roots of the N-th degree Legendre polynomial. If so we have $M = N$ and therefore $M = 2N$; this is just normal because the rule (4.75) now becomes simply the N-point Gauss–Legendre rule. However, in the following we shall disregard such accidental choices, which are particularly favourable for accuracy, to tacitly assume that only $M = 2N^* = N$ holds.

Odd N. With $N^* := (N + 1)/2$ equation (4.75) becomes

$$\int_{X-h}^{X+h} y(z)dz \approx h \sum_{n=1}^{N^*-1} a_n[(y(X+x_nh)+y(X-x_nh)]+h\,a_{N^*}y(X). \quad (4.85)$$

(Note that now $x_{N^*} = 0$.) Again, only N^* coefficients have to be determined. The $\mathcal{L}[h, \mathbf{a}]$ operator is introduced by

$$\mathcal{L}[h, \mathbf{a}]y(x) := \int_{x-h}^{x+h} y(z)dz - h \sum_{n=1}^{N^*-1} a_n[(y(x+x_nh)+y(x-x_nh)]-h\,a_{N^*}y(x)$$

$$(4.86)$$

and the same procedure as before leads to the following expressions of the relevant quantities:

$$E_0^*(\pm z, \mathbf{a}) = \frac{1}{z}[\exp(z) - \exp(-z)] - \sum_{n=1}^{N^*-1} a_n[\exp(x_n z) + \exp(-x_n z)] - a_{N^*},$$

$$(4.87)$$

$$G^+(Z, \mathbf{a}) = 2[\eta_0(Z) - \sum_{n=1}^{N^*-1} a_n \eta_{-1}(x_n^2 Z)] - a_{N^*}, \quad G^-(Z, \mathbf{a}) = 0, \quad (4.88)$$

$$G^{+(p)}(Z, \mathbf{a}) = 2^{1-p}[\eta_p(Z) - \sum_{n=1}^{N^*-1} a_n x_n^{2p} \eta_{p-1}(x_n^2 Z)], \quad p = 1, 2, \ldots \quad (4.89)$$

and

$$L_{2k}^*(\mathbf{a}) = 2[\frac{1}{2k+1} - \sum_{n=1}^{N^*-1} a_n x_n^{2k}] - \delta_{k0} a_{N^*}, \quad L_{2k+1}^*(\mathbf{a}) = 0, \quad k \geq 0. \quad (4.90)$$

Step *ii*. The linear system of N^* equations $L_{2k}^*(\mathbf{a}) = 0$, $k = 0, 1, \ldots, N^* - 1$ has a unique solution. As before, the determination of M would imply the examination of the moments $L_{2k}^*(\mathbf{a}) = 0$, $k \geq N^*$ to find the first k such that $L_{2k}^*(\mathbf{a}) \neq 0$. The typical answer will be $k = N^*$ and then $M = 2N^* = N+1$. This is actually the case for the Newton–Cotes rule and the difference with respect to the case of even N is that now M is greater by one unit than N. This is often mentioned in the literature as a reason for giving preferrence to Newton–Cotes formulae with odd N, see e.g. [6]. It is also worth noticing that, as before, some special choice of the abscissas may lead to $k > N^*$, and then to $M > N+1$. However, in the following we assume that the typical relation $M = 2N^* = N+1$ holds.

With the mentioned assumptions on M the cases of even and odd N share the common feature that $M = 2N^*$, i.e. that M is even and uniquely determined by N^*. From this and by virtue of the selfconsistency condition it follows that K is always odd. The following steps are then common.

Step *iv*. If only the pair P, K is considered, then the following $N^* + 1$ possibilities consistent with $M = 2N^*$ exist: $P = N^* - s$, $K = 2s - 3$, with $s = 1, 2, \ldots, N^* + 1$.

Situations with a bigger number of P-s, as in (3.58), are also allowed where the maximal value admitted for I is N^* and this opens the possibility of approaching linear combinations of functions of the form (3.38) or (3.39). For example triplets P_1, P_2, K may be considered if $N^* \geq 2$. There are $(N^*+1)(N^*+2)/2$ such triplets, namely $P_1 = -1, 0, \ldots, N^* - s$, $P_2 = N^* - P_1 - s - 1$, $K = 2s - 3$, $(s = 1, 2, \ldots, N^* + 1)$. However, not all of these are genuine triplets. For example, the subset corresponding to $P_1 = -1$ consists only of pairs.

Step *v*. For fixed I the system to be solved is in general (3.60). Since in our case K is odd and $G^-(Z, \mathbf{a})$ is identically vanishing the system reads

$$G^{+(k)}(0, \mathbf{a}) = 0, \ 0 \le k \le (K-1)/2,$$
$$G^{+(p)}(Z_i, \mathbf{a}) = 0, \ 0 \le p \le P_i, \ i = 1, 2, \ldots, I. \qquad (4.91)$$

(To derive this form equation (3.52) was also used.) This is a nonhomogeneous linear system of N^* equations with N^* unknowns. Its solution, if it exists, is unique and then the condition in the lemma 4.1 is fulfilled. As a matter of fact, since the system is linear the solution does exist and it is unique except for a discrete set of critical values, much similar with what happened for the formulae considered in the previous sections.

The analytic form of the solution can be written only for small N^*. For example, for the Newton–Cotes rule with $N = 2$ (this is also called the trapezoidal rule) we have $N^* = 1$ and the only coefficient a_1 consistent with the best tuned version (that is with $P = 0$ and $K = -1$) is given by solving the linear equation $G^+(Z, a_1) = 0$, with the solution

$$a_1(Z) = \frac{\eta_0(Z)}{\eta_{-1}(Z)}. \qquad (4.92)$$

For functions of the form (3.38) the latter can be written as a function of $\theta := \omega h$ (note that $Z = -\theta^2$), viz.:

$$a_1(\theta) = \frac{\sin(\theta)}{\theta \cos(\theta)}. \qquad (4.93)$$

This expression has been mentioned before in the Introduction. It is consistent with the expression obtained by Vanden Berghe *et al.* in [39], as an application of Ehrenmark's technique.

In general only the numerical values of the coefficients can be calculated. The subroutine EFQS is used for this purpose.

Step *vi*. If $\mathbf{a}(\mathbf{Z})$ is the solution of (4.91) then the *lte* is given by (3.61) with $l = 1$ and $M = N$ or $M = N + 1$ for even or odd N, respectively.

In particular, for the just mentioned trapezoidal rule this is

$$lte_{ef} = -h^3 \frac{L_0^*(a_1(Z))}{Z}(\mu^2 + D^2)y(X) = -2h^3 \frac{\eta_1(Z)}{\eta_{-1}(Z)}(\mu^2 + D^2)y(X). \qquad (4.94)$$

In [7] and [39] it has been shown that, for the oscillatory case, if $\theta < \pi/2$ then some $\eta \in [X - h, X + h]$ does exist such that the last member represents the whole error if X is replaced by this η. This result is also confirmed in the frame of Coleman's procedure, [5].

2.2.2 Nonsymmetric abscissas

In some applications the values of $y(x)$ are known only at certain previously established points whose distribution may exhibit no regularity. For example, this is the case when $y(x)$ represents the numerical solution of some ordinary differential equation: the points were fixed by the solver chosen for that equation in terms of its internal needs.

The $\mathcal{L}[h, \mathbf{a}]$ operator associated to approximation (4.75) is

$$\mathcal{L}[h, \mathbf{a}]y(x) := \int_{x-h}^{x+h} y(z)dz - h \sum_{n=1}^{N} a_n y(x + x_n h). \qquad (4.95)$$

By applying the same procedure as before we find:

$$E_0^*(\pm z, \mathbf{a}) = \frac{1}{z}[\exp(z) - \exp(-z)] - \sum_{n=1}^{N} a_n \exp(\pm x_n z), \qquad (4.96)$$

$$G^+(Z, \mathbf{a}) = 2\eta_0(Z) - \sum_{n=1}^{N} a_n \eta_{-1}(x_n^2 Z), \quad G^-(Z, \mathbf{a}) = -\sum_{n=1}^{N} a_n x_n \eta_0(x_n^2 Z) \qquad (4.97)$$

$$G^{+(p)}(Z, \mathbf{a}) = 2^{-p}[2\eta_p(Z) - \sum_{n=1}^{N} a_n x_n^{2p} \eta_{p-1}(x_n^2 Z)],$$

$$G^{-(p)}(Z, \mathbf{a}) = -2^{-p} \sum_{n=1}^{N} a_n x_n^{2p+1} \eta_p(x_n^2 Z), \quad p = 1, 2, \ldots \qquad (4.98)$$

$$L_{2k}^*(\mathbf{a}) = \frac{2}{2k+1} - \sum_{n=1}^{N} a_n x_n^{2k}, \quad L_{2k+1}^*(\mathbf{a}) = -\sum_{n=1}^{N} a_n x_n^{2k+1}, \quad k = 0, 1, \ldots \qquad (4.99)$$

Step *ii*. The linear system of N equations $L_k^*(\mathbf{a}) = 0$, $k = 0, 1, \ldots, N-1$ has a unique solution and, after finding it, one must search for the first $k \geq N$ such that $L_k^*(\mathbf{a}) \neq 0$. The typical answer will be $k = N$ although for some special choices of the abscissa points it may well happen that $k > N$.

We assume that $k = N$. If so, we have $M = N$ no matter whether N is even or odd, and this is different from what was typically encountered for symmetrically placed nodes.

Step *iv*. Since the discussion is restricted only to the situation when $M = N$, the selfconsistency condition implies that N and K have opposite parities; if N is even/odd then K is odd/even. This is also different with respect to the case of the symmetric nodes where K was always odd. With $N^* := \lfloor (N-1)/2 \rfloor$, $N^* + 1$ pairs P, K consistent with $M = N$ are allowed for even N, viz.: $P = N^* - s$, $K = 2s - 3$ ($1 \leq s \leq N^* + 1$), but N^* pairs for odd N, viz.:

$P = N^* - s - 1$, $K = 2s - 2$ $(1 \leq s \leq N^*)$. Triplets, quadruplets and so on can be also considered for reasonable big values of N.

Step v. For fixed I, and for P_i and K taken such that the selfconsistency condition with $M = N$ holds, the system (3.60) is

$$G^{\pm(k)}(0, \mathbf{a}) = 0, \ 0 \leq k \leq (K - 1)/2, \ G^{\pm(p)}(Z_i, \mathbf{a}) = 0, \qquad (4.100)$$

for even N, and

$$G^{\pm(k)}(0, \mathbf{a}) = 0, \ 0 \leq k \leq K/2 - 1, \ G^{+(K/2)}(0, \mathbf{a}) = 0, \ G^{\pm(p)}(Z_i, \mathbf{a}) = 0, \qquad (4.101)$$

for odd N. In both cases $0 \leq p \leq P_i$ and $i = 1, 2, \ldots, I$.

Step vi. If $\mathbf{a}(\mathbf{Z})$ is the solution then the lte is given by (3.61) with $l = 1$ and $M = N$.

2.2.3   Subroutines EFQS and EFQNS

The subroutine EFQS computes the coefficients of the rule (4.75) with symmetrically distributed (but not necessarily equidistant) nodes; it can be then used for the trapezoidal, Simpson and the Newton–Cotes rules but not only for them. The computation of these coefficients implies solving the linear system (4.91) for given $N \geq 2$, $K \geq -1$ (K must be odd), $I \geq 1$, $P_i \geq -1$ and Z_i $(1 \leq i \leq I)$ such that the selfconsistency condition with $M = 2N^*$ is satisfied.

With the regularization procedure in mind and since only the function G^+ is involved, we consider the system $G^+(u_k, \mathbf{a}) = 0$ for $1 \leq k \leq N^*$. This is a system of N^* equations of the form (3.63) with $k_{line} = N^*$, that is with one block only. The block label is then superfluous and (3.63) reads

$$f_1(u_k)a_1 + f_2(u_k)a_2 + \cdots + f_{N^*}(u_k)a_{N^*} = g(u_k) \text{ for } 1 \leq k \leq N^* , \quad (4.102)$$

where

$$f_m(u) = \eta_{-1}(x_m^2 u), \ 1 \leq m \leq N^*, \ g(u) = \eta_0(u) ,$$

for even N^* and

$$f_m(u) = 2\eta_{-1}(x_m^2 u), \ 1 \leq m \leq N^* - 1, \ f_{N^*}(u) = 1, \ g(u) = 2\eta_0(u) ,$$

for odd N^*. As for the specific values of the argument which are consistent with our concern, with $K^* = (K - 1)/2$ and

$$k_1 = K^* + 1, \ k_{i+1} = k_i + P_i + 1, \ 1 \leq i \leq I, \qquad (4.103)$$

these are

$$u_k = 0 \text{ for } 1 \leq k \leq k_1, \text{ and } u_k = Z_i \text{ for } k_i + 1 \leq k \leq k_{i+1} \ (1 \leq i \leq I). \qquad (4.104)$$

Note that this set of assignments correctly discards the cases when $k_{i+1} < k_i+1$ which appear when $P_i = -1$.

The subroutine first checks whether the input data satisfy the selfconsistency condition. If this does not hold a message is written and the execution is stopped. Otherwise the execution goes on to the regularization of the system by subroutine REGSOLV2 with ad-hoc versions of INFF and of INFG, named as INFFNC1 and INFGNC1, respectively. To compute the derivatives of the coefficients with respect to u we used

$$\eta_s^{(i)}(x_n^2 u) = \frac{1}{2^i} x_n^{2i} \eta_{i+s}(x_n^2 u), \ i = 0, 1, \ldots, \ (s = -1, 0),$$

see (3.32).

Calling this subroutine is

$$\text{CALL EFQS(N, XMESH, K, I, IP, Z, A)}$$

The arguments are:
Input:

- N - number of the mesh points.

- K, I - the values of K and I.

- XMESH - double precision vector for the abscissa weights x_n, $1 \leq n \leq N$; these must satisfy $x_n = -x_{N+1-n}$ for all n involved.

- IP - integer vector for P_i, $1 \leq i \leq I$,

- Z - double precision vector for Z_i, $1 \leq i \leq I$.

Output:

- A - double precision vector for the coefficients $a_n(\mathbf{Z})$, $1 \leq n \leq N$.

The linear system is affected by ill conditioning which increases with N. For this reason only values of N up to twenty or so should be used.

For the illustration we take the case $N = 6$, $I = 2$, $K = -1$, $P_1 = 1$, $P_2 = 0$, $Z_1 = -1$, $Z_2 = -2$ for the equidistant partition (4.76). The driving program and the output are

```
implicit double precision (a-h,o-z)
dimension z(2),xmesh(20),ip(2),a(20)
data n/6/, i/2/, k/-1/, ip/1,0/, z/-1,-2/
do j=0,n-1
   xmesh(j+1)=-1.d0+2.d0*j/(n-1)
enddo
```

```
      call efqs(n,xmesh,k,i,ip,z,a)
      do j=1,n
        write(*,50)j,a(j)
      enddo
50    format(2x,'a(',i2,') = ',f15.12)
      stop
      end
```

and

```
  a( 1) =   0.134965518520
  a( 2) =   0.512374495027
  a( 3) =   0.352698370536
  a( 4) =   0.352698370536
  a( 5) =   0.512374495027
  a( 6) =   0.134965518520
```

The subroutine EFQNS computes the coefficients of the rule (4.75) with arbitrary nodes. For fixed N, I, and P_i, K chosen such that $K + 2(P_1 + P_2 + \cdots + P_I) = N - 2I - 1$ the system of N equations consists of two blocks, one with the G^+ based and the other with the G^- based expressions of the coefficients, see (4.100) and (4.101). When N is even the two blocks have the same number of equations $N/2$, while when N is odd their number is $(N+1)/2$ and $(N-1)/2$, respectively. In short, the system is of the form (3.63) with

$$f_m^1(u) = \eta_{-1}(x_m^2 u), \; f_m^2(u) = x_m \eta_0(x_m^2 u), \; 1 \le m \le N,$$
$$g^1(u) = 2\eta_0(u), \; g^2(u) = 0,$$

while k_{line} is $N/2$ or $(N+1)/2$ for even or odd N, respectively. The values of the arguments are assigned in terms of the given Z_i, $i = 1, 2, \ldots, I$ in the following way:

Even N. The arguments of the equations in the first block are fixed by equations (4.103) and (4.104) for $K^* = (K-1)/2$. The arguments in the second block are

$$u_{N/2+k} = u_k, \; k = 1, 2, \ldots, N/2.$$

Odd N. For the first and second block we use (4.103) and (4.104) for $K^* = K/2$, and

$$u_{(N+1)/2+k} = u_{k+1}, \; k = 1, 2, \ldots (N-1)/2,$$

respectively.

The system is regularized and solved by REGSOLV2 to which ad-hoc versions INFFNC2 and INFGNC2 of INFF and INFG are attached.

Calling it is

$$\text{CALL EFQNS}(N, XMESH, K, I, IP, Z, A)$$

where the arguments are precisely the same as in EFQS except that at least one value of n exists for which the condition $x_n = -x_{N+1-n}$ does not hold.

For the illustration we take the same case as for EFQS except that now the nodes are nonsymmetrically distributed by the rule

$$x_n = -1 + \frac{2(n-1)}{N-1} + \frac{1}{10n}, \quad n = 1, 2, \ldots, N.$$

The driving program is

```
      implicit double precision (a-h,o-z)
      dimension z(2),xmesh(20),ip(2),a(20)
      data n/6/, i/2/, k/-1/, ip/1,0/, z/-1,-2/
      do j=0,n-1
        xmesh(j+1)=-1.d0+2.d0*j/(n-1)+1.d0/(10*(j+1))
      enddo
      call efqns(n,xmesh,k,i,ip,z,a)
      do j=1,n
        write(*,50)j,a(j)
      enddo
50    format(2x,'a(',i2,') = ',f15.12)
      stop
      end
```

with the output

```
a( 1) =   0.256505876379
a( 2) =   0.361479715907
a( 3) =   0.468464117154
a( 4) =   0.260814125064
a( 5) =   0.540963196191
a( 6) =   0.111808392590
```

2.3 Extended quadrature rules with predetermined abscissa points

The extension consists in admitting that not only the pointwise values of the integrand are available but also the values of a number of its derivatives. The extended form of rule (4.75) is

$$\int_a^b y(z)dz = \int_{X-h}^{X+h} y(z)dz \approx \sum_{j=0}^J h^{j+1} \sum_{n=1}^N a_n^{(j)} y^{(j)}(X + x_n h), \quad (4.105)$$

where $y^{(0)}$ is simply y. The extended rule is then identified by the number of nodes N, the weights x_n of their position, and the order J of the highest derivative of the integrand accepted in the rule. The standard rule (4.75) is the particular case $J = 0$. When the points are distributed as in (4.76) the rule will be called an extended Newton–Cotes rule. For earlier versions of such rules see Section 2.7 of [6] and references therein.

The extended rules are of use in several practical problems, and the Sturm-Liouville problem is one of them; for a detailed description of this see [34]. In essence, this consists in searching for the eigenvalues and the associated normalized eigenfunctions of a boundary value problem. The determination of the eigenvalues is typically achieved by shooting, with usual numerical solvers for the generation of the solutions to be matched. In this way the user will dispose not only of the pointwise solution but also of those of the first and second derivative and then an extended rule is appropriate for the computation of the normalization integral.

A recent investigation on the extended rules is that of Kim, Cools and Ixaru. In papers [26] and [27] these authors construct polynomial and ef-based versions for the case when the abscissa points are distinct and symmetrically distributed. The following lemma concerns the symmetry properties of the coefficients of these rules:

LEMMA 4.2 *If the nodes of the quadrature rule (4.105) are symmetrically distributed around X, that is $x_n = -x_{N+1-n}$, and if the coefficients $a_n^{(j)}$ are the only ones which ensure that the rule is exact for the functions (3.58), then the coefficients $a_n^{(0)}$ and $a_n^{(j)}$ (even j) are symmetric while $a_n^{(j)}$ (odd j) are antisymmetric.*

Proof This closely follows the proof of lemma 4.1. As there we take $X = 0$ and use the fact that the RDE is of form (4.77). If $y(x)$ is a solution of this RDE then $y(-x)$ also is, and therefore the rule is exact for both functions. We have

$$\int_{-h}^{h} y(-z)dz = \sum_{j=0}^{J} h^{j+1} \sum_{n=1}^{N} a_n^{(j)} (-1)^j y^{(j)}(-x_n h)$$

$$= \sum_{j=0}^{J} h^{j+1} \sum_{n=1}^{N} a_{N+1-n}^{(j)} (-1)^j y^{(j)}(x_n h) \quad (4.106)$$

and

$$\int_{-h}^{h} y(z)dz = \sum_{j=0}^{J} h^{j+1} \sum_{n=1}^{N} a_n^{(j)} y^{(j)}(x_n h). \quad (4.107)$$

Since the two integrals are equal and the quadrature rule is assumed unique we must have

$$a_n^{(j)} = (-1)^j a_{N+1-n}^{(j)}. \quad (4.108)$$

Q. E. D.

For the classical, polynomial fitting version of the rule the existence and uniqueness is guaranteed by Hermite interpolation theory (Chapter 3 in [29]). Tables 4.2, 4.3 and 4.4, taken from [26] and [27], list the coefficients of the extended N-point Newton–Cotes rule for $J = 1$, 2 and 3, respectively, and for N between 2 and 6.

Let us assume that the rule, if it exists, is uniquely defined also for ef versions; indeed, as it will be seen later on, this holds because the coefficients result by solving a linear system of equations whose determinant is nonvanishing except for the critical values. If so, the lemma can be applied and this allows halving the number of coefficients to be determined. Again, we distinguish two cases:

Table 4.2. Extended Newton–Cotes rule: classical coefficients for $J = 1$.

coefficient	$N = 2$	$N = 3$	$N = 4$	$N = 5$	$N = 6$
$a_1^{(0)}$	1	$\frac{7}{15}$	$\frac{465}{1680}$	$\frac{1601}{8505}$	$\frac{446719}{3193344}$
$a_2^{(0)}$		$\frac{16}{15}$	$\frac{1215}{1680}$	$\frac{4096}{8505}$	$-\frac{968625}{3193344}$
$a_3^{(0)}$				$\frac{5616}{8505}$	$\frac{1778000}{3193344}$
$a_1^{(1)}$	$\frac{1}{3}$	$\frac{1}{15}$	$\frac{38}{1680}$	$\frac{87}{8505}$	$\frac{17748}{3193344}$
$a_2^{(1)}$		0	$-\frac{54}{1680}$	$-\frac{384}{8505}$	$-\frac{150900}{3193344}$
$a_3^{(1)}$				0	$-\frac{130800}{3193344}$

Even N. With $N^* := N/2$ the extended rule reads

$$\int_{X-h}^{X+h} y(z)\,dz \approx \sum_{n=1}^{N^*} \sum_{j=0}^{J} h^{j+1} a_n^{(j)} [y^{(j)}(X + x_n h) + (-1)^j y^{(j)}(X - x_n h)],$$

(4.109)

The total number of coefficients to be determined is $N_t = (J + 1)N^*$. We denote $\mathbf{a} := [a_n^{(j)},\ 1 \le n \le N^*,\ 0 \le j \le J]$ and define the operator $\mathcal{L}[h, \mathbf{a}]$ by

$$\mathcal{L}[h, \mathbf{a}]y(x) := \int_{x-h}^{x+h} y(z)\,dz$$

Table 4.3.   Extended Newton–Cotes rule: classical coefficients for $J = 2$.

coefficient	$N = 2$	$N = 3$	$N = 4$	$N = 5$	$N = 6$
$a_1^{(0)}$	1	$\frac{41}{105}$	$\frac{1283}{4928}$	$\frac{628741}{3648645}$	$\frac{3964974041}{29035708416}$
$a_2^{(0)}$		$\frac{128}{105}$	$\frac{3645}{4928}$	$\frac{311296}{331695}$	$\frac{89828125}{119488512}$
$a_3^{(0)}$				$-\frac{256}{1155}$	$\frac{101328125}{907365888}$
$a_1^{(1)}$	$\frac{2}{5}$	$\frac{2}{35}$	$\frac{311}{12320}$	$\frac{13138}{1216215}$	$\frac{572685527}{84687482880}$
$a_2^{(1)}$		0	$-\frac{243}{12320}$	$\frac{72704}{1216215}$	$\frac{20025625}{513257472}$
$a_3^{(1)}$				0	$\frac{51299375}{1058593536}$
$a_1^{(2)}$	$\frac{1}{15}$	$\frac{1}{315}$	$\frac{17}{18480}$	$\frac{43}{173745}$	$\frac{2580673}{21171870720}$
$a_2^{(2)}$		$\frac{16}{315}$	$\frac{81}{6160}$	$\frac{15872}{1216215}$	$\frac{22625}{3290112}$
$a_3^{(2)}$				$-\frac{848}{45045}$	$-\frac{1666625}{264648384}$

$$-\sum_{n=1}^{N^*}\sum_{j=0}^{J} h^{j+1} a_n^{(j)} [y^{(j)}(x + x_n h) + (-1)^j y^{(j)}(x - x_n h)], \quad (4.110)$$

to obtain

$$E_0^*(\pm z, \mathbf{a}) = \frac{1}{z}[\exp(z) - \exp(-z)]$$

$$-\sum_{n=1}^{N^*}\sum_{j=0}^{J} z^j a_n^{(j)} [\exp(x_n z) + (-1)^j \exp(-x_n z)], \quad (4.111)$$

whence

$$G^+(Z, \mathbf{a}) = 2\eta_0(Z) - 2\sum_{n=1}^{N^*} \left[\sum_{j=0\,(e)}^{J} a_n^{(j)} Z^{j/2}\eta_{-1}(x_n^2 Z) \right.$$

$$\left. + \sum_{j=1\,(o)}^{J} a_n^{(j)} Z^{(j+1)/2} x_n \eta_0(x_n^2 Z) \right], \quad (4.112)$$

Table 4.4. Extended Newton–Cotes rule: classical coefficients for $J = 3$.

coefficient	$N = 2$	$N = 3$	$N = 4$	$N = 5$	$N = 6$
$a_1^{(0)}$	1	$\frac{103}{231}$	$\frac{9635}{36608}$	$\frac{174876341}{964237365}$	$\frac{1496215546227913}{10962954355212288}$
$a_2^{(0)}$		$\frac{256}{231}$	$\frac{26973}{36608}$	$\frac{186646528}{623918295}$	$-\frac{562928515625}{1670927351808}$
$a_3^{(0)}$				$\frac{1679616}{1616615}$	$\frac{6425836328125}{5353005056256}$
$a_1^{(1)}$	$\frac{3}{7}$	$\frac{19}{231}$	$\frac{18059}{640640}$	$\frac{2752961}{207972765}$	$\frac{47565699805097}{6395056707207168}$
$a_2^{(1)}$		0	$-\frac{21627}{640640}$	$-\frac{2244608}{24724035}$	$-\frac{343630689921875}{2131685569069056}$
$a_3^{(1)}$				0	$-\frac{10445050703125}{49961380525056}$
$a_1^{(2)}$	$\frac{2}{21}$	$\frac{26}{3465}$	$\frac{703}{480480}$	$\frac{5986}{12950685}$	$\frac{51220899809}{266460696133632}$
$a_2^{(2)}$		$\frac{128}{3465}$	$\frac{1539}{160160}$	$-\frac{2203648}{1178512335}$	$-\frac{674253921875}{88820232044544}$
$a_3^{(2)}$				$\frac{228224}{14549535}$	$\frac{47311421875}{4163448377088}$
$a_1^{(3)}$	$\frac{1}{105}$	$\frac{1}{3465}$	$\frac{1}{32032}$	$\frac{7703}{1178512335}$	$\frac{1326980021}{666151740334080}$
$a_2^{(3)}$		0	$-\frac{9}{12320}$	$-\frac{151552}{235702467}$	$-\frac{22458216875}{44410116022272}$
$a_3^{(3)}$				0	$-\frac{11872639375}{8326896754176}$

where e or o means that the sum is running only over even or odd values of j, respectively. It also results that $G^-(Z, \mathbf{a}) = 0$ for any Z and $\mathbf{a}$. Only G^+ and its derivatives are therefore involved in the computation of the coefficients; for the derivatives of G^+ with respect to Z use (3.31). Also only the starred classical moments with even indices appear. Their expressions, which result directly by (3.49), are

$$L_{2k}^*(\mathbf{a}) = 2[\frac{1}{2k + 1} - \sum_{n=1}^{N^*} \sum_{j=0}^{J^*} \frac{(2k)!}{(2k - j)!} a_n^{(j)} x_n^{2k-j}], \quad k = 0, 1, \cdots \quad (4.113)$$

where $J^* = \min\{2k, J\}$.

Odd N. With $N^* := (N+1)/2$ the extended rule reads

$$\int_{X-h}^{X+h} y(z)dz \approx \sum_{n=1}^{N^*-1} \sum_{j=0}^{J} h^{j+1} a_n^{(j)} [y^{(j)}(X + x_n h) + (-1)^j y^{(j)}(X - x_n h)]$$

$$+ \sum_{j=0\,(e)}^{J} h^{j+1} a_{N^*}^{(j)} y^{(j)}(X). \tag{4.114}$$

The number of coefficients N_t to be collected in a is no longer $(J+1)N^*$, as before, but $(J+1)N^* - (J+1)/2$ for odd J and $(J+1)N^* - J/2$ for even J. The reason is that the number of coefficients with a fixed j is N^* if j is even and $N^* - 1$ if j is odd. The operator $\mathcal{L}[h, \mathbf{a}]$ is introduced by

$$\mathcal{L}[h, \mathbf{a}] y(x) := \int_{x-h}^{x+h} y(z)dz - \sum_{n=1}^{N^*-1} \sum_{j=0}^{J} h^{j+1} a_n^{(j)} [y^{(j)}(x + x_n h)$$

$$+ (-1)^j y^{(j)}(x - x_n h)] - \sum_{j=0\,(e)}^{J} h^{j+1} a_{N^*}^{(j)} y^{(j)}(x), \tag{4.115}$$

and then

$$E_0^*(\pm z, \mathbf{a}) = \frac{1}{z}[\exp(z) - \exp(-z)]$$

$$- \sum_{n=1}^{N^*-1} \sum_{j=0}^{J} z^j a_n^{(j)} [\exp(x_n z) + (-1)^j \exp(-x_n z)] - \sum_{j=0\,(e)}^{J} z^j a_{N^*}^{(j)}, \tag{4.116}$$

whence

$$G^+(Z, \mathbf{a}) = 2\eta_0(Z) - 2 \sum_{n=1}^{N^*-1} \left[\sum_{j=0\,(e)}^{J} a_n^{(j)} Z^{j/2} \eta_{-1}(x_n^2 Z) \right.$$

$$\left. + \sum_{j=1\,(o)}^{J} a_n^{(j)} Z^{(j+1)/2} x_n \eta_0(x_n^2 Z) \right] - \sum_{j=0\,(e)}^{J} a_{N^*}^{(j)} Z^{j/2}, \tag{4.117}$$

$$L_{2k}^*(\mathbf{a}) = 2 \left[\frac{1}{2k+1} - \sum_{n=1}^{N^*-1} \sum_{j=0}^{J^*} \frac{(2k)!}{(2k-j)!} a_n^{(j)} x_n^{2k-j} \right]$$

$$- (2k)! \sum_{j=0}^{J^*} \delta_{j,2k} a_{N^*}^{(j)}, \quad k = 0, 1, \cdots \tag{4.118}$$

where $J^* = \min\{2k, J\}$.

Step *ii*. First of all, it is useful to observe that

PROPOSITION 4.1 *The value of N_t is given by the compact formula $N_t = \lfloor Q/2 \rfloor$, where $Q := (J+1)N$, irrespective of whether N is even or odd.*

The determination of the N_t unknowns for the classical version requires the solution of the linear system $L^*_{2k}(\mathbf{a}) = 0$ $k = 0, 1, \cdots, N_t - 1$. The system matrix is nonsingular and then the solution exists and it is unique. As a matter of fact, the coefficients listed in Tables 4.2, 4.3 and 4.4 for $1 \leq J \leq 3$, were obtained by the analytic solution of this system.

To fix M we have to search for the first $k \geq N_t$ for which $L^*_{2k}(\mathbf{a}) = 0$. As a rule this will be just $k = N_t$ but, similarly with the rules discussed in the previous section, some special sets of abscissas may lead to $k > N_t$. However, in the following we assume that $k = N_t$ or, equivalently, that $M = 2N_t$; in particular, this holds for the extended Newton -Cotes rule. If so, it follows that $M = Q$ irrespective of J if N is even, and also when both J and N are odd, but $M = Q + 1$ when J is even and N is odd. This can be formulated compactly in terms of Q alone:

PROPOSITION 4.2 *M is an even number. Its value is $M = Q$ or $M = Q+1$ if Q is even or odd, respectively.*

A direct consequence is that the advantage of the standard rule with an odd number of symmetric nodes over those with an even number is conserved only for the extended versions with even J; otherwise the two are equivalent.

The considerations made with respect to the steps *iv*, *v* and *vi* for the standard rule with symmetric nodes remain valid here provided that N^* is replaced by N_t. Specifically:

Step *iv*. K is always odd and if only the pair P, K is considered, then the following $N_t + 1$ possibilities exist: $P = N_t - s$, $K = 2s - 3$, $s = 1, 2, \ldots, N_t + 1$. Triplets, quadruplets and so on can be also considered for suitable N_t.

Step *v*. The system to be solved is

$$G^{+(k)}(0, \mathbf{a}) = 0, \; 0 \leq k \leq (K-1)/2,$$
$$G^{+(p)}(Z_i, \mathbf{a}) = 0, \;\; 0 \leq p \leq P_i, \; i = 1, 2, \ldots, I \,, \quad (4.119)$$

where the selfconsistency condition is satisfied with $M = 2N_t$. This is a nonhomogeneous linear system of N_t equations with N_t unknowns and then the solution, if it exists, is unique such that the condition in the lemma 4.2 is satisfied. The experimental evidence indicates that this happens always, with the exception of a discrete set of critical values. As for the way of calculating the solution we are again opting only for the numerical approach. To this aim we use subroutine EFEXTQS.

Step *vi.* If $a(Z)$ is the solution then the *lte* is given by (3.61) with $l = 1$ and $M = 2N_t$.

Numerical experiments

Extensive numerical tests with various ef-based versions of the extended Newton–Cotes rule were carried out for $2 \leq N \leq 6$ and $0 \leq J \leq 3$, see [27].

We select only the versions generated upon the requirement of being maximally adapted for functions of the form (3.38) or (3.39), which means using $P = N_t - 1$ and $K = -1$ as the basis for the construction of the coefficients, to actually concentrate the attention to the case of the oscillatory functions; remember that in such a case the associated Z is $Z = -\omega^2 h^2$.

In the following we examine the ωh dependence of various quantities of interest.

A first issue regards the behaviour of the coefficients. It is found experimentally that the parity of J plays an important role in this context. Indeed:

1. No critical value appears when J is odd. Moreover, all coeficients tend to zero as ωh tends to infinity.

2. In contrast, critical values do exist for even J (or for $J = 0$, as in the standard rule). In the vicinity of each such value the coefficients exhibit a pole behaviour. Technically the reason is that the determinant of the system, seen as a function of ωh, is an oscillatory function whose zeros are the critical values. However, if the regions around the critical values are disregarded, then the tendency of the coefficients to tend to zero when ωh increases holds too.

These two distinct behaviours are visible on Figures 4.7, 4.8, 4.9 and 4.10, for $N = 2$ and $J = 0, 1, 2, 3$. As for the shape of variation for odd J, each coefficient exhibits an oscillatory behaviour with damped amplitude. As a matter of fact, the coefficients do not change the sign on Figures 4.8 and 4.10 but for bigger values of N the oscillatory behaviour of the coefficients may well involve alternations in sign.

Another issue of practical interest is the dependence of the error with respect to ω when h is fixed. The forms of the *lte* are of help in this context. We recall that the *lte* of the ef version of the extended Newton–Cotes rule with given N and J and which is optimally tuned on functions of the form (3.38) is

$$
\begin{aligned}
lte_{ef} &= (-1)^{N_t} h^{2N_t+1} \frac{L_0^*(a(Z))}{Z^{N_t}} (\omega^2 + D^2)^{N_t} y(X) \\
&= (-1)^{N_t} h^{2N_t+1} \frac{2 - \sum_{n=1}^{N} a_n(Z)}{Z^{N_t}} (\omega^2 + D^2)^{N_t} y(X), \quad (4.120)
\end{aligned}
$$

(use (3.61) for $l = 1$, $M = 2N_t$ and $Z = -\omega^2 h^2$) while the *lte* of the classical version is

$$lte_{class} = h^{2N_t+1} \frac{L^*_{2N_t}(\mathbf{a})}{(2N_t)!} D^{2N_t} y(X). \tag{4.121}$$

A comparison of the formulae (4.120) and (4.121) allows deriving some qualitative conclusions. Each of the two consists of a product of three significant factors. The first factor, h^{2N_t+1}, is the same in both expressions. To compare the third factors we use the following

LEMMA 4.3 *If $y(x)$ is of the form (3.38) then, for big ω, $y''(x)$ and $(\omega^2 + D^2)y(x)$ behave dominantly as ω^2 and as ω, respectively.*

Proof The direct differentiation gives

$$y''(x) = -\omega^2 y(x)$$

$$+2\omega[f'_1(x)\cos(\omega x) - f'_2(x)\sin(\omega x)] + f''_1(x)\sin(\omega x) + f''_2(x)\cos(\omega x) .$$

Q.E.D.

The consequence is that $D^{2N_t} y(x)$ and $(\omega^2 + D^2)^{N_t} y(x)$ behave like ω^{2N_t} and like ω^{N_t} only, respectively.

The factor in the middle is a constant in lte_{class} but a function of $Z = -\omega^2 h^2$ in lte_{ef}. The numerator in the latter is two minus the sum of the coefficients $a_n^{(0)}(Z)$. Since for odd J these coefficients tend to zero when $\omega h \to \infty$, it follows that $L^*_0(\mathbf{a}(Z))$ tends to 2 and then the whole middle factor, seen as a function of ω, damps out as ω^{-2N_t}. Altogether, lte_{class} *increases* as ω^{2N_t} while lte_{ef} corresponding to odd J *decreases* as ω^{-N_t}. As a matter of fact, the integral itself also decreases but only as ω^{-1} such that for $N_t > 1$ the relative error from the ef version is expected to decrease when ω tends to infinity.

As a numerical illustration we take the integral

$$\int_{-1}^{1} \cos[(\omega + 1/2)x]\,dx = \frac{2\sin(\omega + 1/2)}{\omega + 1/2}, \tag{4.122}$$

which is of the form (3.38) with $f_1(x) = -\sin(x/2)$ and $f_2(x) = \cos(x/2)$. On Figure 4.11 we display the scaled error $(\omega + 1/2)(I - I^{comput})$, which mimics the relative error, for the extended Newton–Cotes formulae with $N = 2$, and $J = 1$ and 3. For the latter version the scaled error is presented on the figure after being multiplied by 10. The prediction is nicely confirmed.

However, it is important to notice that in general a theoretical result of this type should be taken with some caution because it is relying on the belief that *lte* alone is sufficient for a realistic description of the whole error, a situation which may hold in some cases but not in the others. The experimental tests with $N_t > 1$ have confirmed the existence of a general, systematic decrease of both absolute and relative error when ω increases. However, quite often the experimental rate of the decrease was somewhat smaller than predicted.

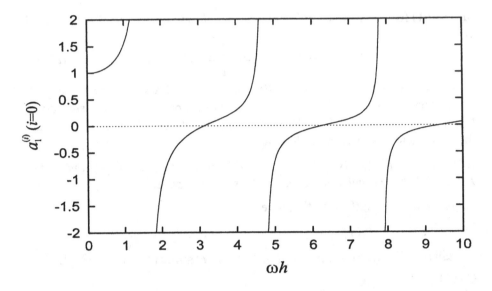

Figure 4.7. The ωh dependence of coefficient $a_1^{(0)}$ of the extended Newton–Cotes rule with $N = 2$ and $J = 0$.

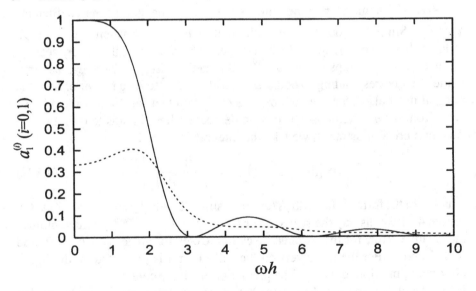

Figure 4.8. The ωh dependence of the coefficients $a_1^{(0)}$ (solid) and $a_1^{(1)}$ (dashed) of the extended Newton–Cotes rule with $N = 2$ and $J = 1$.

2.3.1   Subroutine EFEXTQS

This subroutine computes the coefficients of the extended rule (4.105) with symmetrically distributed (but not necessarily equidistant) nodes. The scheme

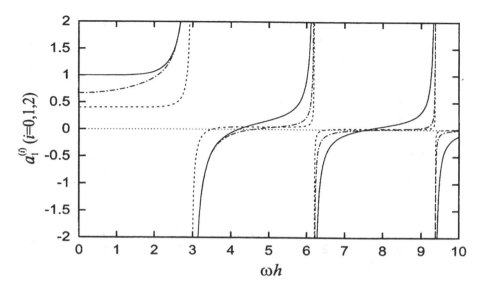

Figure 4.9. The ωh dependence of the coefficients $a_1^{(0)}$ (solid), $a_1^{(1)}$ (dashed) and $10 \times a_1^{(2)}$ (dash–and–dots) of the extended Newton–Cotes rule with $N = 2$ and $J = 2$.

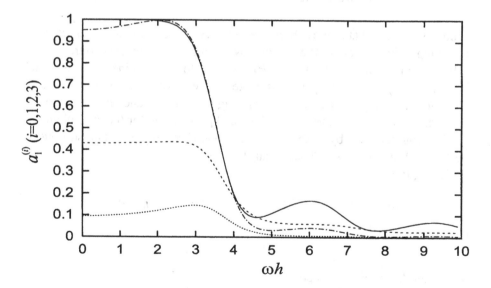

Figure 4.10. The ωh dependence of the coefficients $a_1^{(0)}$ (solid), $a_1^{(1)}$ (dashed), $10 \times a_1^{(2)}$ (dash–and–dots) and $10 \times a_1^{(3)}$ (dots) of the extended Newton–Cotes rule with $N = 2$ and $J = 3$.

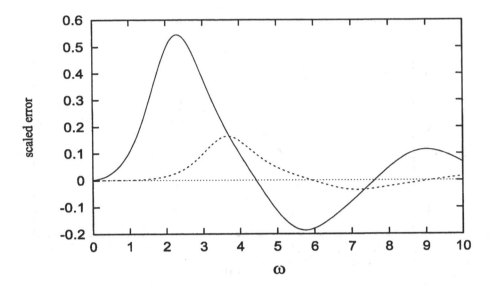

Figure 4.11. The ω dependence of the scaled error $(\omega + 0.5)(I - I^{comput})$ for the integral (4.122) calculated by the Newton–Cotes rule with $N = 2$ and $J = 1$ (solid), and with $N = 2$ and $J = 3$ (dashed). For the latter rule the scaled error multiplied by 10 is plotted.

is much similar to that of the subroutine EFQS. As there, the central part consists in solving a linear system of equations by the regularization procedure.

For given N, J, K, I and P_1, P_2, $\cdots$, P_I the subroutine first computes $N_t = \lfloor (J+1)N + 1 \rfloor$ and performs the test of selfconsistency with $M = 2N_t$. If the test is not passed a warning message is written and the execution is stopped. Otherwise the system (4.119) of N_t equations for the N_t unknowns is solved numerically by REGSOLV2 in the one block regime. The subroutine works for N and $0 \leq J \leq 3$ such that $N_t \leq 20$.

Calling it is

```
CALL EFEXTQS(N, XMESH, J, K, I, IP, Z, A0, A1, A2, A3)
```

The arguments are:
Input:

- N - number of the mesh points.

- XMESH - double precision vector for the mesh point weights x_n, $1 \leq n \leq N$.

- J, K, I - the values of J, K and I.

- IP - integer vector for P_i, $1 \leq i \leq I$,

- Z - double precision vector for Z_i, $1 \leq i \leq I$.

Output:

- A0, A1, A2, A3 - double precision vectors for the coefficients $a_n^{(0)}(\mathbf{Z})$, $a_n^{(1)}(\mathbf{Z})$, $a_n^{(2)}(\mathbf{Z})$, $a_n^{(3)}(\mathbf{Z})$, $1 \le n \le N$.

For the illustration we take $N = 3$ with the equidistant partition (4.76), $I = 1$, $K = -1$ and $J = 0, 1, 2$ and 3. $P_1 = N_t - 1$ is taken for each J and $Z = -1$. The driving program and the output are

```
      implicit double precision (a-h,o-z)
      dimension z(1),xmesh(20),ip(1)
      dimension a0(20),a1(20),a2(20),a3(20)
      data n/3/, i/1/, k/-1/,z/-1/
      do m=0,n-1
        xmesh(m+1)=-1.d0+2.d0*m/(n-1)
      enddo
      do j=0,3
        iq=n*(j+1)
        nt=(iq+1)/2
        ip(1)=nt-1
        call efextqs(n,xmesh,j,k,i,ip,z,a0,a1,a2,a3)
        write(*,*)' j = ',j
        write(*,50)(a0(m),m=1,n)
        if(j.ge.1)write(*,51)(a1(m),m=1,n)
        if(j.ge.2)write(*,52)(a2(m),m=1,n)
        if(j.ge.3)write(*,53)(a3(m),m=1,n)
      enddo
      stop
50    format(2x,'a0: ', 3(1x,f12.9))
51    format(2x,'a1: ', 3(1x,f12.9))
52    format(2x,'a2: ', 3(1x,f12.9))
53    format(2x,'a3: ', 3(1x,f12.9))
      end
```

```
  j =  0
  a0:    0.357907384  1.296185600  0.357907384
  j =  1
  a0:    0.473839352  1.052093043  0.473839352
  a1:    0.070600139  0.000000000 -0.070600139
  j =  2
  a0:    0.392771970  1.214456068  0.392771970
  a1:    0.058010498  0.000000000 -0.058010498
  a2:    0.003272582  0.050037235  0.003272582
```

```
j =  3
a0:    0.447428579   1.105142841   0.447428579
a1:    0.082951886   0.000000000  -0.082951886
a2:    0.007625172   0.036558180   0.007625172
a3:    0.000296796   0.000000000  -0.000296796
```

2.4   Integration of a product of functions

Integrals of products of functions of the form (3.38) or (3.39) appear quite often in the current practice. For example, this happens with the integral (3.20). In the subintervals where E_i, $E_j > \bar{V}$ both $y_i(x)$ and $y_j(x)$ exhibit an oscillatory variation. When E_i, $E_j < \bar{V}$ they behave like (3.39) while when $E_i < \bar{V}$, $E_j > \bar{V}$ or viceversa, one function is oscillatory while the other has a hyperbolic variation. The problem was considered by Kim in [25] for the case when both functions are of the same type but his approach can be extended easily.

To fix the ideas let us assume that

$$y(x) = y_a(x) \cdot y_b(x) \tag{4.123}$$

where y_a and y_b are characterized by the pairs of individual frequencies $\pm\mu_a$ and $\pm\mu_b$, respectively. The whole product is then characterized by the pairs $\pm\mu_1$, $\pm\mu_2$ where $\mu_1 = \mu_a + \mu_b$ and $\mu_2 = \mu_a - \mu_b$. In other words, the set of quadruplets

$$\exp(\pm\mu_i x), \ i = 1, \ 2; \ x\exp(\pm\mu_i x), \ i = 1, \ 2; \ x^2\exp(\pm\mu_i x), \ i = 1, \ 2; \ \cdots \tag{4.124}$$

is appropriate for the whole y. The treatment of this case is then covered by equations (3.58) and (3.60) with $I = 2$ and with $P_1 = P_2$.

The values of the two Z-s are $Z_i = \mu_i^2 h^2$. Both are real when y_a and y_b are of the same type, i.e. when μ_a and μ_b are either purely real or purely imaginary, but they have complex conjugate values when y_a and y_b have different behaviours (for example when $\mu_a = i\omega$ and $\mu_b = \lambda$).

2.4.1   Subroutine CENC1

To ensure a uniform treatment of all these situations by the extended rule (4.105) with symmetric nodes the subroutine EFEXTQS for the numerical calculation of the coefficients has to be adapted to accept pairs of mutually complex conjugate values of Z.

Subroutine CENC1 does this. Its main special ingredients are the use of the subroutine CGEBASEV which computes sets of η functions of complex arguments, and of a complex extension of the pair LUDEC and LUSBS for the solution of linear systems of algebraic equations. It is important to mention that, since $P_1 = P_2$ and since the Z-s are chosen in the mentioned way, the resultant

coefficients are real. Similar to EFEXTQS, this subroutine works only if N and $0 \le J \le 3$ are such that $N_t \le 20$.
Calling it is

$$\text{CALL CENC1(N, XMESH, J, K, I, IP, Z, A0, A1, A2, A3)}$$

Again, the arguments are the same as in EFEXTQS. However, pay attention to the restrictions on I, IP and Z :
Input:

- N - number of the mesh points.

- XMESH - double precision vector for the mesh point weights x_n, $1 \le n \le N$.

- J, K, I - the values of J, K and I. I must be even.

- IP - integer vector for P_i, $1 \le i \le I$. The IP-s must be equal pairwise: IP(1)=IP(2), IP(3)=IP(4) etc.

- Z - double complex vector for Z_i, $1 \le i \le I$. The Z-s must be complex conjugate pairwise: Z(1)=Z*(2), Z(3)=Z*(4) etc.

Output:

- A0, A1, A2, A3 - double precision vectors for the coefficients $a_n^{(0)}(\mathbf{Z})$, $a_n^{(1)}(\mathbf{Z})$, $a_n^{(2)}(\mathbf{Z})$, $a_n^{(3)}(\mathbf{Z})$, $1 \le n \le N$.

The following driving program calculates the coefficients of the rule with $N = 4$ nodes and with $J = 1$ and $J = 3$. Since for these values of J one has $M = 4(J + 1)$, equation (3.59) with $K = -1$ gives $P_1 = P_2 = 1$ for $J = 1$ and $P_1 = P_2 = 3$ for $J = 3$. The input frequencies for the components of the product are $\mu_a = i\omega$ and $\mu_b = \lambda$ with $\omega = \lambda = 1$.

```
implicit double precision (a-h,o-z)
double complex z(2)
dimension xmesh(20),ip(2)
dimension a0(20),a1(20),a2(20),a3(20)
data n/4/, i/2/, k/-1/,omega/1.d0/,rl/1.d0/
h=1.d0
z(1)=(dcmplx(rl,omega)*h)**2
z(2)=(dcmplx(rl,-omega)*h)**2
do m=0,n-1
   xmesh(m+1)=-1.d0+2.d0*m/(n-1)
enddo
do j=1,3,2
   if(j.eq.1)ip(1)=1
```

```
            if(j.eq.3)ip(1)=3
            ip(2)=ip(1)
            call cenc1(n,xmesh,j,k,i,ip,z,a0,a1,a2,a3)
            write(*,*)' j = ',j
            write(*,50)(a0(m),m=1,n)
            write(*,51)(a1(m),m=1,n)
            if(j.eq.3)write(*,52)(a2(m),m=1,n)
            if(j.eq.3)write(*,53)(a3(m),m=1,n)
         enddo
         stop
50       format(2x,'a0: ', 4(1x,f10.7))
51       format(2x,'a1: ', 4(1x,f10.7))
52       format(2x,'a2: ', 4(1x,f10.7))
53       format(2x,'a3: ', 4(1x,f10.7))
         end
```

The output is:

```
 j =  1
 a0:    0.2766276  0.7233664  0.7233664  0.2766276
 a1:    0.0226041 -0.0323092  0.0323092 -0.0226041
 j =  3
 a0:    0.2631870  0.7368130  0.7368130  0.2631870
 a1:    0.0281875 -0.0337642  0.0337642 -0.0281875
 a2:    0.0014630  0.0096089  0.0096089  0.0014630
 a3:    0.0000312 -0.0007307  0.0007307 -0.0000312
```

To illustrate the accuracy of a rule adapted for a product of functions we take the integral

$$I = \int_{-1}^{1} \cos(\alpha x) \exp(-\beta x) dx$$

$$= \frac{1}{\alpha^2 + \beta^2} [-\beta \cos(\alpha x) + \alpha \sin(\alpha x)] \exp(-\beta x)|_{-1}^{1}. \quad (4.125)$$

where $\alpha = \omega + 1/2$ and $\beta = \lambda + 1/2$. We calculate this integral by the versions with $N = 2$ and $J = 1, 3$ where $\mu_a = i\omega$ and $\mu_b = \lambda$ are taken for input. (If $\mu_a = i\alpha$ and $\mu_b = \beta$ are taken for input the computed integral would be exact except for the round-off.)

On Figure 4.12 we represent the error $I - I^{comput}$ in terms of $0 \leq \omega \leq 10$, for fixed $\lambda = 1$. The curve displayed for the version with $J = 3$ represents the error multiplied by ten. The two main features, i.e. that the error damps out when ω increases, and than the bigger J the more accurate is the result, are confirmed.

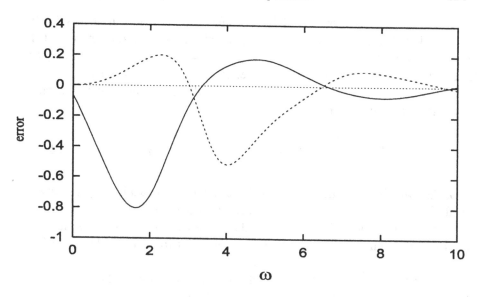

Figure 4.12. The ω dependence of the error when computing the integral (4.125) with fixed $\lambda = 1$ by the extended Newton–Cotes rule with $N = 2$ and $J = 1$ (solid) and $J = 3$ (broken). The curve displayed for the latter version represents actually the error multiplied by ten.

2.5   Gauss–Legendre quadrature rule

This rule allows the calculation of the integral of $y(x)$ on $[a, b]$ in terms of the values of the function at a set of points. With $X = (a+b)/2$ and $h = (b-a)/2$ the N-point Gauss–Legendre rule is

$$\int_a^b y(z)dz = \int_{X-h}^{X+h} y(z)dz \approx h \sum_{n=1}^N w_n y(X + x_n h), \qquad (4.126)$$

which is formally the same as the formula (4.75). The main difference is that, when constructing the rule, the abscissas are fixed and then only the weights have to be determined in the formula (4.75), while in the Gauss–Legendre rule both are open for determination. In general, the latter will be then much more accurate than the former.

The classical, polynomial-fitted representative is that where the values of the $2N$ parameters (that is the weights and the abscissas) are expressed on the basis of the Legendre polynomials. In particular, the abscissas are the zeros of the N-th degree Legendre polynomial. The parameters for various values of N are presented in many books in tabulated form (see, e.g. [6]) and quadrature subroutines based on these parameters are available in any library of scientific programs.

The theory behind this rule is based on the properties of the orthogonal polynomials, see e.g. Section 2.7 of [6]. Simple and elegant, this theory is able to furnish a direct derivation of some general features, among them that the weights are positive and symmetric (i.e. $w_1 = w_N$, $w_2 = w_{N-1}, \ldots$) and that the abscissas are antisymmetric (i.e. $x_1 = -x_N$, $x_2 = -x_{N-1}, \ldots$) and inside $[-1, 1]$. It also allows writting the error of the N-point Gauss–Legendre formula in a compact form,

$$err_{GL}^{(N)} = \frac{2^{2N+1}(N!)^4}{(2N+1)[(2N)!]^3} y^{(2N)}(\eta), \quad -1 < \eta < 1, \tag{4.127}$$

a formula which shows that the Gauss–Legendre rule is exact if $y(x)$ is a polynomial of class $\mathcal{P}_{2N-1}$. The rule is then accurate enough for oscillatory integrands of the form (3.38) only if ω is small, while a drastic deterioration of the accuracy is expected when ω is increased. The same holds for integrands of the form (3.39).

A way to improve the accuracy of the approximation (4.126) consists in trying to determine weights and abscissas which will be no longer constants but integrand dependent. The specific problem consists in constructing w_n and x_n ($n = 1, 2, \ldots, N$), depending on ωh or on λh, in such a way that the approximation is best tuned for integrands of the forms (3.38) or (3.39), respectively, where the functions $f_1(x)$ and $f_2(x)$ are assumed smooth enough to be well approximated by low degree polynomials on the interval of interest. The new rule is expected to work well also when the value of μ ($\mu = i\omega$ or λ) is either not known very accurately or it is slightly x-dependent. For example, if the actual frequency in the integrand is x-dependent, of the form $\bar{\mu}(x) = \mu + \Delta\mu(x)$ where $\Delta\mu(x)$ is a small perturbation, the structure of $y(x)$ remains as in (3.38) or (3.39), with new $f_1(x)$ and $f_2(x)$ which are as smooth as the original ones.

This problem was considered by Ixaru and Paternoster, [24]. Following that paper we introduce the operator

$$\mathcal{L}[\mathbf{a}]y(x) := \int_{x-h}^{x+h} y(z)dz - h \sum_{n=1}^{N} w_n y(x + x_n h), \tag{4.128}$$

where a is the set of the $2N$ parameters, viz.: $\mathbf{a} := [w_1, w_2, \ldots, w_N, x_1, x_2, \ldots, x_N]$. We apply this operator on $\exp(\pm\mu x)$. With $z := \mu h$, as usual, we get

$$\mathcal{L}[h, \mathbf{a}] \exp(\pm\mu x) = h \exp(\pm\mu x) E_0^*(\pm z, \mathbf{a}), \tag{4.129}$$

where

$$E_0^*(z, \mathbf{a}) = 2\eta_0(Z) - \sum_{n=1}^{N} w_n \exp(\pm x_n z) \tag{4.130}$$

with $Z = z^2$. On this basis we form $G^{\pm}(Z)$,

$$G^{+}(Z, \mathbf{a}) = 2\eta_0(Z) - \sum_{n=1}^{N} w_n \eta_{-1}(x_n^2 Z), \quad G^{-}(Z, \mathbf{a}) = -\sum_{k=1}^{N} w_n x_n \eta_0(x_n^2 Z).$$

$$(4.131)$$

Their partial derivatives with respect to Z read

$$G^{+(p)}(Z, \mathbf{a}) = \frac{1}{2^p} (2\eta_p(Z) - \sum_{n=1}^{N} w_n x_n^{2p} \eta_{p-1}(x_n^2 Z)),$$

$$G^{-(p)}(Z, \mathbf{a}) = -\frac{1}{2^p} \sum_{n=1}^{N} w_n x_n^{2p+1} \eta_p(x_n^2 Z), \quad (4.132)$$

whence

$$L_{2k}^{*}(\mathbf{a}) = \frac{2}{2k+1} - \sum_{n=1}^{N} w_n x_n^{2k}, \quad L_{2k+1}^{*}(\mathbf{a}) = -\sum_{n=1}^{N} w_n x_n^{2k+1}, \quad k = 0, 1, \ldots$$

$$(4.133)$$

In this way the steps *i* and *iii* of the flow chart are completed.

Step *ii* implies the determination of the maximal M for which the algebraic system (3.53) is compatible. The unusual feature is that now the system is nonlinear and then more difficult to investigate, but the required information is available from the general theory of the Gauss–Legendre rule. In fact, equation (4.127) indicates that $M = 2N$ i.e. the maximal system consists of exactly $2N$ equations for the $2N$ unknowns,

$$L_{2k}^{*}(\mathbf{a}) = 0, \quad L_{2k+1}^{*}(\mathbf{a}) = 0, \quad k = 0, 1, \ldots, N - 1 \quad (4.134)$$

which in detailed form reads

$$\frac{2}{2k-1} - \sum_{n=1}^{N} w_n x_n^{2(k-1)} = 0, \quad \sum_{n=1}^{N} w_n x_n^{2k-1} = 0, \quad k = 1, 2, \ldots, N. \quad (4.135)$$

The mentioned symmetry / antisymmetry properties of the weights / abscissas established in the frame of the general theory are also of help in so much they allow reducing the computational effort for the solution of this system. It is easy to see that, if these properties are accounted for, the second equation in (4.135) is automatically satisfied for all k and therefore the system to be solved is reduced to only N equations for the subset α with N components of the original $\mathbf{a}$, defined by

$$\alpha := [w_1, w_2, \ldots, w_{(N+1)/2}, x_1, x_2, \ldots, x_{(N-1)/2}]$$

for odd N (note that $x_{(N+1)/2} = 0$ by default, i.e. it is no longer an unknown) and by

$$\alpha := [w_1, w_2, \ldots, w_{N/2}, x_1, x_2, \ldots, x_{N/2}]$$

for even N.

The system which remains to be solved then is

$$\frac{1}{2k-1} - \sum_{n=1}^{(N-1)/2} w_n x_n^{2(k-1)} - \frac{1}{2}\delta_{k\,1} w_{(N+1)/2} = 0, \ k = 1, 2, \ldots, N,$$

$$(4.136)$$

for odd N, and

$$\frac{1}{2k-1} - \sum_{n=1}^{N/2} w_n x_n^{2(k-1)} = 0, \ k = 1, 2, \ldots, N, \qquad (4.137)$$

for even N.

The evaluation of the parameters by solving such a system can be carried out in the frame of the so-called algebraic approach, see Section 2.7.2 of [6]. This approach consists in first associating a linear algebraic system of equations whose solution furnishes the values of the coefficients of some polynomial. Finally, the roots of the latter give the required parameters of the rule. The practical inconvenience of such an approach is that when N is increased the linear system becomes more and more ill conditioned and therefore multiple length arithmetic is needed at big N.

The parameters resulting in this way are of course identical with the ones from the general theory based on the orthogonal polynomials. Is then any reason for considering this way when the other is already available and it is much simpler? The answer is that the construction of the parameters through orthogonal polynomials is affected by a severe restriction: it asks for a positive weight function. The restriction holds for the classical rule but not for the ef extension and then only the first way will be applicable in such a case.

Step *iv*. Since $M = 2N$ is even it results that K must be odd. Moreover, since we want to derive a version maximally tuned on integrands of the form (3.38) or (3.39), we must take $K = -1$ and therefore $P = N - 1$.

Step *v*. The algebraic system to be solved is

$$G^{\pm\,(p)}(Z, \mathbf{a}) = 0, \ p = 0, 1, \ldots, N - 1$$

but we shall accept that the symmetry / antisymmetry properties for the weights / abscissas of the standard Gauss–Legendre rule (that is when $Z = 0$) continue to be satisfied also when $Z \neq 0$. With this assumption the equations involving $G^-(Z, \mathbf{a})$ and its derivatives are automatically satisfied. The system is then reduced to $G^{+\,(k-1)}(Z, \alpha) = 0, \ k = 1, 2, \ldots, N$, that is

$$\eta_{k-1}(Z) - \sum_{n=1}^{(N-1)/2} w_n x_n^{2(k-1)} \eta_{k-2}(x_n^2 Z) - \frac{1}{2}\delta_{k\,1} w_{(N+1)/2} = 0,$$

$$k = 1, 2, \ldots, N \qquad (4.138)$$

for odd N, and

$$\eta_{k-1}(Z) - \sum_{n=1}^{N/2} w_n x_n^{2(k-1)} \eta_{k-2}(x_n^2 Z) = 0, \quad k = 1, 2, \ldots, N \qquad (4.139)$$

for even N.

Each of these two algebraic systems for the set $\alpha(Z)$ is nonlinear. Solutions exist but they are not unique. One of them, call it the principal set, has the property that it tends to the classical set when $Z \to 0$; only this is computed by subroutine EFGAUSS described in Section 2.5.1. Also, all considerations below have the principal set for reference.

Step *vi*. The leading term of the error when calculating the integral by (4.126) results by particularization of (3.57). The standard (classical) rule is identified by $K = 2N - 1$ and $P = -1$ such that

$$lte_{class} = \frac{h^{2N+1}}{(2N)!} \left(\frac{2}{2N+1} - \sum_{n=1}^{N} w_n x_n^{2N} \right) y^{(2N)}(X). \qquad (4.140)$$

The numerical values of the factors multiplying $y^{(2N)}$ in (4.127) and (4.140) are equal, of course. As for the optimally tuned rule, we take $K = -1$ and $P = N - 1$ with the result

$$lte_{ef} = (-1)^N h^{2N+1} \frac{2 - \sum_{n=1}^{N} w_n(Z)}{Z^N} (D^2 - \mu^2)^N y(X), \qquad (4.141)$$

where $Z = \mu^2 h^2$.

Numerical experiments

The following experiments cover only the (trigonometric) case of nonpositive $Z = -\omega^2 h^2$, $1 \leq N \leq 6$, $K = -1$ and $P = N - 1$. The parameters are calculated by subroutine EFGAUSS. The variation with ωh of the weights and of the abscissas for ωh between 0 and 50 is presented in Figure 4.13. For odd N the central abscissa $x_{(N+1)/2} = 0$ is not drawn. These data enable some conclusions.

1. The system of equations for the parameters has a solution for all the values of ωh. In other words, there are no critical values of ωh, that is similar with the case of the extended Newton–Cotes rule with odd J.

2. In all cases the abscissas are placed inside $[-1, 1]$, thus preserving a well known property of the classical Gauss–Legendre rule.

3. The properties of the classical weights are preserved only partially. Thus, all weights are positive for even N. When N is odd the weights are also positive but with one notable exception. This is the weight $w_{(N+1)/2}$ corresponding to the central abscissa, which exhibits oscillations around zero.

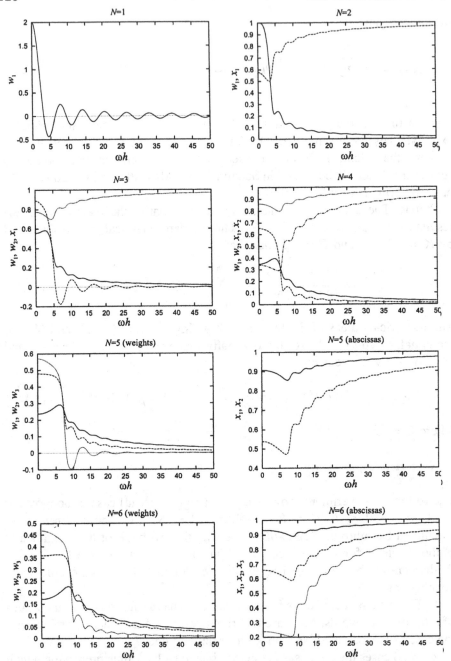

Figure 4.13. ωh dependence of the weights w_n and the abscissas x_n for the N-point Gauss–
Legendre rule with $P = N - 1$ and $K = -1$. $N = 1$: w_1 (solid); $N = 2$: w_1 (solid) and
x_1 (broken); $N = 3$: w_1 (solid), w_2 (broken) and x_1 (dots); $N = 4$: w_1 (solid), w_2 (broken),
x_1 (dots) and x_2 (dash and dots); $N = 5$ (weights): w_1 (solid), w_2 (broken) and w_3 (dots);
$N = 5$ (abscissas): x_1 (solid) and x_2 (broken); $N = 6$ (weights): w_1 (solid), w_2 (broken) and
w_3 (dots); $N = 6$ (abscissas): x_1 (solid), x_2 (broken) and x_3 (dots).

4. There are clearly two distinct regimes in the variation of the parameters with respect to the product $\theta = \omega h$. For small θ, from 0 up to some $\bar{\theta}$ which typically increases with N, a smooth variation is exhibited but the picture changes when $\theta > \bar{\theta}$. In the latter case the variation consists of a succession of oscillations with decaying amplitude around some average curves which tend to zero for the weights and to one for the abscissas. In other words, there is a tendency that, when ωh is increased, the weights become smaller and smaller while the abscissas tend to occupy more and more eccentric positions on $[-1, 1]$.

We stress that all these conclusions are of experimental nature and that some separate mathematical consideration is needed in the future for a better understanding of the mentioned behaviours.

As for the error, equations (4.120) and (4.141) suggest that for one and the same number of points N the ef versions of the Gauss–Legendre rule and of the Newton–Cotes rule with one derivative ($J = 1$) should show similar accuracies, and this expectation is reasonably well confirmed by experiments. The choice for practice then depends mainly on the input environment of the actual problem: the first rule requires only the values of the integrand, but at very special nodes, while both the values of the integrand and of its first derivative are needed by the second rule, but no serious restriction has to be imposed on the position of the nodes. As a matter of fact, a version of the latter rule which is valid on nonsymmetric partitions can be easily formulated.

For a numerical illustration we take the integral (4.122). The ω dependence of the scaled error is displayed on Figure 4.14 for $1 \leq N \leq 6$. Particularly interesting is the behaviour of this in the asymptotic region. It is seen that the amplitude of the scaled error is constant only for $N = 1$, which is fully consistent with the asymptotic behaviour $lte_{ef} \sim \omega^{-1}$ for this N. For bigger N the amplitude of the scaled error decreases but for $N \geq 3$ the rate of the decrease is not as fast as suggested by the expression of the lte_{ef}. The qualitative reason is the same as for the extended Newton–Cotes rule. However, additional work is needed in the future on both rules for a better understanding of the asymptotic behaviour of the error.

For completeness we mention that the quadrature rules whose exponential fitting versions have been considered in this chapter are only the most popular ones, except perhaps for the extended rules. As a matter of fact, rules that generalize the standard and extended rules with predetermined nodes or the Gauss–Legendre rule also exist, but only with polynomial forms of the integand in mind. In essence, the existing generalizations accept that a specified number of derivatives of the integrand is available and also that the positions of some nodes are fixed, to determine the position of the other nodes and the values of the weights such that the error is minimized. Examples of such generalizations are the Turan and the Chakalov–Popoviciu rules, [4, 32, 36, 37]; see also [30]. To our knowledge, exponential fitting extensions of such rules have

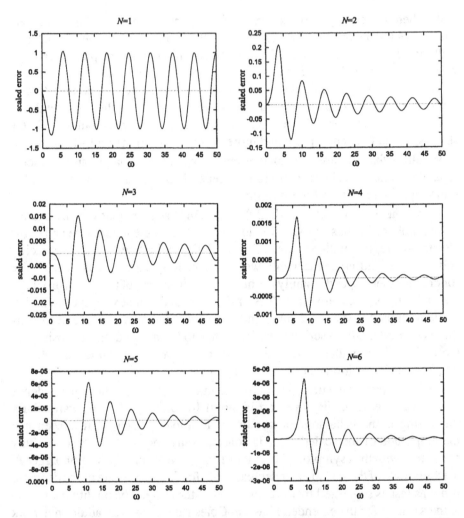

Figure 4.14. The ω dependence of the scaled error when the integral (4.122) is computed by the N–point Gauss–Legendre rule with $P = N - 1$ and $K = -1$ for $1 \leq N \leq 6$.

not been published but these can be constructed by means of the same flow chart.

A special case

It has been mentioned in Section 3 of Chapter 3 that the flow chart may not work in some accidental situations; a similar warning appeared in Section 4.3 of Chapter 2. The following example represents an illustration of such a case.

We consider the standard rule (4.75) for $N = 2$ and with the symmetrically distributed abscissas $x_1 = -x_2 = 1/\sqrt{3}$. The two coefficients are equal, $a_1 = a_2$. For the classical version we have $a_1 = 1$ (this is the solution of

$L_0^*(a_1) = 0$, see equation (4.84)) and, since $L_1^*(1) = L_2^*(1) = L_3^*(1) = 0$ but $L_4^*(1) = 8/45 \neq 0$, it results that $M = 4$.

The special feature is that this M is bigger than the value $M = 2$ which normally results from the general treatment of the classical rule with $N = 2$ and symmetrically distributed abscissas. However, this special value of M is a direct consequence of the fact that the abscissas were chosen in a particular way: they are the roots of the second degree Legendre polynomial and then the discussed rule is simply the classical two-point Gauss–Legendre rule whose *lte* is

$$lte_{class} = \frac{1}{135}h^5 y^{(4)}(X), \qquad (4.142)$$

see (4.140).

This raises a problem with the flow chart because the assumption of the latter that M resulting from the classical version is preserved also when $Z \neq 0$ is violated. In fact, for the pair $K = -1, P$ the selfconsistency condition $K + 2P = M - 3$ would imply that $P = 1$, i. e. that $a_1(Z)$ will satisfy simultaneously two equations, $G^+(Z, a_1(Z)) = 0$ and $G^{+(1)}(Z, a_1(Z)) = 0$. However, this is not true. The first equation (use the expression (4.82)) gives

$$a_1(Z) = \frac{\eta_0(Z)}{\eta_{-1}(Z/3)}, \qquad (4.143)$$

and then

$$G^{+(1)}(Z, a_1(Z)) = \frac{\eta_1(Z)\,\eta_{-1}(Z/3) - \eta_0(Z)\,\eta_0(Z/3)}{\eta_{-1}(Z/3)}. \qquad (4.144)$$

This is a non identically vanishing function which tends to zero when $Z \to 0$. It follows that $M = 2$ is the correct value when $Z \neq 0$ but, since this is not the same as the one resulting from the classical case, the form (3.57) of the lte_{ef} is no longer valid.

An appropriate treatment of such a case relies on the approach presented in Chapter 2 after equation (2.83). The discussed case is covered by that equation for $I = M = 2$, $m_1 = m_2 = 1$, $\mu_1 = -\mu_2 = \mu$ (and then $z_1 = -z_2 = z$ where $z = \mu h$). Also the RDE is of the second order with $c_1 = 0$ and $c_2 = -\mu^2$. The equation (2.84) becomes

$$\mathcal{L}[h, a_1(z)]y(x) = h^3 \sum_{k=0}^{\infty} h^k T_k^*(z, a_1(z))D^k(D^2 - \mu^2)y, \qquad (4.145)$$

where $a_1(z)$ is given by (4.143), that is it depends on the compressed argument $Z = z^2$. Equations (2.86), (2.87) and (2.88) with $c_2^* = -Z$ allow expressing $T_k^*(z, a_1(Z))$ in terms of the starred classical moments (4.84). All T_k^* with odd k are vanishing while the first nonvanishing ones have the expressions

$$T_0^*(z, a_1(Z)) = \frac{2[a_1(Z) - 1]}{Z},$$

and

$$T_2^*(\mathbf{z}, a_1(Z)) = \frac{[a_1(Z) - 1](Z + 6)}{3Z^2}.$$

Equation (4.145) then becomes

$$\mathcal{L}[h, a_1(\mathbf{z})]y(x) = \frac{a_1(Z) - 1}{3Z^2}[6Zh^3 + (Z + 6)h^5 D^2](D^2 - \mu^2)y + \mathcal{O}(h^7),$$

$$(4.146)$$

The series expansion of the η components in $a_1(Z)$ give that $a_1(Z) - 1 = Z^2(1/270 + \mathcal{O}(Z))$ and then the front factor tends to $1/810$ when $\mu \to 0$. The first term in the brackets can be written as $6\mu^2 h^5$ and then its h dependence is the same as in the second term. It folows that the *lte* of the discussed case is not of the form (3.57) but it is a sum of two terms,

$$lte_{ef} = \frac{a_1(Z) - 1}{3Z^2}h^5[6\mu^2 + (\mu^2 h^2 + 6)D^2](D^2 - \mu^2)y(X). \qquad (4.147)$$

As expected, it tends to the classical expression (4.142) when $\mu \to 0$.

2.5.1   Subroutines EFGAUSS and EFGQUAD

The systems (4.138) and (4.139) are nonlinear and then the procedures described before cannot be used for the computation of the parameters of the Gauss–Legendre rule. This is why a procedure specially adapted for this case must be formulated.

The procedure is based on the Newton–Raphson iteration. If the left hand side of each equation in (4.138) or (4.139) is denoted $R_k(\alpha(Z))$ then any of the two systems can be written compactly as

$$R_k(\alpha(Z)) = 0, \quad k = 1, 2, \ldots, N.$$

For a numerical approach it is convenient to first re-scale each equation in an advantagous way. Since the η functions satisfy some hierarchy each equation is multiplyed by a numerical factor $f_k(Z)$ which approximates the inverse of the amplitude of the free term $\eta_{k-1}(Z)$ in the expression of $R_k(Z)$. We take $f_k(Z) = \max\{(1 + \sqrt{|Z|})^k, 1 \times 3 \times \cdots \times (2k - 1)\}$ and define $Q_k(Z) := f_k(Z) \cdot R_k(Z)$. The system to be solved becomes

$$Q_k(\alpha(Z)) = 0, \quad k = 1, 2, \ldots, N. \qquad (4.148)$$

On defining some set of coefficients $\alpha^0(Z)$ for start, increasingly better values of the N dimensional vector $\alpha(Z)$ are given by the following matrix iteration for the transposed $\alpha(Z)$:

$$\alpha^{i+1^T}(Z) = \alpha^{i^T}(Z) - \mathbf{D}^{-1}(\alpha^i(Z))\mathbf{Q}(\alpha^i(Z)), \quad i = 0, 1, 2, \ldots, \qquad (4.149)$$

Table 4.5. Values of the parameters g_n, v_n, p_n and s for $1 \leq N \leq 6$.

		$n = 1$	$n = 2$	$n = 3$	$n = 4$	$n = 5$	$n = 6$
$N = 1$	g_n	0.00					
	v_n	−100.00					
	p_n	1.00					
$s = -6.0$							
$N = 2$	g_n	2.25	.33				
	v_n	12.50	−3.33				
	p_n	1.00	.55				
$s = -4.0$							
$N = 3$	g_n	1.09	−2.25	2.50			
	v_n	4.70	−12.50	−7.50			
	p_n	.60	.90	.75			
$s = -3.5$							
$N = 4$	g_n	.38	.85	.58	.19		
	v_n	−1.25	2.20	−4.20	−.62		
	p_n	.40	.60	.80	.30		
$s = -2.5$							
$N = 5$	g_n	1.12	2.30	−2.00	.58	.18	
	v_n	6.25	16.70	−7.00	−4.20	−.62	
	p_n	.27	.48	.55	.87	.50	
$s = 0.0$							
$N = 6$	g_n	.56	.85	3.00	.81	.25	.15
	v_n	0.00	4.00	19.00	−4.00	−1.00	0.00
	p_n	.15	.35	.45	.90	.60	.20
$s = 0.0$							

up to convergence in the desired number of figures. Here $\mathbf{Q}(\alpha(Z))$ is the column vector $[Q_1(\alpha(Z)), Q_2(\alpha(Z)), \ldots, Q_N(\alpha(Z))]^T$ and $\mathbf{D}$ is the N by N matrix with the elements

$$D_{nm}(\alpha) = \frac{\partial Q_n(\alpha)}{\partial \alpha_m}, \quad n, m = 1, 2, \ldots, N. \tag{4.150}$$

To obtain these matrix elements in analytic form use is made of the differentiation properties of the η functions.

The number of iterations depends on how close the initial set $\alpha^{(0)}(Z)$ is to the true solution. The following empirical rule works well for the determination of the principal set when Z is negative and $1 \leq N \leq 6$. For given input ω and h we form $\theta = \omega h$; this means that $Z = -\theta^2$. Let N be given and $N^* := \lfloor (N + 1)/2 \rfloor$. The first N^* components of the vector α represent the weights, $w_n = \alpha_n$ ($n = 1, 2, \ldots, N^*$), while the others are the nonvanishing abscissas, $x_n = \alpha_{N^*+n}$, ($n = 1, 2, \ldots, N - N^*$). For some N dependent s values between -6 and 0, see Table 4.5, three regions in θ are delimited and the following simple formulae are used:

$$\alpha_n^0(Z) = p_n \quad \text{for} \quad 1 \leq n \leq N \tag{4.151}$$

if $0 \leq \theta \leq 7 + s$,

$$\alpha_n^0(Z) = \begin{cases} \dfrac{1}{g_n\theta - v_n} & \text{for } 1 \leq n \leq N^* \\ 1 - \dfrac{1}{g_n\theta - v_n} & \text{for } N^* + 1 \leq n \leq N \end{cases} \tag{4.152}$$

if $\theta \geq 10$, and the linear interpolation formula

$$\alpha_n^0(Z) = \frac{1}{3 - s}\{[(\alpha_n(10) - \alpha_n(7 + s)]\theta$$
$$+ 10\alpha_n(7 + s) - (7 + s)\alpha_n(10)\}, \quad \text{for} \quad 1 \leq n \leq N \tag{4.153}$$

if $7 + s < \theta < 10$. The numerical values of the parameters g_n, v_n and p_n are listed in Table 4.5. If this is used for $\alpha^0(Z = -\theta^2)$ the number of iterations to obtain the parameters with fourteen significant figures ranges between 3 and 7 for $0 \leq \theta \leq 50$.

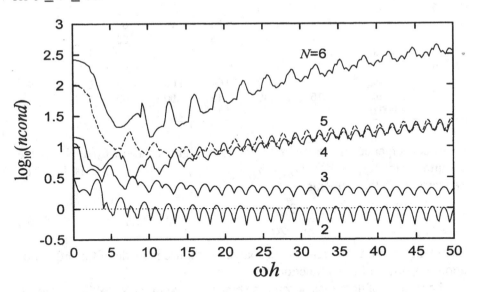

Figure 4.15. The ωh dependence of $\log_{10}(ncond)$ for the $N \times N$ matrix $D(\alpha^i(-\omega^2 h^2))$ corresponding to the last iteration, $2 \leq N \leq 6$. The curve with broken line corresponds to $N = 5$. The condition number $ncond$ is defined as $ncond = \max\limits_{1 \leq i, j \leq N} |D_{ij}| \times \max\limits_{1 \leq i, j \leq N} |D_{ij}^{-1}|$.

One of the reasons for the rescaling of the system was the reduction of the ill conditioning effects in the matrix D. However, such effects cannot be removed completely. In fact, as noticed, it is well known that they exist and increase with N for the clasical case ($\omega = 0$), and the experimental evidence indicates that the same holds also when $\theta \neq 0$, see Figure 4.15. The extension of the approach to higher N will then face more and more accuracy problems.

The subroutine EFGAUSS computes the parameters of the Gauss–Legendre rule for $1 \leq N \leq 6$. Calling it is

$$\text{CALL EFGAUSS(N, OMH, PAR)}$$

Input arguments:

- N - the number of nodes.

- OMH - the value of ωh (double precision); this must be nonnegative.

Output argument:

- PAR - a double precision vector with N components for the parameters. The identification is as follows: if N is even then $w_i = \text{PAR(i)}$ and $x_i = \text{PAR(N/2 + i)}$ for $i = 1, 2, \ldots N/2$, while if N is odd then $w_i = \text{PAR(i)}$ for $i = 1, 2, \ldots (N+1)/2$ and $x_i = \text{PAR((N + 1)/2 + i)}$ for $i = 1, 2, \ldots (N-1)/2$. Note that in the latter case the central abscissa, which is zero by default, does not appear in the vector PAR.

Subroutine EFGQUAD computes the integral of a function with the Gauss–Legendre rule whose parameters were generated by EFGAUSS. Calling it is

$$\text{CALL EFGQUAD(N, A, B, PAR, FUNC, QUAD)}$$

Input arguments:

- N - the number of nodes.

- A, B - the ends a and b ($a < b$) of the integration interval (double precision).

- PAR - double precision vector with the N values resulting from EFGAUSS.

- FUNC - the function to be integrated (external). It is of the form DOUBLE PRECISION Y(X).

Output argument:

- QUAD - the computed value of the integral (double precision)

The following driving program and the written form of function $y(x)$ enable the evaluation of the integral (4.122) for $\omega = 10$ by the Gauss–Legendre rule with $N = 5$. The output consists in the list of the parameters of the rule, the value of the computed integral and the absolute error.

```
      implicit double precision (a-h,o-z)
      dimension par(6)
      common/comega/omref
      external y
      data n/5/,a/-1.d0/,b/1.d0/,omega/10.d0/
      h=0.5d0*(b-a)
      omref=omega
      omh=omega*h
      call efgauss(n,omh,par)
      write(*,50)(par(i),i=1,(n+1)/2)
      write(*,51)(par(i),i=(n+1)/2+1,n)
      qex=2.d0*sin(omega+.5d0)/(omega+.5d0)
      call efgquad(n,a,b,par,y,quad)
      err=qex-quad
      write(*,52)quad,err
50    format(2x,'w_i : ',3(2x,f10.8))
51    format(2x,'x_i : ',3(2x,f10.8))
52    format(2x,'I_comp = ',f10.8,2x,'I-I_comp = ',e9.2)
      stop
      end


      double precision function y(x)
      implicit double precision (a-h,o-z)
      common/comega/omega
      y=cos((omega+.5d0)*x)
      return
      end
```

The output is:

```
w_i :    0.18500934  0.14994084  -.08437731
x_i :    0.89223531  0.62722994
I_comp = -.16756458  I-I_comp =  0.35E-05
```

3.   Interpolation

In its most familiar form, the interpolation problem consists in finding a formula which allows the calculation of an approximation of y at some arbitrary x ($a < x < b$) when the values of y at the mesh points $x_0 = a < x_1 < x_2 < \ldots < x_n = b$ are given. The classical formulae rely on the assumption that $y(x)$ is smooth enough to be approximated satisfactorily by a polynomial. This is the case with the Lagrange interpolation formula. When $y(x)$ is not of this type, in particular when it behaves as (3.38) or as (3.39), the formula has to be adapted, and the exponential fitting technique is a convenient tool for such an adaptation.

The quality of the interpolation depends on the available input information. For one and the same partition the interpolated values will be better if both $y(x_i)$ and its first derivative $y'(x_i)$ are available (this is known as the Hermite interpolation) and, even better, if we additionally use the second derivative $y''(x_i)$. Again, the classical formulae refer to the case when $y(x)$ is assumed smooth enough, while the exponential fitting technique allows extending them to become tuned on the forms (3.38) or (3.39).

3.1 A simple scheme

The aim of this section consists in just illustrating the way in which the flow chart can be used for the interpolation problems. The things are best seen on the simplest situation when the numerical information is available at only two points. As for the number of data available at these points, we consider three cases, to treat in detail only the most difficult of them. The other two result by suitable particularization.

Let a and b be given. We denote $x_0 = (a + b)/2$ and $h = (b - a)/2$ and assume that the values at the end points $a = x_0 - h$ and $b = x_0 + h$ of y, of its first derivative y' and of its second derivative y'' are known. As said, we want to obtain an approximation to y at some arbitrary point x $(x_0 - h < x < x_0 + h)$. One direct application consists in calculating the values of the wave function inbetween the mesh points after applying a one-step solver for the Schrödinger equation (Runge-Kutta-Nyström or CP methods, for example; the latter are based on the piecewise perturbation theory, see [18], [20], [21], [22]). The mentioned data are known indeed since these solvers give y and y' at the mesh points, while the values of the second derivative at the same points are available through the very equation, viz. $y'' = (V(x) - E)y$.

As in [23], the following interpolation formula is considered for the calculation of $y(x_0 + th)$, $-1 \leq t \leq 1$:

$$
\begin{aligned}
y(x_0 + th) \approx\ & a_0(t)y(x_0 - h) + a_1(t)y(x_0 + h) \\
& + h[b_0(t)y'(x_0 - h) + b_1(t)y'(x_0 + h)] \\
& + h^2[c_0(t)y''(x_0 - h) + c_1(t)y''(x_0 + h)].
\end{aligned} \tag{4.154}
$$

Note that the parameter t which identifies the point x at which the interpolated value is needed and the six t dependent coefficients a_i, b_i, c_i, $(i = 0, 1)$ which have to be determined, are dimensionless. The associated operator $\mathcal{L}$ is introduced by

$$
\begin{aligned}
\mathcal{L}[h, t, \mathbf{a}]y(x) :=\ & y(x + th) - [a_0 y(x - h) + a_1 y(x + h)] \tag{4.155} \\
& - h[b_0 y'(x - h) + b_1 y'(x + h)] - h^2[c_0 y''(x - h) + c_1 y''(x + h)].
\end{aligned}
$$

where vector a collects the six coefficients. A formal novelty is that now the operator depends on one more parameter, t, but this does not affect the use of the flow chart.

To construct the moments the operator $\mathcal{L}[h, t, \mathbf{a}]$ is applied on $\exp(\pm\mu x)$. This gives

$$E_0^*(t, \pm z, \mathbf{a}) = \exp(\pm tz) - [a_0 \exp(\mp z) + a_1 \exp(\pm z)] \quad (4.156)$$
$$\mp z[b_0 \exp(\mp z) + b_1 \exp(\pm z)] - z^2[c_0 \exp(\mp z) + c_1 \exp(\pm z)].$$

The expressions of G^+ and G^- result without difficulty. With

$$a^\pm := a_0 \pm a_1, \quad b^\pm := b_0 \pm b_1, \quad c^\pm := c_0 \pm c_1 \quad (4.157)$$

they read

$$G^+(t, Z) = \eta_{-1}(Zt^2) - a^+\eta_{-1}(Z) + b^- Z\eta_0(Z) - c^+ Z\eta_{-1}(Z),$$
$$G^-(t, Z) = t\eta_0(Zt^2) + a^-\eta_0(Z) - b^+\eta_{-1}(Z) + c^- Z\eta_0(Z). \quad (4.158)$$

Their derivatives are

$$G^{+(1)}(t, Z) = \frac{1}{2}[t^2\eta_0(Zt^2) - a^+\eta_0(Z) + b^-(\eta_{-1}(Z) + \eta_0(Z))$$
$$-c^+(2\eta_{-1}(Z) + Z\eta_0(Z))], \quad (4.159)$$
$$G^{-(1)}(t, Z) = \frac{1}{2}[t^3\eta_1(Zt^2) + a^-\eta_1(Z) - b^+\eta_0(Z) + c^-(\eta_{-1}(Z) + \eta_0(Z))],$$

and

$$G^{+(p)}(t, Z) = \frac{1}{2^p}[t^{2p}\eta_{p-1}(Zt^2) - a^+\eta_{p-1}(Z) + b^-(\eta_{p-2}(Z) + \eta_{p-1}(Z))$$
$$-c^+(\eta_{p-3}(Z) + 3\eta_{p-2}(Z))], \quad (4.160)$$
$$G^{-(p)}(t, Z) = \frac{1}{2^p}[t^{2p+1}\eta_p(Zt^2) + a^-\eta_p(Z) - b^+\eta_{p-1}(Z)$$
$$+c^-(\eta_{p-2}(Z) + \eta_{p-1}(Z))], \quad p = 2, 3, \ldots$$

whence the classical moments are

$$L_{2k}^*(t, \mathbf{a}) = t^{2k} - a^+ + 2kb^- - 2k(2k - 1)c^+, \quad (4.161)$$
$$L_{2k+1}^*(t, \mathbf{a}) = t^{2k+1} + a^- - (2k + 1)b^+ + 2k(2k + 1)c^-, \quad k = 0, 1, \ldots$$

The coefficients of the classical, polynomial interpolation rule result by solving the linear system

$$L_m^*(t, \mathbf{a}) = 0, \quad m = 0, 1, 2, 3, 4, 5 \quad (4.162)$$

for the six unknowns. They are:

$$a_0(t) = -\frac{1}{16}(t-1)^3(3t^2 + 9t + 8),$$

$$a_1(t) = \frac{1}{16}(t+1)^3(3t^2 - 9t + 8),$$

$$b_0(t) = -\frac{1}{16}(t^2 - 1)(t-1)^2(3t + 5),$$

$$b_1(t) = \frac{1}{16}(t^2 - 1)(t+1)^2(-3t + 5),$$

$$c_0(t) = \frac{1}{16}(t^2 - 1)^2(1-t),$$

$$c_1(t) = \frac{1}{16}(t^2 - 1)^2(1+t).$$
$$(4.163)$$

Since $L_6^*(t, \mathbf{a}(t)) \neq 0$ it follows that $M = 6$.

For functions of the forms (3.38) or (3.39) the best tuned version is the one with $K = -1$ and $P = 2$. To determine its coefficients the linear systems

$$G^+(t, Z) = G^{+(1)}(t, Z) = G^{+(2)}(t, Z) = 0,$$
$$(4.164)$$

with the unknowns $a^+(t, Z)$, $b^-(t, Z)$ and $c^+(t, Z)$, and

$$G^-(t, Z) = G^{-(1)}(t, Z) = G^{-(2)}(t, Z) = 0,$$
$$(4.165)$$

with the unknowns $a^-(t, Z)$, $b^+(t, Z)$ and $c^-(t, Z)$, have to be correspondingly solved and their solutions will lead to the required individual coefficients through (4.157). Analytic expressions can be written but we have opted for a numerical solution (subroutine EFINT).

Finally, on using (3.57) the leading term of the error reads

$$lte_{ef} = -h^6 \frac{L_0^*(t, \mathbf{a}(t, Z))}{Z^3} (D^2 - \mu^2)^3 y(x_0),$$
$$(4.166)$$

where $L_0^*(t, \mathbf{a}(Z)) = 1 - [a_0(t, Z) + a_1(t, Z)]$, see (4.161). Critical values are present for negative $Z = -\omega^2 h^2$ at $\omega h = 0.9438\pi$, 1.4647π, 1.9741π, 2.4794π etc.

The expressions of the coefficients for the simpler cases when (*i*) the values of y at the end points are only given, and (*ii*) the values of both y and y' are given, can be evaluated without difficulty. We give only the final results.

Case (i)

$$y(x_0 + th) \approx a_0 y(x_0 - h) + a_1 y(x_0 + h).$$
$$(4.167)$$

Classical $(M = 2)$:

$$a_0(t) = \frac{1}{2}(1 - t), \ a_1(t) = \frac{1}{2}(1 + t). \qquad (4.168)$$

Tuned $(K = -1, P = 0)$:

$$a_0(t, Z) = \frac{1}{2\eta_{-1}(Z)\eta_0(Z)}[\eta_{-1}(Zt^2)\eta_0(Z) - t\eta_0(Zt^2)\eta_{-1}(Z)],$$

$$a_1(t, Z) = \frac{1}{2\eta_{-1}(Z)\eta_0(Z)}[\eta_{-1}(Zt^2)\eta_0(Z) + t\eta_0(Zt^2)\eta_{-1}(Z)]. \quad (4.169)$$

For the case of the oscillatory functions these expressions have simple trigonometric forms,

$$a_0(t, -\omega^2 h^2) = \frac{\sin[(1 - t)\omega h]}{\sin(2\omega h)}, \ a_1(t, -\omega^2 h^2) = \frac{\sin[(1 + t)\omega h]}{\sin(2\omega h)}. \quad (4.170)$$

– The leading term of the error:

$$lte_{ef} = -h^2 \frac{\eta_{-1}(Z) - \eta_{-1}(Zt^2)}{Z\eta_{-1}(Z)} (D^2 - \mu^2)y(x_0). \qquad (4.171)$$

– Critical values: $\omega h = n\pi/2, \ n = 1, 2, \ldots$

Case (ii)

$$y(x_0 + th) \approx a_0 y(x_0 - h) + a_1 y(x_0 + h) + h(b_0 y'(x_0 - h) + b_1 y'(x_0 + h)). \tag{4.172}$$

Classical $(M = 4)$:

$$a_0(t) = \frac{1}{4}(t - 1)^2(t + 2), \ a_1(t) = \frac{1}{4}(t + 1)^2(-t + 2), \quad (4.173)$$

$$b_0(t) = \frac{1}{4}(t - 1)^2(t + 1), \ b_1(t) = \frac{1}{4}(t + 1)^2(t - 1).$$

Tuned $(K = -1, P = 1)$:

$$a^+(t, Z) = \frac{\eta_{-1}(Zt^2)(\eta_{-1}(Z) + \eta_0(Z)) - Zt^2\eta_0(Zt^2)\eta_0(Z)}{1 + \eta_{-1}(Z)\eta_0(Z)},$$

$$a^-(t, Z) = \frac{t[-\eta_0(Zt^2)\eta_0(Z) + t^2\eta_1(Zt^2)\eta_{-1}(Z)]}{\eta_0^2(Z) - \eta_{-1}(Z)\eta_1(Z)}, \qquad (4.174)$$

$$b^+(t, Z) = \frac{t[-\eta_0(Zt^2)\eta_1(Z) + t^2\eta_1(Zt^2)\eta_0(Z)]}{\eta_0^2(Z) - \eta_{-1}(Z)\eta_1(Z)},$$

$$b^-(t, Z) = \frac{\eta_{-1}(Zt^2)\eta_0(Z) - t^2\eta_0(Zt^2)\eta_{-1}(Z)}{1 + \eta_{-1}(Z)\eta_0(Z)}.$$

from which the individual $a_0(t, Z)$, $a_1(t, Z)$, $b_0(t, Z)$ and $b_1(t, Z)$ can be extracted without difficulty.

– The leading term of the error:

$$lte_{ef} = h^4 \frac{1 + D(t, Z)}{N(t, Z)} (D^2 - \mu^2)^2 y(x_0), \qquad (4.175)$$

where

$$D(t, Z) = \eta_{-1}(Z)\eta_0(Z) - \eta_{-1}(Zt^2)(\eta_{-1}(Z) + \eta_0(Z)) + Zt^2\eta_0(Zt^2)\eta_0(Z),$$

$$N(t, Z) = Z^2(1 + \eta_{-1}(Z)\eta_0(Z)).$$

– Critical values: none

A numerical illustration

The function

$$y(x) = (1 + 2x + 3x^2 + 4x^3)\sin(\omega x + 1), \qquad (4.176)$$

is interpolated on the interval $0 < x < 1$ by the discussed tuned formulae with $P = 0$, 1 and 2 for $\omega = 20$. We take $x_0 = h = 1/2$ in all these formulae. The results are displayed on Figure 4.16. It is seen that the quality of the interpolation increases with P, as expected.

3.1.1 Subroutine EFINT

This subroutine computes the interpolated value of $y(x)$ of the form (3.38) in terms of the values of y and of its derivatives at two points a and b by means of the three tuned versions $P = 0$, 1, 2. $y(a)$ and $y(b)$ should be given for $P = 0$, $y(a)$, $y'(a)$ and $y(b)$, $y'(b)$ for $P = 1$, and $y(a)$, $y'(a)$, $y''(a)$ and $y(b)$, $y'(b)$, $y''(b)$ for $P = 2$. Calling it is:

$$\text{CALL EFINT(IP, A, B, X, OMEGA, YA, YB, YINT)}$$

The arguments are:
Input:

- IP - the integer value of P ($P = 0$, 1 or 2).

- A, B - the points a and b, $a < b$ (double precision).

- X - the point x ($a < x < b$) where the interpolated value is required (double precision).

- OMEGA - the value of ω (double precision).

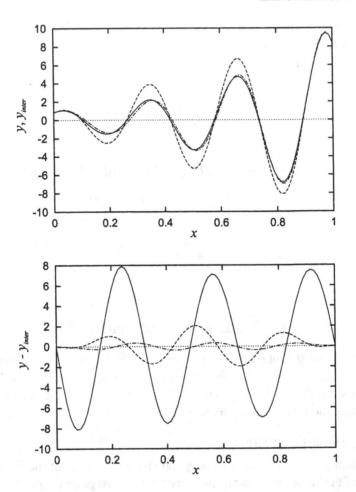

Figure 4.16. Up: the function $y(x)$ (solid) and its interpolation by the formulae with $P = 1$ (broken) and $P = 2$ (dash and dots). Down: error $y(x) - y_{inter}(x)$ for $P = 0$ (solid), $P = 1$ (broken) and $P = 2$ (dash and dots).

- YA, YB - double precision vectors with three components each, for the data at the two end points. YA(1), YA(2), YA(3) should contain $y(a)$, $y'(a)$ and $y''(a)$, respectively, while YB(1), YB(2), YB(3) are for $y(b)$, $y'(b)$ and $y''(b)$. As said, only subsets of these data are effectively used for IP < 2. For example, only YA(1) and YB(1) have to be furnished for IP=0.

Output:

- YINT - the interpolated value of $y(x)$ (double precision).

The following driving program computes the interpolated values for the function (4.176) for $a = 0$, $b = 0.5$, $\omega = 40$, at points $x = 0.1$, 0.2, 0.3, and 0.4 by each of the three versions. In output it publishes the exact value of $y(x)$, the interpolated values produced by the versions with $P = 0$, 1, 2, respectively, and also the errors. The subroutine YTEST, also listed, is used for the generation of the data at the two end points and for the evaluation of the error.

```
      implicit double precision (a-h,o-z)
      dimension ya(3),yb(3),yint(0:2)
      data a/0.d0/,b/.5d0/,omega/40.d0/
      do ix=1,4
        x=ix*(b-a)/5.d0
        call ytest(omega,a,ya(1),ya(2),ya(3))
        call ytest(omega,b,yb(1),yb(2),yb(3))
        call efint(0,a,b,x,omega,ya,yb,yint(0))
        call efint(1,a,b,x,omega,ya,yb,yint(1))
        call efint(2,a,b,x,omega,ya,yb,yint(2))
        call ytest(omega,x,y,yp,ypp)
          write(*,50)x,y,(yint(k),k=0,2),((y-yint(k)),k=0,2)
      enddo
50    format(' x=',f3.1,' y(x)=',f7.3,' y_int: ',3(f6.3,1x)
     ,/ !,20x,'err  : ',3(f6.3,1x),/)
      stop
      end

      subroutine ytest(om,x,y,yp,ypp)
      implicit double precision(a-h,o-z)
      y0=1.d0+x*(2.d0+x*(3.d0+4.d0*x))
      yp0=2.d0+x*(6.d0+12.d0*x)
      ypp0=6.d0+24.d0*x
      sinox=sin(om*x+1.d0)
      cosox=cos(om*x+1.d0)
      y=sinox*y0
      yp=om*cosox*y0+sinox*yp0
      ypp=-om*om*sinox*y0+2*om*cosox*yp0+sinox*ypp0
      return
      end
```

The output is:

```
x=0.1 y(x)= -1.183 y_int: -2.519 -1.356 -1.174
                   err  :  1.336  0.173 -0.009

x=0.2 y(x)=  0.640 y_int:  2.452  0.769  0.675
                   err  : -1.813 -0.129 -0.035
```

```
x=0.3 y(x)=  0.831 y_int: -0.686  0.956  0.796
                   err  :  1.517 -0.125  0.036


x=0.4 y(x)= -2.438 y_int: -1.555 -2.646 -2.447
                   err  : -0.883  0.208  0.009
```

References

[1] Blatt, J. M. (1967). Practical points concerning the solution of the Schrödinger equation. *Comput. Phys. Commun.*, 1: 382–396.

[2] Bocher, P. , De Meyer, H. and Vanden Berghe, G. (1994). On Gregory and modified Gregory–type corrections to Newton–Cotes quadrature. *J. Comput. Appl. Math.*, 50: 145–158.

[3] Burden, R. L. and Faires, J. D. (1981). *Numerical Analysis*. Prindle, Weber and Schmidt, Boston.

[4] Chakalov, L. (1954). General quadrature formulae of Gaussian type. *Bulgar. Akad. Izv. Mat. Inst.*, 1: 67–84. (English transl.: (1995) *East J. Approx.*, 1: 261–276.)

[5] Coleman, J. P. (2003). Private communication.

[6] Davis, P. J. and Rabinowitz, P. (1984). *Methods of Numerical Integration*. Academic Press, New York.

[7] De Meyer, H. , Vanthournout, J. , Vanden Berghe, G. and Vanderbauwhede, A. (1990). On the error estimation for a mixed type of interpolation. *J. Comput. Appl. Math.*, 32: 407–415.

[8] Duris, C. S. (1976). Generating and compounding product-type Newton–Cotes quadrature formulas. *TOMS*, 2: 50–58.

[9] Ehrenmark, U. T. (1988). A three–point formula for numerical quadrature of oscillatory integrals with variable frequency. *J. Comput. Appl. Math.*, 21: 87–99.

[10] Ehrenmark, U. T. (1996). On the error and its control in a two-parameter generalised Newton–Cotes rule. *J. Comput. Appl. Math.*, 75: 171–195.

[11] Ehrenmark, U. T. (2001). A note on a recent study of oscillatory integration rules. *J. Comput. Appl. Math.*, 131: 493–496.

[12] Evans, G. (1993). *Practical Numerical Integration*. John Wiley & Sons Ltd., Chichester.

[13] Evans, G. A. and Webster, J. R. (1997). A high order, progressive method for the evaluation of irregular oscillatory integrals. *Appl. Num. Math.*, 23: 205-218.

[14] Evans, G. A. and Webster, J. R. (1999). A comparison of some methods for the evaluation of highly oscillatory integrals *J. Comput. Appl. Math.*, 112: 55-69.

[15] Gerald, G. F. and Wheatley, P. O. (1994). *Applied Numerical Analysis*. Addison-Wesley, Reading.

[16] Ghizzetti, A. and Ossicini, A. (1970) *Quadrature formulae*. Birkhaüser, Basel.

[17] Fröberg, K. -E. (1970) *Introduction to numerical analysis*. Addison–Wesley, Reading.

[18] Ixaru, L. Gr. (1984). *Numerical Methods for Differential Equations and Applications*. Reidel, Dordrecht - Boston - Lancaster.

[19] Ixaru, L. Gr. (1997). Operations on oscillatory functions. *Comput. Phys. Commun.*, 105: 1–19.

[20] Ixaru, L. Gr. , De Meyer, H. and Vanden Berghe, G. (1997). CP methods for the Schrödinger equation revisited. *J. Comput. Appl. Math.*, 88: 289–314.

[21] Ixaru, L. Gr. , De Meyer, H. and Vanden Berghe, G. (1999). SLCPM12 – a program for solving regular Sturm–Liouville problems. *Comput. Phys. Comm.*, 118: 259–277.

[22] Ixaru, L. Gr. (2000). CP methods for the Schrödinger equation. *J. Comput. Appl. Math.*, 125: 347-357.

[23] Ixaru, L. Gr. (2001). Numerical operations on oscillatory functions. *Computers and Chemistry*, 25: 39-53.

[24] Ixaru, L. Gr. and Paternoster, B. (2001). A Gauss quadrature rule for oscillatory integrands. *Comput. Phys. Comm.*, 133: 177-188.

[25] Kim, K. J. (2003). Quadrature rules for the integration of the product of two oscillatory functions with different frequencies. *Comput. Phys. Comm.*, 153: 135-144.

[26] Kim, K. J., Cools, R. and Ixaru, L. Gr. (2002). Quadrature rules using first derivatives for oscillatory integrands. *J. Comput. Appl. Math.*, 140: 479-497.

[27] Kim, K. J., Cools, R. and Ixaru, L. Gr. (2003). Extended quadrature rules for oscillatory integrands. *Appl. Num. Math.*, 46: 59-73.

[28] Köhler, P. (1993). On the error of parameter-dependent compound quadrature formulas. *J. Comput. Appl. Math.*, 47: 47-60.

[29] Krylov, V. I. (1962) *Approximate Calculation of Integrals*. Macmillan, New York.

[30] Milovanovic, G. V. and Spalevic, M. M. (2002). Quadrature formulae connected to σ–orthogonal polynomials. *J. Comput. Appl. Math.*, 140: 619-637.

[31] Phillips, J. M . and Taylor, P. J. (1973). *Theory and Applications of Numerical Analysis*. Academic Press. London and New York.

[32] Popoviciu, T. (1955). Sur une generalisation de la formule d'integration numerique de Gauss. *Acad. R.P. Romane Fil. Iasi Stud. Cerc. Sti.*, 6: 29–57.

[33] Press, W. H., Teukolsky, S. A., Vetterling, W. T. and Flannery, B. P. (1986). *Numerical Recipes, The Art of Scientific Computing*. Cambridge University Press, Cambridge.

[34] Pryce, J. D. (1993). *Numerical Solution of Sturm-Liouville Problems*. Oxford University Press, Oxford.

[35] Stroud, A. and Secrest, D. (1966). *Gaussian Quadrature Formulas*, Prentice Hall, Englewood Cliffs, N.J.

[36] Stroud, A. and Stancu, D. D. (1965). Quadrature formulas with multiple Gaussian nodes. *J. SIAM Numer. Anal. Ser. B*, 2: 129–143.

[37] Turan, T. (1950). On the theory of the mechanical quadrature. *Acta Sci. Math. Szeged*, 12: 30–37.

[38] Vanden Berghe, G. , De Meyer, H. and Vanthournout, J. (1990). On a new type of mixed interpolation. *J. Comput. Appl. Math.*, 30: 55–69.

[39] Vanden Berghe, G. , De Meyer, H. and Vanthournout, J. (1990). On a class of modified Newton–Cotes quadrature formulae based upon mixed-type interpolation. *J. Comput. Appl. Math.*, 31: 331–349.

Chapter 5

LINEAR MULTISTEP SOLVERS FOR ORDINARY DIFFERENTIAL EQUATIONS

The solution of the initial value problem for ordinary differential equations is one of the main topics in numerical analysis. The linear multistep methods (algorithms) form a class of methods which benefitted from much attention over the years. The theory of these methods is basically due to Dahlquist, [14], and a series of well-known books which cover both theoretical and practical aspects are available, to mention only the books of Henrici, [22], Lambert, [40], Hairer, Nørsett and Wanner, [20], Shampine, [52], Hairer and Wanner, [21], and Butcher, [4].

Seen from the perspective of the present book the methods discussed in the quoted references are classical methods. They were constructed on the assumption that the solution $y(x)$ is smooth enough on the current multistep subinterval. Their coefficients are then constants, as in any classical approximation formula.

There is no technical difficulty in extending such algorithms to become tuned on other forms of the solution. In particular the six-point flow chart can be applied for forms (3.38) or (3.39). However, the capacity of computing the coefficients of the extended form of a classical algorithm is not enough for applications. Two additional issues have to be considered adequately:

1 Operations like numerical differentiation or quadrature are local operations while the numerical solution of an initial value problem (IVP) for ordinary differential equations (ODE) implies a propagation process: the approximate solution y_n at the mesh point x_n is calculated in terms of the values of the approximate solution at a number of previous mesh points. Since the latter are affected by the errors introduced at all previous applications of the procedure, the whole error will exhibit a cumulative effect. Even if it is known that this effect is kept under control by the classical version, it is not obvious that the same will hold for the extended version. Expressed in

more technical terms, the new issue consists in investigating how important may be the modification of the stability properties of a classical algorithm when this is extended via exponential fitting.

2 When deriving approximation formulae for differentiation and quadrature it is assumed that the shape of the function $y(x)$ on which these formulae will be applied is known. In particular, the values of the frequencies (or some good approximation of them, at least) are assumed available, and they are used for the determination of the appropiate coefficients of the formula.

In contrast, when solving an ODE the shape of the solution $y(x)$ is in general not known in advance. [There are only a few exceptions of practical importance; the solution of the radial Schrödinger equation is one of them.] The solution may be smooth enough on some subranges but with a dramatic variation (oscillatory or hyperbolic, but not only) on other subranges. An efficient computation must keep the track of such different potential behaviours and one set of frequencies may be suited on one subrange but a different set on another subrange. The problem of how the suited set of frequencies can be determined in terms of the local behaviour of the solution is therefore another specific problem to be answered adequately.

1.   First order equations

We consider the initial value problem

$$y' = f(x, y), \ x \in [a, b], \ y(a) = y_0, \tag{5.1}$$

where $f(x, y)$, $f : \Re \times \Re \to \Re$, the ends a and b of the interval and the initial value y_0 are given.

It is assumed that the solution exists and is unique. As a matter of fact, the following standard theorem gives sufficient conditions for a unique solution to exist (see, e.g., Lambert, [40]):

THEOREM 5.1 *If $f(x, y)$, $f : \Re \times \Re \to \Re$ is defined and continuous on all $x \in [a, b]$ and $-\infty < y < \infty$ and a constant L exists such that*

$$|f(x, y) - f(x, y^*)| < L|y - y^*| \tag{5.2}$$

for every pair (x, y) and (x, y^*) in the quoted region then, for any $y_0 \in \Re$ the stated initial value problem admits a unique solution which is continuous and differentiable on $[a, b]$.

The stated condition is called the Lipschitz condition.

For a numerical solution we introduce a partition of $[a, b]$: $x_0 = a$, $x_n = x_0 + nh$, $(n = 1, 2, \ldots, n_{max})$ such that $x_{n_{max}} = b$ which means that n_{max} and h are linked, $h = (b - a)/n_{max}$.

1.1 Exponential fitting versions of the two-step bdf algorithm

Let n be fixed ($0 \leq n \leq n_{max} - 2$) and assume that y_n and y_{n+1} are known approximations to the exact solution $y(x)$ at the points x_n and x_{n+1}, respectively. We want to construct an approximation y_{n+2} to $y(x_{n+2})$ by a formula of the form

$$a_0 y_n + a_1 y_{n+1} + y_{n+2} = h b_2 f(x_{n+2}, y_{n+2}). \tag{5.3}$$

The formula can be used to propagate the solution in the usual way, that is by taking $n = 0, 1, \ldots$. The algorithm based on formula (5.3) is called a two-step bdf algorithm, where bdf is the abbreviation for *backwards difference formula*. The rational behind such a denomination is that if we consider $y(x)$ as a function with known $y_{n+j} = y(x_{n+j})$, $j = 0, 1, 2$, equation (5.3) allows determining an approximation to $y'(x_{n+2})$,

$$y'(x_{n+2}) \approx \frac{1}{h b_2} [a_0 y(x_n) + a_1 y(x_{n+1}) + y(x_{n+2})], \tag{5.4}$$

which is nothing else than the three point backwards formula for the first derivative.

To determine the three coefficients a_0, a_1 and b_2 we follow the $\mathcal{L}$ operator scheme. We denote $\mathbf{a} := [a_0, a_1, b_2]$ and define $\mathcal{L}[h, \mathbf{a}]$ by

$$\mathcal{L}[h, \mathbf{a}] y(x) := a_0 y(x - h) + a_1 y(x) + y(x + h) - h b_2 y'(x + h). \tag{5.5}$$

$\mathcal{L}$ is dimensionless, that is $l = 0$.
To construct $E_0^*(z, \mathbf{a})$, where $z = \mu h$, we apply $\mathcal{L}$ on $y(x) = \exp(\mu x)$ (see (2.61) and (2.69)) to obtain

$$E_0^*(z, \mathbf{a}) = a_0 \exp(-z) + a_1 + (1 - z b_2) \exp(z), \tag{5.6}$$

while for obtaining $E_k^*(z, \mathbf{a})$, $k = 1, 2, \ldots$, we use (2.70) with the result

$$E_k^*(z, \mathbf{a}) = (-1)^k a_0 \exp(-z) + [1 - b_2(k + z)] \exp(z). \tag{5.7}$$

Classical L_k^* moments result by taking $z = 0$:

$$L_k^*(\mathbf{a}) = (-1)^k a_0 + a_1 \delta_{k0} + 1 - k b_2, \quad k = 0, 1, \ldots \tag{5.8}$$

As for the functions $G^{\pm}(Z, \mathbf{a})$ and their derivatives with respect to $Z = z^2$, we have

$$
\begin{aligned}
G^+(Z, \mathbf{a}) &= a_0 \eta_{-1}(Z) + a_1 - b_2 Z \eta_0(Z) + \eta_{-1}(Z), \\
G^-(Z, \mathbf{a}) &= -a_0 \eta_0(Z) - b_2 \eta_{-1}(Z) + \eta_0(Z), \\
G^{+(m)}(Z, \mathbf{a}) &= 2^{-m} [a_0 \eta_{m-1}(Z) - b_2(3 \eta_{m-1}(Z) - \eta_m(Z)) + \eta_{m-1}(Z)], \\
G^{-(m)}(Z, \mathbf{a}) &= 2^{-m} [-a_0 \eta_m(Z) - b_2 \eta_{m-1}(Z) + \eta_m(Z)], \quad m = 1, 2, \ldots
\end{aligned}
\tag{5.9}
$$

This set of expressions allows determining the three coefficients of the bdf algorithm under various assumptions on the three functions required for making $\mathcal{L}[h, \mathbf{a}]y(x)$ identically vanishing in x and in h. In [35] and [36] three such versions are considered, and they are called A0, A1 and A2.

A0. This is the classical version. The functions are $y(x) = 1$, x and x^2. The system to be solved is $L_k^*(\mathbf{a}) = 0$, $k = 0, 1, 2$, with the well-known solution

$$a_0 = \frac{1}{3}, \; a_1 = -\frac{4}{3}, \; b_2 = \frac{2}{3}. \tag{5.10}$$

A1. Here $\mathcal{L}[h, \mathbf{a}]y(x)$ is required being identically vanishing for $y(x) = 1$, x and $\exp(\mu x)$. The system is $L_0^*(\mathbf{a}) = L_1^*(\mathbf{a}) = E_0^*(z, \mathbf{a}) = 0$, where $z = \mu h$, as usual, with the solution

$$a_0(z) = \frac{(1 - z)\exp(z) - 1}{-z\exp(z) - \exp(-z) + 1}, \tag{5.11}$$
$$a_1(z) = -(1 + a_0(z)), \; b_2(z) = 1 - a_0(z).$$

$a_0(z)$ can be written in power series, viz.:

$$a_0(z) = \frac{\displaystyle\sum_{n=0}^{\infty} \frac{1}{(n+2)!}(n+1)z^n}{\displaystyle\sum_{n=0}^{\infty} \frac{1}{(n+2)!}[(-1)^n + n + 2]z^n}.$$

This form is useful for the computation at small $|z|$. For double precision computation $a_0(z)$ has to be evaluated by (5.11) for $|z| > 10^{-2}$, and by the truncated power series (with six terms retained) otherwise.

A2. The three functions are $y(x) = 1$, $\exp(\pm\mu x)$ and then, according to the flow chart, we solve the system $L_0^*(\mathbf{a}) = G^+(Z, \mathbf{a}) = G^-(Z, \mathbf{a}) = 0$, where $Z = z^2 = \mu^2 h^2$, to obtain

$$a_0(Z) = \frac{1}{1 + 2\eta_{-1}(Z)}, \; a_1(Z) = -1 - a_0(Z), \; b_2(Z) = 2\eta_0(Z)a_0(Z). \tag{5.12}$$

The functions used to generate each of these versions are the linearly independent solutions of the following RDE of order $M = 3$:

$$(D - \mu_1)(D - \mu_2)(D - \mu_3)y = 0, \tag{5.13}$$

where $\mu_1 = \mu_2 = \mu_3 = 0$ for A0, $\mu_1 = \mu_2 = 0$, $\mu_3 = \mu$ for A1, and $\mu_1 = 0$, $\mu_2 = -\mu_3 = \mu$ for A2. Thus, no adjustable frequency is present in A0, while there is one (μ) in A1, and two ($\mu^{\pm} = \pm\mu$) in A2.

The expressions of the *lte* are

$$lte = \frac{1}{3!}h^3 L_3^*(\mathbf{a})y^{(3)}(x_{n+1}) = -\frac{2}{9}h^3 y^{(3)}(x_{n+1}), \qquad (5.14)$$

for A0,

$$lte = h^3 T_0(z, \mathbf{a}(z))[y^{(3)}(x_{n+1}) - \mu\, y''(x_{n+1})], \qquad (5.15)$$

where

$$
\begin{aligned}
T_0(z, \mathbf{a}(z)) &= -\frac{L_2^*(\mathbf{a}(z))}{2!z} = \frac{-a_0(z) + 2b_2(z) - 1}{2z} \\
&= \frac{(2z - 3)\exp(z) - \exp(-z) + 4}{2z(-z\exp(z) - \exp(-z) + 1)}
\end{aligned}
\qquad (5.16)
$$

for A1, and

$$lte = h^3 Q_0(Z, \mathbf{a}(Z))[y^{(3)}(x_{n+1}) - \mu^2\, y'(x_{n+1})], \qquad (5.17)$$

where

$$
\begin{aligned}
Q_0(Z, \mathbf{a}(Z)) &= -\frac{L_1^*(\mathbf{a}(Z))}{Z} = \frac{a_0(Z) + b_2(Z) - 1}{Z} \\
&= -\frac{2[\eta_{-1}(Z) - \eta_0(Z)]}{Z(1 + 2\eta_{-1}(Z))} = -\frac{2\eta_1(Z)}{1 + 2\eta_{-1}(Z)},
\end{aligned}
\qquad (5.18)
$$

for A2; the formula (3.57) has been used.

The most general ef extension of the two-step bdf algorithm will be the one in which $\mathcal{L}[h, \mathbf{a}]y(x)$ is required to be identically vanishing for $y(x) = \exp(\mu_1 x)$, $\exp(\mu_2 x)$ and $\exp(\mu_3 x)$ with arbitrary complex μ_1, μ_2 and μ_3. This will be consistent with an RDE with complex coefficients and then it will also lead to complex values for the coefficients of the method. However, since the differential equation to be solved by the algorithm is real, it is sufficient to consider only the case when the coefficients of the RDE are real, an assumption which will be tacitly admitted throughout the whole chapter. Versions A1 and A2 are consistent with this assumption if μ in A1 and μ^2 in A2 are real. In the latter case this implies that the two frequencies $\mu^\pm$ are either real or purely imaginary.

The versions A1 and A2 do not exhaust the possibilities available for the generation of ef versions with real coefficients. For example, a version of this type is also the one based on real μ_1 and on $\mu_2 = \lambda + i\omega$, $\mu_3 = \lambda - i\omega$, where λ and ω are real, but the reference to only A1 and A2 is sufficient for making an idea on the advantages which may be offered by the exponential fitting.

A few specific comments follow:

The existence of different versions of the two-step bdf algorithm represents a simple illustration of a general feature: the use of the exponential fitting allows making any multistep algorithm flexible. Indeed, the three versions can be regarded as the three components of a flexible algorithm, call it AEF (see [35], [36] and also Section 1.4), whose coefficients depend on the set $\mathbf{z} = [z_1, z_2, z_3]$ where $z_1 = 0$ by default and some specific correlations exist between z_2 and z_3. If $z_2 = z_3 = 0$ AEF is simply A0 while if $z_2 = 0$ and $z_3 = \mu h$ (real μ) it is A1 with the argument $z = \mu h$. Finally, if $z_2 = -z_3 = \mu h$ (μ is either real or purely imaginary) AEF becomes A2 with the real argument $Z = \mu^2 h^2$. The use of the classical component A0 is convenient in the regions where the solution is smooth enough but the other two are still available otherwise. The problems of how to choose between different components in practice and of how to select the convenient value of the real parameter μ or μ^2 will be discussed later on.

As expected, when $z, Z \rightarrow 0$ the coefficients of A1 and of A2 tend to the classical coefficients (5.10). Also $T_0(z, \mathbf{a}(z))$ and $Q_0(Z, \mathbf{a}(Z))$ tend to the classical value $-2/9$. It is also important to notice that the values of these functions are also reasonably close to $-2/9$ when $-1 < z, Z < 1$, see Figure 5.1. It follows that for z or Z in this range the comparison only of the different forms of the last factor in the lte is sufficient to draw conclusions on the relative merits of each version for accuracy.

When introducing the $\mathcal{L}$ operator by (5.5) we used the midpoint for x but this choice had only a technical reason behind it: the expressions of E^*, L^* and of $G^{(\pm)}$ resulting in this way are particularly simple and easy to manipulate. Other options are also possible, for example by taking the first point for x, which means introducing $\mathcal{L}$ by

$$\mathcal{L}[h, \mathbf{a}]y(x) := a_0 y(x) + a_1 y(x + h) + y(x + 2h) - h b_2 y'(x + 2h). \quad (5.19)$$

The expressions of E^*, L^* and $G^{(\pm)}$ will look differently but the important point is that the final expressions of the coefficients and of the relevant components in lte are unchanged; the replacement of x_{n+1} by x_n in the lte will be the only (minor) change. This is not surprising at all. It reflects the fact that the complete set of the linearly independent solutions of an RDE was used in the whole construction which, as explained in Section 2.1, is closed with respect to translation.

1.2   Theoretical issues

Extensions of the classical multistep methods for solving r-th order ordinary differential equations of the form $y^{(r)} = f(x, y)$ (that is with f depending only on x and on y) have been considered by Lyche, [43], and called there Chebyshevian multistep methods. The extensions based on the exponential fitting belong to this class and in that paper it is shown that the theory of the convergence for Chebyshevian methods can be approached in the frame of the standard theory for classical multistep methods provided some basic definitions in the latter are conveniently adapted.

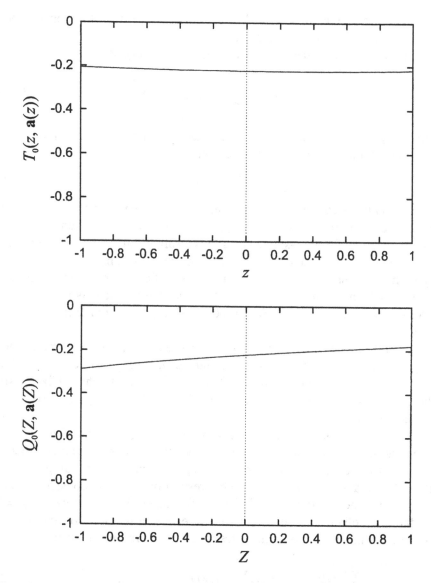

Figure 5.1. Behaviour of the functions $T_0(z, \mathbf{a}(z))$ (up) and $Q_0(Z, \mathbf{a}(Z))$ (down) corresponding to the exponential fitting versions A1 and A2, respectively, of the two-step bdf algorithm (5.3) for $-1 < z, Z < 1$.

Hereinafter we briefly review the main concepts of the standard theory and indicate the modifications needed for the treatment of the exponential fitting versions. These concepts hold both for single differential equations and for systems of such equations but for reasons of simplicity we shall consider only the former case, equation (5.1). We examine two issues: the convergence theory, and the linear stability theory.

The classical J-step method is a method with the algorithm

$$\sum_{j=0}^{J} a_j y_{n+j} = h \sum_{j=0}^{J} b_j f_{n+j}, \, n = 0, 1, \ldots \tag{5.20}$$

where $f_{n+j} = f(x_{n+j}, y_{n+j})$. The coefficients a_j and b_j are constants which do not depend on n subject to the conditions

$$a_J = 1, \, |a_0| + |b_0| \neq 0. \tag{5.21}$$

At each n it allows computing y_{n+J}, which is an approximation to $y(x_{n+J})$, in terms of the approximations y_{n+j} to $y(x_{n+j})$, $0 < j < J-1$, assumed known.

An exponential fitting J-step algorithm is an algorithm of the same form but we shall consider only the case when the coefficients a_j and b_j are allowed to vary with n in a special way. With the two-step bdf algorithm in mind, that is with $J = 2$ and with $b_0 = b_1 = 0$ by default, the coefficients may be as in A0 for some values of n, as in A1 for other values, and as in A2 for still other values.

In general, for given n the coefficients will depend on M local frequencies μ_i, $i = 1, 2 \ldots, M$ (distinct or not) and on h, through the corresponding $\mathbf{z} = [z_1, z_2, \ldots, z_M]$. Functions $a_j(\mathbf{z})$ and $b_j(\mathbf{z})$ are well-behaved except for critical values but, since $\mathbf{z} = \mathbf{0}$ is never a critical value, $a_j(\mathbf{0})$ and $b_j(\mathbf{0})$ exist and they are finite.

We shall also assume that $a_j(\mathbf{0})$ and $b_j(\mathbf{0})$ are independent of n. This is convenient because in such a case only a slight adaptation of the standard theory will be needed to cover the convergence theory of the exponential fitting version. With a $J = 2$ algorithm in mind, such a restriction rules out considering bdf versions for some n and central diffference formulae (that is with $b_0 = b_2 = 0$ by default) for other n because the two will lead to different $a_j(\mathbf{0})$ and $b_j(\mathbf{0})$.

Altogether the discussion will be restricted to algorithms of the form

$$\sum_{j=0}^{J} a_j(\mathbf{z}) y_{n+j} = h \sum_{j=0}^{J} b_j(\mathbf{z}) f_{n+j}, \, n = 0, 1, \ldots \tag{5.22}$$

with $a_J(\mathbf{z}) = 1$, whose classical counterpart

$$\sum_{j=0}^{J} a_j(\mathbf{0}) y_{n+j} = h \sum_{j=0}^{J} b_j(\mathbf{0}) f_{n+j}, \, n = 0, 1, \ldots \tag{5.23}$$

has coefficients independent of n, i.e. it is of a form which is directly covered by the standard theory.

Convergence theory

To start the computation ($n = 0$), $y_0, y_1, \ldots, y_{J-1}$ are needed but only y_0 is known. The generation of the other starting values thus represents a separate task and in practice this is covered by means of a one-step method (Runge–Kutta, for example). The values determined in this way will be normally affected by errors which will propagate and combine with the additional errors introduced at each new step $n = 1, 2, \ldots$. It follows that the algorithm will be reliable only if it is convergent which, loosely speaking, means that if the accuracy in the starting values is kept under control then, when the partition becomes finer and finer the approximate value y_n is expected to tend to the true $y(x_n)$ on all mesh points of the partition.

For a rigorous definition of the convergence the following process is considered. We fix some $x \in [a, b]$ and focus on the interval $[a, x]$. We take some integer n_{max} and define $h = (x - a)/n_{max}$ to examine the behaviour of $y_{n_{max}}$, which is an approximation to $y(x)$, when $n_{max} \to \infty$ (and then $h \to 0$); this is called a *fixed station limit*, see [40]. It is also assumed that the starting data depend on h, $y_j = \phi_j(h)$, $0 \leq j \leq J - 1$, in such a way that when $h \to 0$ one has $\phi_j(h) \to y_0$. This assumption is just normal because when $h \to 0$ the involved mesh points tend to $x_0 = a$, where the initial condition was imposed.

DEFINITION 5.1 *The method defined by (5.22) and (5.23) is said to be convergent if, for all initial value problems satisfying the hypotheses of the theorem 5.1, the fixed station limit*

$$\lim_{\substack{h \to 0 \\ x = a + n_{max}h}} y_{n_{max}} = y(x) \tag{5.24}$$

holds for all $x \in [a, b]$ and for all solutions of the equation (5.22) satisfying starting conditions $y_j = \phi_j(h)$, $0 \leq j \leq J-1$, for which $\lim_{h \to 0} \phi_j(h) = y_0$. A method which is not convergent is said to be divergent.

In essence this definition is the same as the one in [40] for a classical method.

The direct determination of the coefficients in terms of the conditions for the convergence is difficult but it becomes transparent and easier to manipulate if the whole problem is divided into two separate problems. On one hand, the intuition says that a necessary condition for convergence is that, at each n, the resultant y_{n+J} should be indeed a reasonable approximation to the desired $y(x_{n+J})$ if the input values for y_{n+j} are close enough to the true $y(x_{n+j})$ ($0 \leq j \leq J - 1$). On the other hand, the computation of the numerical solution is a propagation process. The input values at each n are affected by errors accumulated all the way until that n and it is important to choose the coefficients such that the propagation of these errors from one n to another be kept under control. The two problems are therefore a local problem, which will lead to the consistency condition, and a propagation problem, which will lead to the zero-stability condition.

This separation of the whole problem in two distinct problems lies at the basis of the convergence theory of Dahlquist. The following theorem holds for classical methods:

THEOREM 5.2 *The necessary and sufficient conditions for the method (5.20) (or (5.23)) to be convergent is that it be both consistent and zero-stable.*

1.2.1 Local problem: order of the method and consistency

Let n be fixed. The operator $\mathcal{L}[h, \mathbf{a}]$ is introduced by

$$\mathcal{L}[h, \mathbf{a}]y(x) := \sum_{j=0}^{J} [a_j y(x + jh) - hb_j y'(x + jh)], \qquad (5.25)$$

where $\mathbf{a} := [a_j \ (0 \leq j \leq J - 1), \ b_j \ (0 \leq j \leq J)]$ is a set with $2J + 1$ components. $\mathcal{L}$ is dimensionless, that is $l = 0$.

Let us take some M and ask that $\mathcal{L}[h, \mathbf{a}]y(x) = 0$ holds identically in x and in h if and only if $y(x)$ satisfies an RDE of the form

$$(D - \mu_1)(D - \mu_2) \cdots (D - \mu_M)y = 0, \qquad (5.26)$$

with any given set of frequencies.

The 'if' requirement means that the coefficients must satisfy the following system of M equations

$$E_0^*(z_i, \mathbf{a}) = 0, \ 0 \leq i \leq M - 1. \qquad (5.27)$$

Strictly speaking, this is of use only if all frequencies are distinct but suited systems are also available otherwise. If, for example, only μ_i, $1 \leq i \leq I$ are distinct and the multiplicity of μ_i is m_i, then the system of M equations is

$$E_k^*(z_i, \mathbf{a}) = 0, \ k = 0, 1, \ldots, m_i - 1, \qquad (5.28)$$
$$i = 1, 2, \ldots, I, \ m_1 + m_2 + \cdots + m_I = M,$$

(see (2.83)), while if the frequencies are clustered in pairs with opposite values, the system will involve the functions $G^\pm$ and their derivatives; see e.g. (3.60).

Since the number of unknowns is $2J + 1$ one may think that the maximal M should be $2J + 1$. However, the first Dahlquist barrier (see [40]) imposes the severe restrictions $M \leq J + 2$ for odd J, and $M \leq J + 3$ for even J. It follows that $2J + 1 - M$ coefficients have to be fixed in advance (or taken as free parameters) and then the remaining M coefficients will depend on these data. A situation of this type was met for two-step bdf method where the values $b_0 = b_1 = 0$ were taken by default.

To construct $E_0^*(z, \mathbf{a})$ we use lemma 2.1 and (2.69) to get

$$E_0^*(z, \mathbf{a}) = \sum_{j=0}^{J} (a_j - zb_j) \exp(jz), \qquad (5.29)$$

while for $E_k^*(z, \mathbf{a})$, $k = 1, 2, \ldots$ we use (2.70) with the result

$$E_k^*(z, \mathbf{a}) = \sum_{j=0}^{J} [j^k a_j - j^{k-1}(jz + k)b_j] \exp(jz). \qquad (5.30)$$

The classical moments result by taking $z = 0$:

$$L_k^*(\mathbf{a}) = \sum_{j=0}^{J} (j^k a_j - j^{k-1} k b_j), \qquad (5.31)$$

and the coefficients of the classical version must satisfy

$$L_k^*(\mathbf{a}) = 0, \ 0 \le i \le M - 1. \qquad (5.32)$$

The 'if' requirement thus means that

$$\mathcal{L}[h, \mathbf{a}(z)]y(x) = h^M \sum_{k=0}^{\infty} h^k T_k^*(z, \mathbf{a}(z)) D^k (D - \mu_1) \cdots (D - \mu_M) y(x), \qquad (5.33)$$

see (2.84).

The 'only if' requirement implies that $T_0^*(z, \mathbf{a}(z))$ is not identically vanishing in z and that, in particular, $T_0^*(0, \mathbf{a}(0)) \ne 0$. It follows that the *lte* of the method whose coefficients are constructed in this way is

$$lte = h^M T_0^*(\mathbf{z}, \mathbf{a}(\mathbf{z}))(D - \mu_1)(D - \mu_2) \cdots (D - \mu_M) y(x_n), \qquad (5.34)$$

which reduces to

$$lte = \frac{1}{M!} h^M L_M^*(\mathbf{a}(\mathbf{0})) y^{(M)}(x_n), \qquad (5.35)$$

for the classical counterpart of the method.

DEFINITION 5.2 *The method defined by (5.22) and (5.23) with the coefficients constructed as described before is said of order* $p = M - 1$. *It is said to be consistent if* $p \ge 1$.

This definition contains the standard definition of the order as a particular case; the latter is : the classical method (5.23) is of order p if $L_k^*(\mathbf{a}(\mathbf{0})) = 0$ for $0 \le k \le p$ and $L_{p+1}^*(\mathbf{a}(\mathbf{0})) \ne 0$, see [40]. As for the consistency conditions, their standard formulation is given in terms of the first and second characteristic polynomials

$$\rho(d) = \sum_{j=0}^{J} a_j(0)d^j, \text{ and } \sigma(d) = \sum_{j=0}^{J} b_j(0)d^j, \qquad (5.36)$$

and the method (5.23) is said to be consistent if $\rho(1) = 0$ and $\rho'(1) = \sigma(1)$. However, this is equivalent to saying that $L_0^*(\mathbf{a}(0)) = L_1^*(\mathbf{a}(0)) = 0$ and then $p = 1$, at least.

Local truncation error

As everywhere in this book, the leading term of the error *lte* of an approximation formula is the first nonvanishing term in the series expansion of $\mathcal{L}[h, \mathbf{a}]y(x)$ evaluated at a particular value of x. In our case this is x_n. What link may exist between this *lte* and the error produced by the algorithm when evaluating y_{n+J}? There is not a direct answer to this problem because the entrance data in the algorithm are affected by the errors accumulated up to the current n. A simpler, but also relevant question may be still addressed: what about the error in y_{n+J} when all the entrance data are assumed exact, that is $y_{n+j} = y(x_{n+j})$, $0 \le j \le J - 1$? The latter is called the localizing assumption. If so, the algorithm reads

$$\sum_{j=0}^{J-1}[a_j y(x_{n+j}) - hb_j f(x_{n+j}, y(x_{n+j}))] + y_{n+J} - hb_J f(x_{n+J}, y_{n+J}) = 0,$$

$$(5.37)$$

while (5.25) is

$$\sum_{j=0}^{J-1}[a_j y(x_{n+j}) - hb_j f(x_{n+j}, y(x_{n+j}))] \qquad (5.38)$$

$$+y(x_{n+J}) - hb_J f(x_{n+J}, y(x_{n+J})) = \mathcal{L}[h, \mathbf{a}]y(x_n).$$

According to the usual terminology in the context of the multistep methods for ODE (see again [40]), $\mathcal{L}[h, \mathbf{a}]y(x_n)$ is called the *local truncation error* at x_{n+J} and denoted LTE.

By subtraction we get

$$y(x_{n+J}) - y_{n+J} - hb_J[f(x_{n+J}, y(x_{n+J})) - f(x_{n+J}, y_{n+J})] = LTE, \quad (5.39)$$

while by using the mean value theorem we have

$$f(x_{n+J}, y(x_{n+J})) - f(x_{n+J}, y_{n+J}) = \frac{\partial f(x_{n+J}, \xi)}{\partial y}[y(x_{n+J}) - y_{n+J}],$$

$$(5.40)$$

where ξ is in the internal part of the segment joining $y(x_{n+J})$ and y_{n+J}. It follows that

$$y(x_{n+J}) - y_{n+J} = \frac{1}{1 - hb_J \dfrac{\partial f(x_{n+J}, \xi)}{\partial y}} LTE = lte + \mathcal{O}(h^{M+1}), \quad (5.41)$$

and this gives the link between the error $y(x_{n+J}) - y_{n+J}$ produced by the algorithm under the localizing assumption, LTE and lte. Then lte represents just what in the usual terminology adopted in this field is referred to as the *principal local truncation error* and abbreviated $PLTE$, see again [40].

1.2.2 Propagation problem and zero-stability

As seen, the basic concepts of the standard theory for the local problem (order, consistency, LTE etc.) remain valid also for exponential fitting multistep methods provided a few minor adaptations are introduced in some definitions. The situation is comparatively more complicated in the propagation problem. For simplicity we shall assume that one and the same set of frequencies is used in all steps, i.e., $\mathbf{z}$ is independent of n.

One starts with the simplest initial value problem,

$$y' = 0, \ x \in [a, b], \ y(a) = y_0, \qquad (5.42)$$

with the exact solution $y(x) = y_0$. The algorithm (5.22) becomes

$$\sum_{j=0}^{J} a_j(\mathbf{z})y_{n+j} = 0, \ n = 0, 1, \ldots, \qquad (5.43)$$

and, if the set of starting values is properly selected, we should normally expect that

$$\lim_{\substack{h \to 0 \\ x = a + n_{max}h}} y_{n_{max}} = y(x) = y_0 \qquad (5.44)$$

holds for any $x \in [a, b]$, in particular for $x = b$.

The equation (5.43) can be treated in analytic form. We search for a solution of the form $y_m = d^m$ which, when introduced in that equation, leads to

$$d^n \sum_{j=0}^{J} a_j(\mathbf{z})d^j = 0, \ n = 0, 1, \ldots \qquad (5.45)$$

It is appropriate to introduce the polynomials

$$\rho(d, \mathbf{z}) = \sum_{j=0}^{J} a_j(\mathbf{z})d^j, \ \text{and} \ \sigma(d, \mathbf{z}) = \sum_{j=0}^{J} b_j(\mathbf{z})d^j, \qquad (5.46)$$

which for $z = 0$ reduce to the first and the second characteristic polynomials of the associated classical method. Equation (5.45) reads

$$d^n \rho(d, z) = 0, \ n = 0, 1, \ldots \tag{5.47}$$

The behaviour of y_m, in particular whether the fixed station limit (5.44) does or does not hold, depends on the values of the roots of $\rho(d, z)$. The following definition is pertinent for the classical version:

DEFINITION 5.3 [40] *The method (5.23) is said to satisfy the root condition if all of the roots of the polynomial $\rho(d, 0)$ have modulus less than or equal to unity, and those of modulus unity are simple. It is said to be zero-stable if it satisfies the root condition.*

According to the theorem 5.2 the root condition and the requirements that $\rho(1, 0) = 0$ and that $\rho'(1, 0) = \sigma(1, 0)$ are the minimal conditions for the convergence of the method (5.23).

Let us assume that this is convergent indeed. Is its ef-based extension (5.22) also convergent ? Expressed in different words, the roots of $\rho(d, z)$ are z-dependent and, as assumed, for $z = 0$ they satisfy the root condition but what about their behaviour when at least one of the frequencies is different from zero ?

We investigate these issues on the case when the roots $d_1(z)$, $d_2(z)$, $\ldots$, $d_J(z)$ of $\rho(d, z)$ are distinct, where $d_1(0) = 1$ is the only root of $\rho(d, 0)$ which equals 1 in modulus. We show below that the ef-based version is convergent even if some of the roots of $\rho(d, z)$ may happen to be bigger than 1 in modulus for some $h > 0$.

The analytic solution of the recurrence relation (5.43) is

$$y_m = \sum_{j=1}^{J} C_j d_j^m(z), \ m = 0, 1, \ldots, n_{max}, \tag{5.48}$$

where the coefficients C_j are fixed in terms of the starting values. The roots $d_j(0)$ are constants which satisfy the root condition and then

$$\left| \lim_{n_{max} \to \infty} d_j(0)^{n_{max}} \right| = \delta_{1j},$$

but for $d_j(z)$ we have to consider the fixed station limit and to check whether

$$\left| \lim_{\substack{h \to 0 \\ b - a = n_{max}h}} d_j(z)^{n_{max}} \right| = \delta_{1j} \tag{5.49}$$

holds as well. Since each z_i is linear in h we have $\lim_{h \to 0} z = 0$. It follows that $\rho(d, z) = \rho(d, 0) + \mathcal{O}(h^r)$ and $\sigma(d, z) = \sigma(d, 0) + \mathcal{O}(h^s)$ where r and s are positive integers.

We assume the roots of $\rho(d, \mathbf{z})$ are of the form $d_j(\mathbf{z}) = d_j(0) + \alpha_j h^{p_j} + \mathcal{O}(h^{p_j+1})$, where α_j depends on the frequencies, and want to determine the value of p_j.

The behaviour of the root $d_1(\mathbf{z})$ is particularly important. We apply the operator $\mathcal{L}[h, \mathbf{a}(\mathbf{z})]$ on the function $y(x) = 1$. From the definition of $L_0^*(\mathbf{a}(\mathbf{z}))$ we have $\mathcal{L}[h, \mathbf{a}(\mathbf{z})]1 = L_0^*(\mathbf{a}(\mathbf{z}))$ while from comparison of (5.31) and (5.46) we get $L_0^*(\mathbf{a}(\mathbf{z})) = \rho(1, \mathbf{z})$. Also, from (5.33) we have

$$\mathcal{L}[h, \mathbf{a}(\mathbf{z})]1 = (-h)^M T_0^*(\mathbf{z}, \mathbf{a}(\mathbf{z}))\mu_1\mu_2\cdots\mu_M$$

where $T_0^*(\mathbf{z}, \mathbf{a}(\mathbf{z}))$ is not identically vanishing. It follows that

$$\rho(1, \mathbf{z}) = (-h)^M T_0^*(\mathbf{z}, \mathbf{a}(\mathbf{z}))\mu_1\mu_2\cdots\mu_M$$

If at least one frequency is vanishing, we have $\rho(1, \mathbf{z}) = 0$ and then the first root is $d_1(\mathbf{z}) = 1$. If all $\mu_i \neq 0$ we search for an approximation of the form $d_1(\mathbf{z}) \approx 1 + \Delta d_1$, where Δd_1 is the first correction in a Newton–Raphson iteration process. We have

$$\Delta d_1 = -\frac{\rho(1, \mathbf{z})}{\rho'(1, \mathbf{z})} \approx -\frac{(-h)^M T_0^*(\mathbf{z}, \mathbf{a}(\mathbf{z}))\mu_1\mu_2\cdots\mu_M}{\sigma(1, 0)} \sim \mathcal{O}(h^M).$$

This shows that $p_1 = M$. Since the classical version was assumed convergent we have $M \geq 2$ and then

$$\lim_{\substack{h \to 0 \\ b-a = n_{max}h}} d_1(\mathbf{z})^{n_{max}} = \lim_{n_{max}\to\infty}(1 + \frac{\alpha_1(b-a)^M}{n_{max}^M})^{n_{max}} = 1.$$

Notice the importance of the fact that $M \geq 2$. The value $M = 1$ would also give a finite limit, but this may be bigger than 1.

For $j > 1$ we write $d_j(\mathbf{z}) \approx d_j(0) + \Delta d_j$ where

$$\Delta d_j = -\frac{\rho(d_j(0), \mathbf{z})}{\rho'(d_j(0), \mathbf{z})} = -\frac{\rho(d_j(0), 0) + \mathcal{O}(h)}{\rho'(d_j(0), 0) + \mathcal{O}(h)},$$

and since $\rho(d_j(0), 0) = 0$ we have $\Delta d_j \sim h$, that is $p_j = 1$. As for the fixed station limit, we have

$$\lim_{\substack{h \to 0 \\ b-a = n_{max}h}} d_j(\mathbf{z})^{n_{max}} = \lim_{n_{max}\to\infty} d_j(0)^{n_{max}}$$

$$\times \lim_{n_{max}\to\infty}\left[1 + \frac{\alpha_j(b-a)}{d_j(0)n_{max}}\right]^{n_{max}} = 0.$$

The above results, which say in essence that an ef-based extension of a convergent classical method is also convergent, may seem to have a rather limited relevance for practice since the whole investigation covered only the case when the set of frequencies remains unchanged on all steps. However, the situation is not so dramatic because the stated result allows formulating a way for drawing conclusions on the convergence also when z is n-dependent. The set $\mu = [\mu_i, 1 \leq i \leq M]$ is normally fixed in terms of the specific behaviour of the solution of the given initial value problem in various subintervals of $[a, b]$.

To fix the ideas, let us first introduce an equidistant partition of $[a, b]$ with N_{max} steps, that is with the step width $H = (b - a)/N_{max}$, and let fix some N $(J \leq N \leq N_{max} - J)$ and denote $X = x_N$. We are interested to see what happens when the partition becomes finer and finer by successive halvings of the current step width, that is by taking $n_{max} = 2^k N_{max}$, $h = 2^{-k} H$, $k = 1, 2, \ldots$, if the ef algorithm (5.22) is used with the set μ_1 for all mesh points x_n in the subinterval $I_1 = [a, X)$ and with the set μ_2 in the subinterval $I_2 = [X, b]$. The algorithm used on I_1 will have one and the same z_1 on all n involved, that is $0 \leq n \leq 2^k N - 1$ and then it is of a form whose convergence is secured by the stated results. This means that, if the starting conditions are properly posed, as we assume, the discrete solution on I_1 will tend to the exact solution when $h \to 0$. In particular this is true for $y_{2^k N + j}$, $j = 0, 1, 2 \ldots, J - 1$, which are the starting values for the solution on I_2. The algorithm used on the latter interval is also covered for convergence by the above results and then the discrete solution will also tend to the exact one when $h \to 0$, as desired. The generalization for an arbitrary number of subintervals is obvious.

1.2.3 Linear stability theory

The zero-stability controls the manner in which the errors accumulate when $h \to 0$ but in applications only $h > 0$ is used. It makes sense to get some information on how the errors accumulate in such a case.

The usual way of getting such information consists in selecting some initial value problem whose analytic solution $y(x)$ is known and it tends to 0 when $x \to \infty$, and then checking whether the numerical solution y_n, supposed to approximate $y(a + nh)$ (for fixed $h > 0$) also tends to 0 when $n \to \infty$.

The standard initial value problem used for this purpose is

$$y' = \lambda y, \ x > a, \ y(a) = y_0, \tag{5.50}$$

where λ is a complex constant with $Re \, \lambda < 0$. The exact solution is $y(x) = y_0 \exp(\lambda x)$. It tends to zero when $x \to \infty$.

The advantage of using this problem is that the algorithm (5.23) (for the moment the discussion is restricted to the classical algorithms) reads

$$\sum_{j=0}^{J} [a_j(0) + \lambda h \, b_j(0)] y_{n+j} = 0, \ n = 0, 1, \ldots \tag{5.51}$$

and then it is of a form which can be treated analytically in the same way as it was with (5.43).

We assume that $y_m = d^m$, denote $\hat{h} := \lambda h$, to obtain that d must satisfy the equation

$$\pi(d, \hat{h}) = 0, \tag{5.52}$$

where

$$\pi(d, \hat{h}) := \rho(d, 0) + \hat{h}\sigma(d, 0) \tag{5.53}$$

is called the stability polynomial of the method. It is a J-th degree polynomial with coefficients which depend linearly on the parameter $\hat{h}$. Let the roots of $\pi(d, \hat{h})$ be $d_j(\hat{h})$, $1 \le j \le J$, $|d_1(\hat{h})| \ge |d_2(\hat{h})| \ge \ldots \ge |d_J(\hat{h})|$. If these roots are distinct the general form of y_m is

$$y_m = \sum_{j=1}^{J} C_j d_j^m(\hat{h}), \ m = 0, 1, \ldots, \tag{5.54}$$

[If the roots are not distinct the general form of y_m will look slightly different but the qualitative conclusions which follow will remain the same.]

When $\hat{h} = 0$ (which means that $\lambda = 0$ because $h > 0$ is assumed fixed) $\pi(d, \hat{h})$ becomes the first characteristic polynomial $\rho(d, 0)$ and then its roots are those of the latter. Since the method is assumed to be convergent, all the roots of $\rho(d, 0)$ satisfy $|d_j(0)| \le 1$ but this does not exclude that some of the roots $d_j(\hat{h})$ may be equal to or bigger than 1 in absolute value when $\hat{h} \ne 0$. If the latter happens it follows that $|y_n|$ increases indefinitely when $n \to \infty$, and this is in contrast with the behaviour of the exact solution of the equation although the method was assumed convergent. The numerical solution is also unacceptable when $|d_1(\hat{h})| = 1$ because $|y_n|$ does not tend to 0 when $n \to \infty$, as desired.

The following two definitions are taken from [40]:

DEFINITION 5.4 *The method (5.23) is said to be absolutely stable for given $\hat{h}$ if for that $\hat{h}$ the roots of the stability polynomial (5.53) satisfy $|d_j(\hat{h})| < 1$, $j = 1, 2, \ldots, J$ and to be absolutely unstable otherwise.*

DEFINITION 5.5 *A method is said to have region of stability $\mathcal{R}_A$, where $\mathcal{R}_A$ is a region in the complex $\hat{h}$ plane, if it is absolutely stable for all $\hat{h} \in \mathcal{R}_A$. The intersection of $\mathcal{R}_A$ with the real axis is called the interval of absolute stability.*

Notice that in these definitions the whole complex $\hat{h}$ plane is taken for reference. However, since the whole discussion started from the case when $Re\,\lambda < 0$ the examination of what happens in the left-half $\hat{h}$ plane only will be enough. More than this, the knowledge of the stability properties of a numerical method in the left-half $\hat{h}$ plane is important for the characterization of the quality of such a method when solving systems of stiff equations, see [40].

A simple-minded model problem for the illustration of the stiffness is the system

$$^k y' = \lambda_k \,^k y, \; k = 1, 2, \ldots, N,$$

where the involved λ-s (we tacitly assume that all these have $Re\,\lambda < 0$) differ very much from one equation to another (by orders of magnitude, say). The problem consists in solving all these equations with one and the same h, but with no special attention to the accuracy. What is required is only that the numerical solution mimics the asymptotic behaviour of the analytic solution, that is $\lim_{n \to \infty} \,^k y_n = 0$ for all k.

The following definitions are pertinent for this case:

DEFINITION 5.6 *(Dahlquist [15]) A method is said to be A-stable if $\mathcal{R}_A \supseteq \{\hat{h}|\,Re\,\hat{h} < 0\}$.*

An A-stable method will then impose no restriction on the value of h irrespective of the value of λ ($Re\,\lambda < 0$) and then it would be ideally suited for stiff problems. However, the requirement that the whole left-half $\hat{h}$ plane must belong to the region of absolute stability is too demanding for most of the linear multistep methods. It can be lowered by asking the method be absolutely stable only for the systems whose λ-s are packed in such a way that all $\hat{h} = \lambda h$ lie within an infinite wedge bounded by the rays $\arg \hat{h} = \pi \pm \alpha$:

DEFINITION 5.7 *(Widlund [64]) A method is said to be A(α)-stable, $\alpha \in (0, \pi/2)$, if $\mathcal{R}_A \supseteq \{\hat{h}|\, -\alpha < \pi - \arg \hat{h} < \alpha\}$.*

Finally, if the involved λ are real and negative it is sufficient that the method be A_0-stable:

DEFINITION 5.8 *(Cryer [13]) A method is said to be A_0-stable if $\mathcal{R}_A \supseteq \{\hat{h}|\,Re\,\hat{h} < 0, \, Im\,\hat{h} = 0\}$.*

Up to this moment the discussion regarded the classical multistep algorithms. For the ef version (5.22) we distinguish two situations.

If the set $\mathbf{z}$ is one and the same for all n there is no difficulty in extending the definition 5.4 of the absolute stability.

DEFINITION 5.9 *The method (5.22) with n-independent $\mathbf{z}$ is said to be absolutely stable for given $\hat{h}$ if for that $\hat{h}$ the roots of the stability polynomial*

$$\pi(d, \hat{h}, \mathbf{z}) := \rho(d, \mathbf{z}) + \hat{h}\sigma(d, \mathbf{z}) \tag{5.55}$$

satisfy $|d_j(\hat{h})| < 1$, $j = 1, 2, \ldots, J$ and to be absolutely unstable otherwise.

The region of absolute stability for this version can be then constructed on this basis.

The situation is much more complicated when z depends on n because the coefficients of the polynomial $\pi(d, \hat{h}, z)$ will differ from one n to another. A possible attempt may consist in examining each n separately, to finally accept that the region of absolute stability for the whole (5.22) is the subregion which is common in all individual regions of absolute stability. This view is confirmed in practice although there is no general justification for it. As a matter of fact, the arguments used before for the convergence of algorithm (5.22) with n-dependent z do not hold: the limit $h \to 0$ was examined there while now h is kept fixed.

To see where the problem is and why this view holds in practice we consider the AEF algorithm. When applied to the test problem (5.50) this reads

$$a_0(z_n)y_n + a_1(z_n)y_{n+1} + \alpha(z_n, \hat{h})y_{n+2} = 0, \ n = 0, 1, 2, \ldots \quad (5.56)$$

where $\alpha(z_n, \hat{h}) = 1 - \hat{h}b_2(z_n)$.

This two-step algorithm can be easily re-formulated in a one-step form. We introduce

$$F_{n+1} := y_{n+1} - y_n, \quad (5.57)$$

and evaluate $F_{n+2} - \gamma F_{n+1}$, where γ is some number. We have

$$F_{n+2} - \gamma F_{n+1} = y_{n+2} - y_{n+1} - \gamma(y_{n+1} - y_n) = y_{n+2} - (\gamma+1)y_{n+1} + \gamma y_n,$$

whence

$$F_{n+2} = y_{n+2} - (\gamma + 1)y_{n+1} + y_n + \gamma F_{n+1}.$$

If we choose $\gamma = -a_0(z_n)/\alpha(z_n, \hat{h})$ the latter equation becomes

$$F_{n+2} = -\frac{a_0(z_n) + a_1(z_n) + \alpha(z_n, \hat{h})}{\alpha(z_n, \hat{h})}y_{n+1} + \frac{a_0(z_n)}{\alpha(z_n, \hat{h})}F_{n+1}. \quad (5.58)$$

[The important point is that y_{n+2} and y_n have disappeared from the expression of F_{n+2}.] This is substituted in (5.57) with $n \to n+1$ to obtain

$$y_{n+2} = -\frac{a_0(z_n) + a_1(z_n)}{\alpha(z_n, \hat{h})}y_{n+1} + \frac{a_0(z_n)}{\alpha(z_n, \hat{h})}F_{n+1}.$$

Therefore

$$\begin{pmatrix} y_{n+2} \\ F_{n+2} \end{pmatrix} = M(z_n, \hat{h}) \begin{pmatrix} y_{n+1} \\ F_{n+1} \end{pmatrix}, \ n = 0, 1, 2, \ldots \quad (5.59)$$

is the required one-step form, with

$$M(\mathbf{z}_n, \hat{h}) = \begin{pmatrix} -\dfrac{a_0(\mathbf{z}_n) + a_1(\mathbf{z}_n)}{\alpha(\mathbf{z}_n, \hat{h})} & \dfrac{a_0(\mathbf{z}_n)}{\alpha(\mathbf{z}_n, \hat{h})} \\ -\dfrac{a_0(\mathbf{z}_n) + a_1(\mathbf{z}_n) + \alpha(\mathbf{z}_n, \hat{h})}{\alpha(\mathbf{z}_n, \hat{h})} & \dfrac{a_0(\mathbf{z}_n)}{\alpha(\mathbf{z}_n, \hat{h})} \end{pmatrix}.$$

The characteristic equation of the matrix M is

$$\alpha(\mathbf{z}_n, \hat{h})d^2 + a_1(\mathbf{z}_n)d + a_0(\mathbf{z}_n) = 0,$$

and this is exactly $\pi(d, \hat{h}, \mathbf{z}_n) = 0$. The roots of $\pi(d, \hat{h}, \mathbf{z}_n)$ are then the eigenvalues of $M(\mathbf{z}_n, \hat{h})$ and viceversa.
Equation (5.59) allows writing

$$\begin{pmatrix} y_{n+2} \\ F_{n+2} \end{pmatrix} = P_n \begin{pmatrix} y_1 \\ F_1 \end{pmatrix}, \quad n = 0, 1, 2, \ldots$$

where

$$P_n := M(\mathbf{z}_n, \hat{h}) \cdot M(\mathbf{z}_{n-1}, \hat{h}) \cdots M(\mathbf{z}_0, \hat{h}).$$

Let us assume that the eigenvalues of $M(\mathbf{z}_i, \hat{h})$ are smaller than 1 in modulus for all i involved. Does the behaviour $\lim_{n \to \infty} P_n = 0$ (and then $\lim_{n \to \infty} \{y_n, F_n\} = 0$, as desired) hold ?
There is not a definite answer to this question. In fact, let $T(\mathbf{z}_i, \hat{h})$, denoted T_i for short, be the diagonalization matrix of $M(\mathbf{z}_i, \hat{h})$, that is $M(\mathbf{z}_i, \hat{h}) = T_i^{-1} D_i T_i$, where D_i is diagonal. We have

$$P_n = T_n^{-1} \cdot D_n \cdot T_n \cdot T_{n-1}^{-1} \cdot D_{n-1} \cdot T_{n-1} \cdot T_{n-2}^{-1} \cdots T_1 \cdot T_0^{-1} \cdot D_0 \cdot T_0.$$

This shows that the limit depends not only on the eigenvalues collected in D_i but also on the matrix elements of the interface products $T_i \cdot T_{i-1}^{-1}$; if the latter are big the limit is certainly altered. However, if the set $\mathbf{z}_i$ does not vary substantially over successive i, as is typical in the current practice, each interface product is close enough to the two by two unit matrix and the quoted limit holds, [8]. The subregion which is common in all individual regions of absolute stability will be then a region of absolute stability for the whole algorithm (5.56).

1.3 Frequency evaluation and stability properties for two-step bdf algorithms

In this and in the subsequent sections we concentrate on the solution of stiff problems to show how the exponential fitting is of help in improving the quality of multistep methods.

In the frame of the classical multistep methods it is well-known that the J-step bdf algorithms (that is where the values $b_0 = b_1 = \ldots = b_{J-1} = 0$

are taken by default) are better suited than any other version with the same J for problems where good stability properties are required, see [40]. It is also known that the order p of a classical J-step bdf algorithm is $p = J$ and then it increases with J but, on the opposite, the stability properties deteriorate when J is increased. For example, the two-step bdf algorithm A0 is of the second order and A-stable (and then it is ideally suited for stiff problems), while the three-step bdf algorithm (call it B0),

$$-\frac{2}{11}y_n + \frac{9}{11}y_{n+1} - \frac{18}{11}y_{n+2} + y_{n+3} = \frac{6}{11}hf_{n+3}, \qquad (5.60)$$

is of the third order but it is only A(α)-stable with $\alpha \approx 86°$. It follows that, in the quoted context, the investigation of the exponential fitting extensions of the two-step bdf algorithm must receive particular attention.

Frequency evaluation

Three versions, A0, A1 and A2, are available, see Section 1.1. Each of them has order two but the real parameters μ in A1, and μ^2 in A2, are still at our disposal.

Let us assume that the values of $y'(x_{n+1})$, $y''(x_{n+1})$ and $y^{(3)}(x_{n+1})$, or of some good approximation to these, are available (the determination in practice of such values will be discussed later on). The structure of the last factor in the expressions of the *lte* for A1 and A2 shows how the values of μ or of μ^2 must be selected. Indeed, if we fix

$$\mu = \frac{y^{(3)}(x_{n+1})}{y''(x_{n+1})} \qquad (5.61)$$

in A1, and

$$\mu^2 = \frac{y^{(3)}(x_{n+1})}{y'(x_{n+1})} \qquad (5.62)$$

in A2, the corresponding *lte* will vanish and then each of A1 or A2 will become a third order method. More precisely, if the exact values are taken for the derivatives the effective *lte* with these optimal values is given by the second term in the series expansion (5.33), that is

$$lte^{opt} = h^4 T_1^*(z, \mathbf{a}(z))[y^{(4)}(x_{n+1}) - \mu y^{(3)}(x_{n+1})], \qquad (5.63)$$

where

$$T_1^*(z, \mathbf{a}(z)) = \frac{L_3^*(\mathbf{a}(z)) - 6T_0^*(z, \mathbf{a}(z))}{6z}$$

for A1, and

$$lte^{opt} = h^4 Q_1^*(Z, \mathbf{a}(Z))[y^{(4)}(x_{n+1}) - \mu^2 y^{(2)}(x_{n+1})], \qquad (5.64)$$

where

$$Q_1^*(Z, \mathbf{a}(Z)) = -\frac{L_2^*(\mathbf{a}(Z))}{2Z} = \frac{2\eta_0(Z) - \eta_{-1}(Z) - 1}{[1 + 2\eta_{-1}(Z)]Z}$$

for A2. When z, $Z \to 0$ we have $T_1^*(z, \mathbf{a}(z))$, $Q_1^*(Z, \mathbf{a}(Z)) \to -1/18$. The behaviour *lte* $\sim h^4$ holds also when the involved derivatives of the solution are calculated numerically with errors which are at least linear in h.

The increase by one unit of the order is certainly a gain but it will be achieved at the expense of a number of extra operations such as the separate determination of the higher order derivatives of the solution and the evaluation, on this basis, of the coefficients of the method at each n. Does it make any sense to adopt such a way instead of just using B0, which is exactly of the same order but with the advantage of avoiding such extra computations? The answer will depend on how the stability properties of A1 and of A2 compare with those of B0 in the left-half $\hat{h}$ plane.

Stability properties of B0, A1 *and* A2

A convenient method for finding regions of absolute stability is the *boundary locus technique*, see [40]. According to the definition 5.5, $\mathcal{R}_A$ is the region in the complex $\hat{h}$ plane such that for all $\hat{h} \in \mathcal{R}_A$ all the roots of the stability polynomial π have modulus less than 1. Let us define $\partial \mathcal{R}_A$ as the contour in the complex $\hat{h}$ plane such that for all $\hat{h} \in \partial \mathcal{R}_A$ one of the roots has modulus 1. This means that it is of the form $d = \exp(i\theta)$ with real θ. It follows that identity

$$\pi(\exp(i\theta), \hat{h}, \mathbf{z}) = \rho(\exp(i\theta), \mathbf{z}) - \hat{h}\sigma(\exp(i\theta), \mathbf{z}) = 0 \qquad (5.65)$$

must hold, whence the locus of $\partial \mathcal{R}_A$ is given by

$$\hat{h} = \sigma(\exp(i\theta), \mathbf{z})/\rho(\exp(i\theta), \mathbf{z}), \qquad (5.66)$$

where $0 \le \theta \le 2\pi$.

By its very construction, this contour ensures that one of the roots of π has modulus 1 along it, but this does not exclude that some other roots may exceed 1 in absolute value and, if the latter happens, $\partial \mathcal{R}_A$ does not represent the effective boundary of the region of absolute stability, as we would like. However, such a special situation never appears in the cases to be discussed below; $\partial \mathcal{R}_A$ constructed by the boundary locus technique will be always the true boundary of $\mathcal{R}_A$.

For the method B0 we have

$$\rho(d, \mathbf{0}) = d^3 - \frac{18}{11}d^2 + \frac{9}{11}d - \frac{2}{11}, \quad \sigma(d, \mathbf{0}) = \frac{6}{11}d^3, \qquad (5.67)$$

and the part of the boundary $\partial \mathcal{R}_A$ in the left-half $\hat{h}$ plane is displayed on Figure 5.2. The curve exibits symmetry with respect to the real axis and the origin belongs to it. The latter is just normal because B0 is zero-stable. The region of absolute stability is outside this contour. For example, B0 is absolutely stable for $Re\,\hat{h} = -0.09$, $Im\,\hat{h} = 1$ but it is not for $Re\,\hat{h} = -0.01$, $Im\,\hat{h} = 1$. The two rays tangent to the contour enable characterizing B0 as an $A(\alpha)$-stable method with $\alpha \approx 86°$.

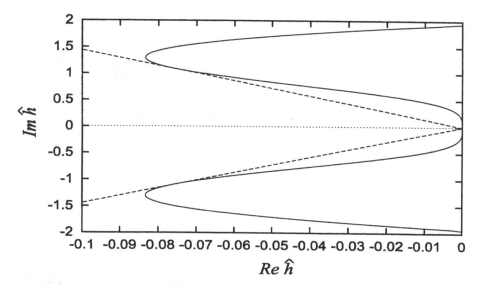

Figure 5.2. The boundary $\partial \mathcal{R}_A$ of the region of absolute stability in the left-half $\hat{h}$ plane for the classical three-step bdf algorithm. The region of absolute stability lies on the left-hand side of the boundary.

We now consider the ef form A1 of the two-step bdf algorithm. Let z be fixed. The two characteristic polynomials are

$$\rho(d,\,z) = d^2 + a_1(z)d + a_0(z), \quad \sigma(d,\,z) = b_2(z)d^2, \qquad (5.68)$$

with the coefficients given in (5.11). The boundary $\partial \mathcal{R}_A$ of the region of absolute stability is presented on Figure 5.3 for $z = 1$, 0.5, 0.25 and 0.125. As before, the region of absolute stability is outside the contour.

It is seen that the position of the boundary depends on z and that the region of absolute stability is progressively enlarged when $z \to 0$. When $z = 0$ this covers the whole left-half plane and then the method is A-stable. The latter is just as expected because in this limit A1 becomes the classical A0. For the quoted values of z, A1 is $A(\alpha)$-stable with $\alpha \approx 82.7°$, $87.2°$, $89°$, and $89.6°$, respectively; the latter three angles are bigger than for B0. Also important, but

not visible from these graphs, is that A1 remains A-stable for any $z < 0$, a fact
first observed by Liniger and Willoughby [41].

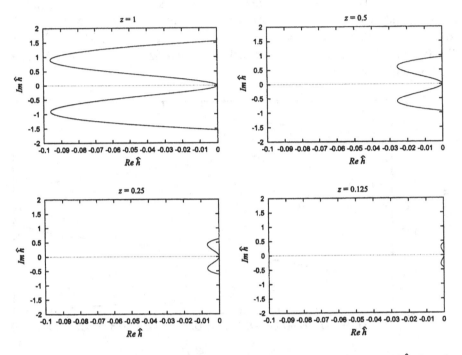

Figure 5.3. The boundary $\partial \mathcal{R}_A$ of the region of absolute stability in the left-half $\hat{h}$ plane for
the ef version A1 of the two-step bdf algorithm at four values for the parameter z. The region
of absolute stability lies on the left-hand side of the boundary.

As for the version A2 let us assume that Z is fixed. The two characteristic
polynomials depend on Z, viz.:

$$\rho(d, Z) = d^2 + a_1(Z)d + a_0(Z), \quad \sigma(d, Z) = b_2(Z)d^2, \qquad (5.69)$$

with the coefficients (5.12). The boundary $\partial \mathcal{R}_A$ of the region of absolute stability is presented on Figure 5.4 for $Z = -1, -0.5, -0.25$ and -0.125. Again,
the region of absolute stability is outside the contour. The shapes are very similar to the ones for A1 and then A2 can be coined as an A(α)-stable method. For
the written four values of Z the angles are $\alpha \approx 82.6°$, $87.8°$, $89.2°$ and $89.7°$,
respectively. The tendency mentioned before for A1 that the region of absolute
stability progressively invades the whole left-half plane when the parameter
tends to zero holds as well; the only difference is that now the parameter is Z
and it tends to zero through negative values. Also, the method A2 is A-stable
for all $Z \geq 0$ in the same way as it was with A1 for all $z \leq 0$.

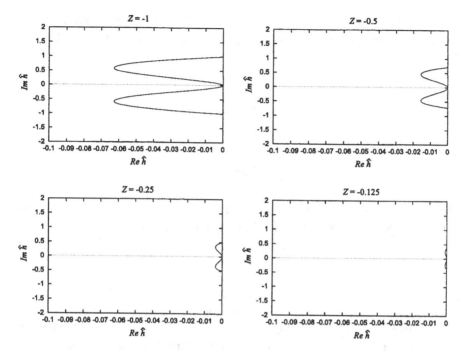

Figure 5.4. The boundary $\partial \mathcal{R}_A$ of the region of absolute stability in the left-half $\hat{h}$ plane for the ef version A2 of the two-step bdf algorithm at four values for the parameter Z. The region of absolute stability lies on the left-hand side of the boundary.

1.4   The flexible two-step bdf algorithm AEF

For each n this algorithm selects the appropriate version in terms of the local behaviour of the solution, see [35] and [36].

1.4.1   Numerical evaluation of the derivatives

The evaluation of μ for A1 by (5.61) and of μ^2 for A2 by (5.62) requires the knowledge of the first, second and third derivative of the exact solution at x_{n+1}. A direct procedure may consist in just using the analytic expressions of $f(x, y)$ and of its total derivatives with respect to x, for instance $y'(x) = f(x, y(x))$ and $y''(x) = \partial f(x, y(x))/\partial x + y'(x)\partial f(x, y(x))/\partial y$, and evaluating these expressions at x_{n+1}. [For notational brevity we shall temporarily denote this x_{n+1} by X.] Such a procedure works well when the exact $y(X)$ is known but not when an approximate value of this is only available, as it is in the usual runs, and the situation is particularly bad for stiff problems.

The reason for such a difficulty is not the stiffness itself but some parallel effect. This effect is connected to the fact that, given functions $y(x)$ and $\Delta y(x)$, the numerical values at X of the derivatives of $y(x)$ and of $y(x) + \Delta y(x)$ may become very different even if $\Delta y(x)$ is practically negligible with respect to

$y(x)$ for any x in the vicinity of X. The link with the stiffness is that it happens that just such pathological forms of $\Delta y(x)$ are typically used to illustrate the stiffness.

To see the things more clearly we take an example. The equation

$$y' = \lambda y + x^2 - \frac{1}{3}\lambda x^3, \; x > 0 \tag{5.70}$$

has the general solution

$$y^{gen}(x) = \frac{1}{3}x^3 + \chi \exp(\lambda x),$$

where χ is an arbitrary constant. If the initial condition is $y(0) = 0$ the exponential component disappears and the solution is simply

$$y(x) = \frac{1}{3}x^3,$$

so that, for $X = 1$, the optimal value of μ for A1 is $\mu = 1$.

The things change when the value assumed as exact is affected by an error. If the distorted $y^*(X) = y(X) + \epsilon$ is taken as exact then the use of the analytic evaluation for the derivatives is equivalent to admitting that the exact solution in some neighbourhood of X is now $y^{gen}(x)$ with $\chi = \epsilon \cdot \exp(-\lambda X)$, that is

$$y^*(x) = y(x) + \epsilon \cdot \exp[\lambda(x - X)],$$

and therefore the value of μ calculated on this basis will be

$$\mu^* = (2 + \lambda^3 \epsilon)/(2 + \lambda^2 \epsilon) \, .$$

Let us assume that $\epsilon = 10^{-3}$. If λ is small, $\lambda = -1$, say, (a nonstiff case) the new μ is in very good agreement with the exact one, but when $\lambda = -1000$ (a stiff case) we get $\mu^* \approx \lambda = -1000$. The accuracy of A1 will be indeed badly affected on the stiff case, but this is because μ has been evaluated so erroneously, not because of stability problems, as one might be first tempted to believe.

This unwanted effect can be removed to a substantial extent if the analytic evaluation of the derivatives is replaced by the usual finite difference formulae. To see this let us concentrate on the first derivative. At the same $X = 1$ and with $\lambda = -1000$ and $\epsilon = 10^{-3}$, the analytic evaluations give $y'(X) = X^2 = 1$ and $y^{*\prime}(X) = X^2 + \lambda\epsilon = 0$, which are significantly different. The situation changes favourably if the central difference formula is adopted. This is because the data are distorted in the specific way in which the errors are accumulating when the solution is propagated by means of the considered algorithm.

By the very meaning of the order of a numerical method, for small h the scaled errors $[y(x_0 + jh) - y_j]/h^p$, $j = 0, 1, 2, \ldots$ from a convergent numerical method of order p for (5.1) will be practically independent on h. When $h \to 0$ they will tend to follow a curve $E(x)$ whose shape depends on the method used and on the form of $f(x, y)$. In the case of the classical J-step method (5.23) the error function $E(x)$ is the solution of the initial value problem

$$E' = \frac{\partial f(x, y(x))}{\partial y} E + C L_{p+1}^*(a(0)) y^{(p+1)}(x), \; E(x_0) = 0, \quad (5.71)$$

where

$$C = \frac{1}{[b_0(0) + b_1(0) + \ldots + b_J(0)](p+1)!},$$

see, e.g., [22]; the form of the inhomogeneous term is closely related with that of the *lte*, compare with (5.35). For A0 we have $p = 2$ and the error equation reads

$$E' = \frac{\partial f(x, y(x))}{\partial y} E - \frac{1}{3} y^{(3)}(x), \; E(x_0) = 0. \quad (5.72)$$

It follows that we can assume that $y_j = y(x_0 + jh) - h^p E(x_0 + jh)$ and, if so, the central difference formula with the numerical values used for input will give the approximation

$$\bar{y}'(X) = \frac{1}{2h}(y_{n+2} - y_n)$$

$$= \frac{1}{2h}[y(X + h) - y(X - h)] - \frac{1}{2h}h^p[E(X + h) - E(X - h)]$$

$$= X^2 + \frac{1}{3}h^2 - h^p[E'(X) + \mathcal{O}(h^2)] = y'(X) + \mathcal{O}(h^2),$$

which is much closer to the desired $y'(X)$ than the one resulting from the analytic evaluation.

For the effective computation of the first three derivatives at x_{n+1} the four-point bdf approximations are appropriate:

$$y^{(m)}(x_{n+1}) \approx h^{-m}[c_0 y_{n-2} + c_1 y_{n-1} + c_2 y_n + c_3 y_{n+1}], \; m = 1, 2, 3 \quad (5.73)$$

with

$$lte = K h^{4-m} y^{(4)}(x_{n+1}). \quad (5.74)$$

The coefficients are:

$$m = 1: \; c_0 = -\frac{1}{3}, \; c_1 = \frac{3}{2}, \; c_2 = -3, \; c_3 = \frac{11}{6}, \; K = \frac{1}{4}$$

$$m = 2: \; c_0 = -1, \; c_1 = 4, \; c_2 = -5, \; c_3 = 2, \; K = \frac{11}{12}$$

$$m = 3: \; c_0 = -1, \; c_1 = 3, \; c_2 = -3, \; c_3 = 1, \; K = \frac{3}{2}$$

1.4.2 Choosing the appropriate version

Let us concentrate on the two-step interval centred at x_{n+1}. The values of $y'(x_{n+1})$, $y''(x_{n+1})$ and $y^{(3)}(x_{n+1})$ are used to compute μ by (5.61) and form $z = \mu h$, and μ^2 by (5.62) to form $Z = \mu^2 h^2$, respectively. If

$$|z| \leq |Z| \text{ and } |z| \leq 1,$$

version A1 with argument z is selected. If

$$|Z| < |z| \text{ and } |Z| \leq 1,$$

A2 with argument Z is selected and, finally, if

$$|Z| > 1 \text{ and } |z| > 1,$$

then the classical A0 is activated. When A0 is never activated the algorithm AEF is a third order method. As a matter of fact, situations when A0 may be forced to appear are the ones when it happens that $y'(x)$ and $y''(x)$ have roots in the close vicinity of x_{n+1}.

The case of systems of differential equations requires a specific treatment. Let

$$^k y' = {}^k f(x, {}^1 y, {}^2 y, \ldots, {}^N y), \quad x > x_0, \quad {}^k y(x_0) = {}^k y_0, \; k = 1, 2, \ldots, N$$

be the initial value problem for such a system. With the column vectors $\mathbf{y} = [{}^1 y, {}^2 y, \ldots, {}^N y]^T$, $\mathbf{f}(x, \mathbf{y}) = [{}^1 f(x, \mathbf{y}), {}^2 f(x, \mathbf{y}), \ldots, {}^N f(x, \mathbf{y})]^T$ this reads compactly:

$$\mathbf{y}' = \mathbf{f}(x, \mathbf{y}), \; \mathbf{y}(x_0) = \mathbf{y}_0. \tag{5.75}$$

The version has to be selected on each equation separately and it may well happen that A1 with some z is selected for the first equation, A2 with some Z for the second equation etc. Expressed differently, while the two-step bdf algorithm (5.3), which now reads

$$a_0 y_n + a_1 y_{n+1} + y_{n+2} = h b_2 \mathbf{f}(x_{n+2}, y_{n+2}), \tag{5.76}$$

has one and the same set a_0, a_1 and b_2 for the classical version A0, the set of coefficients will be k-dependent for the AEF. If the classical version A0 is never activated then the AEF for systems is a third order method, exactly as in the case of a single equation.

The algebraic system (5.76) for the unknown vector y_{n+2} is implicit because y_{n+2} appears also as one of the arguments of $\mathbf{f}$. The solution is direct only if $\mathbf{f}$ is linear in $\mathbf{y}$ but in general an iteration procedure has to be used. Perhaps the most popular of these is the Newton–Raphson procedure,

$$y_{n+2}^{i+1} = y_{n+2}^i - [\mathbf{I} - h b_2 \mathbf{J}(x_{n+2}, y_{n+2}^i)]^{-1}$$
$$\times \; [y_{n+2}^i - h b_2 \mathbf{f}(x_{n+2}, y_{n+2}^i) + a_0 y_n + a_2 y_{n+1}], \; i = 0, 1, 2, \ldots, \tag{5.77}$$

where $J(x, y) = \partial f / \partial y$ and I are the $N \times N$ Jacobian and the unity matrix, respectively. The iteration must be repeated as many times as necessary to reach convergence in the desired accuracy. The starting vector y_{n+2}^0 is generated by the explicit formula

$$y_{n+2}^0 = -\frac{1}{2}y_{n-1} + 3y_n - \frac{3}{2}y_{n+1} + 3f(x_{n+1}, y_{n+1}).$$

As a matter of fact, the approximation

$$y(X + h) \approx -\frac{1}{2}y(X - 2h) + 3y(X - h) - \frac{3}{2}y(X) + 3y'(X), \quad (5.78)$$

on which the former is based, has

$$lte = \frac{1}{4}h^4 y^{(4)}(X).$$

1.4.3 Numerical illustrations

The first test case is

$$y' = 1 - x + \frac{1}{2}x^2, \ x > 0, \ y(0) = 1, \quad (5.79)$$

with the exact solution

$$y(x) = 1 + x - \frac{1}{2}x^2 + \frac{1}{6}x^3. \quad (5.80)$$

Strictly speaking, this is not a genuine differential equation (because y does not appear in its right-hand side) but it is convenient enough to illustrate some interesting points on the relative merits of the various components of the AEF.

Neither the form of the r.h.s. of this equation nor the form of the solution may suggest that a two-step ef version will work better than the classical A0. Yet, since $y'(x) = 1 - x + x^2/2$, $y''(x) = -1 + x$ and $y^{(3)}(x) = 1$, we have all data to construct the optimal μ and μ^2 for the ef versions A1 and A2, respectively. In the two-step interval centred at x_{n+1} the optimal μ for A1 is $\mu = \phi_1(x_{n+1})$, while $\mu^2 = \phi_2(x_{n+1})$ is the appropriate value for A2, where $\phi_1(x) = 1/(x - 1)$ and $\phi_2(x) = 1/(1 - x + x^2/2)$.

The interesting feature is that $\phi_1(x)$ has a pole at $x = \bar{x} = 1$. It follows that A1 is not defined when $x_{n+1} = \bar{x}$ and it is also expected to work badly on a certain number of two-step intervals situated in a larger neighbourhood of this $\bar{x}$. For this reason, on this test case we impose to advance the solution by A1 along all mesh points of the partition with the exception of those situated in the region $0.8 < x < 1.2$, where A2 is used.

In Table 5.1 we collect the absolute errors at $x = 1$, $x = 5$ and $x = 10$, as produced by the classical A0 and by the method based on this combined use of

Table 5.1. Absolute errors $\Delta y(x) = y(x) - y^{comput}(x)$ at $x = 1$, 5 and 10 from the classical two–point bdf algorithm A0 and from its ef extension for equation (5.79).

h	$x = 1$	$x = 5$	$x = 10$
classical A0			
0.0500	$-0.812(-03)$	$-0.415(-02)$	$-0.831(-02)$
0.0250	$-0.206(-03)$	$-0.104(-02)$	$-0.208(-02)$
0.0125	$-0.518(-04)$	$-0.260(-03)$	$-0.521(-03)$
ef method			
0.0500	$-0.161(-04)$	$0.168(-04)$	$0.253(-04)$
0.0250	$-0.208(-05)$	$0.180(-05)$	$0.287(-05)$
0.0125	$-0.273(-06)$	$0.216(-06)$	$0.349(-06)$

the ef algorithms A1 and A2, with the stepwidths $h = 1/20$, $1/40$ and $1/80$. It is seen that, as expected, A0 is a second order method while the alternative ef method is of third order and also that the results of the latter are substantially more accurate.

To produce more evidence that this is not accidental at the mentioned points and also to see better what happens around the critical point $\bar{x} = 1$, the scaled errors $(y(jh) - y_j)/h^3$, $j = 1, 2, 3, \ldots$ from the ef method are displayed on Figure 5.5 for the three mentioned values of h. The three curves displayed on this figure are enough packed together to conclude that the algorithm based exclusively on the ef versions is of the third order, indeed. As for the classical version A0, the error equation (5.72) has the simple form $E' = -1/3$ and then the scaled errors $(y(jh) - y_j)/h^2$, $j = 1, 2, 3, \ldots$ will lie on one and the same straightline $E(x) = -x/3$.

The specific influence of the critical point $\bar{x} = 1$ is also visible on Figure 5.5. The error deteriorates progressively from $x = 0$ up to $x = 0.8$ because the equation is integrated by A1, a version which meets more and more adverse conditions when x is advanced, but the deterioration is supressed between 0.8 and 1.2 because here A2 is used.

The second test case is the system of two equations

$$\begin{cases} {}^1y' = -2\,{}^1y + {}^2y + 2\sin(x), \\ {}^2y' = -(\beta + 2)\,{}^1y + (\beta + 1)({}^2y + \sin(x) - \cos(x)), \ x > 0. \end{cases} \tag{5.81}$$

Its general solution is

$$\begin{aligned} {}^1y(x) &= \chi_1 \exp(-x) + \chi_2 \exp(\beta x) + \sin(x), \\ {}^2y(x) &= \chi_1 \exp(-x) + \chi_2(\beta + 2) \exp(\beta x) + \cos(x), \end{aligned} \tag{5.82}$$

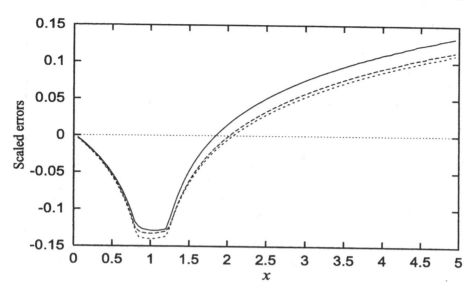

Figure 5.5. Scaled errors $(y(jh) - y_j)/h^3$, $j = 1, 2, 3, \ldots$ for the described ef method at three stepwidths: $h = 0.05$ (solid line), $h = 0.025$ (dashed line) and $h = 0.0125$ (dotted line) for equation (5.79).

Table 5.2. Absolute errors $\Delta^i y(x) = {}^i y(x) - {}^i y^{comput}(x)$ at $x = 10$ for the two components $i = 1, 2$, from the two–point bdf algorithms A0 and AEF for system (5.81).

h	$\beta = -10$		$\beta = -1000$	
	$\Delta^1 y$	$\Delta^2 y$	$\Delta^1 y$	$\Delta^2 y$
classical A0				
0.1000	$-0.212(-02)$	$-0.169(-02)$	$-0.225(-02)$	$-0.225(-02)$
0.0500	$-0.535(-03)$	$-0.425(-03)$	$-0.570(-03)$	$-0.569(-03)$
0.0250	$-0.134(-03)$	$-0.106(-03)$	$-0.143(-03)$	$-0.143(-03)$
ef method				
0.1000	$-0.172(-03)$	$-0.130(-03)$	$-0.241(-03)$	$-0.241(-03)$
0.0500	$-0.150(-04)$	$-0.933(-05)$	$-0.236(-04)$	$-0.236(-04)$
0.0250	$-0.145(-05)$	$-0.708(-06)$	$-0.252(-05)$	$-0.252(-05)$

where χ_1 and χ_2 are arbitrary constants. This system has been used in Section 6.1 of [40] for $\beta = -3$ and $\beta = -1000$, with the aim of illustrating the phenomenon of stiffness and the numerical consequencies of it. If the initial conditions are ${}^1y(0) = 2$ and ${}^2y(0) = 3$, the constants χ_1 and χ_2 get the values $\chi_1 = 2$ and $\chi_2 = 0$ and therefore the exact solution becomes β–independent.

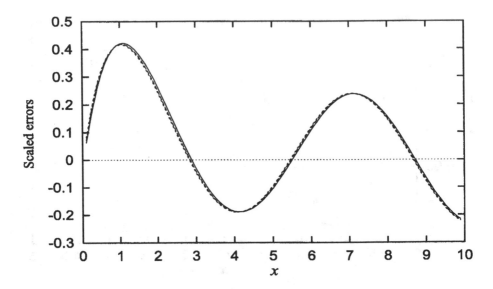

Figure 5.6. Scaled errors $(y(jh) - y_j)/h^2$, $j = 1, 2, 3, \ldots$ for the classical method A0 at three stepwidths: $h = 0.05$ (solid line), $h = 0.025$ (dashed line) and $h = 0.0125$ (dotted line) for the first solution component of the system (5.81).

It is

$$^1y(x) = 2\exp(-x) + \sin(x), \quad ^2y(x) = 2\exp(-x) + \cos(x). \qquad (5.83)$$

We compare the classical A0 and the flexible AEF algorithms for $x \in [0, 10]$ and for $h = 1/10, 1/20$ and $1/40$; the initial conditions $y_1(0) = 2$ and $y_2(0) = 3$ are taken by default in all runs. In Table 5.2 we give the absolute errors from the two methods at $x = 10$, for $\beta = -10$ (a nonstiff case) and $\beta = -1000$ (a stiff case) to see that AEF works much better than A0, irrespective of whether the system is stiff or nonstiff. It is also seen that the order of A0 and of AEF is two and three, respectively.

Some more graphs for $\beta = -1000$ are given for additional evidence. (The graphs corresponding to $\beta = -10$ are similar.) The scaled errors $(y(jh) - y_j)/h^2$, $j = 1, 2, 3, \ldots$ from the classical A0 for the component ^1y, which are displayed on Figure 5.6, confirm that A0 is a second order method. The graph corresponding to the component ^2y is pretty much the same.

Figures 5.7 and 5.8 refer to the AEF. On Figure 5.7 the scaled errors $(y(jh) - y_j)/h^3$, $j = 1, 2, 3, \ldots$ for the first and the second component are given. Again, the behaviours are pretty similar in these two graphs but, as normal, the profile differs from that in Figure 5.6. On Figure 5.8 we present the variation of μ with respect to x for each component. On each of these two graphs one or two curves are alternatively shown. The regions of x where only one curve

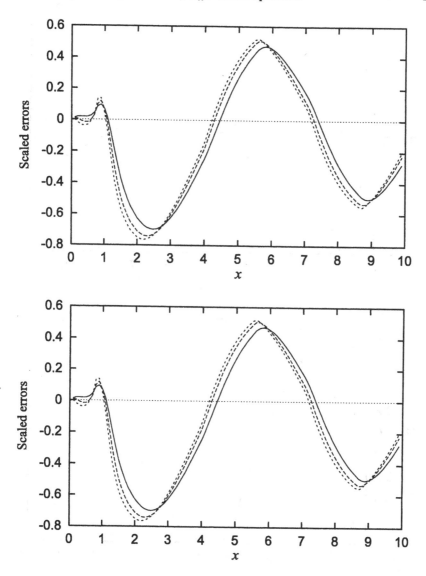

Figure 5.7. Scaled errors $(y(jh) - y_j)/h^3$, $j = 1, 2, 3, \ldots$ for the AEF method at three stepwidths: $h = 0.1$ (solid line), $h = 0.05$ (dashed line) and $h = 0.025$ (dotted line) for the first (up) and the second (down) solution components of the system (5.81).

is shown represent the regions where A1 was chosen while when two curves appear (these are always symmetric) it means that A2 was activated and the two curves give the values of the two frequencies $\mu^\pm$. In the latter case, the solid lines indicate that the two frequencies are real while the dashed lines indicate that they are imaginary.

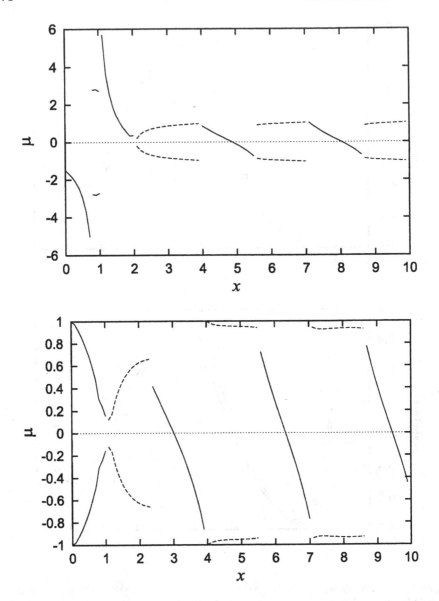

Figure 5.8. Variation with x of the optimal μ in the first equation (up) and the second equation (down) of the system (5.81).

1.5   Variable step form of the AEF

Following [36], we adapt the algorithm (5.3) for a nonequidistant distribution of the mesh points on $[a, b]$. Let us fix x_{n+1} and denote it by X, choose the length of the step on the left as the reference stepsize h of the considered two-step interval, that is $x_n = X - h$, and measure the length of the step on the right in units of h, i.e. $x_{n+2} = X + sh$. The positions of the three mesh points of such a two-step interval are then identified by three parameters: X, h and s. To determine the weights a_0, a_1 and b_2, we define

$$\mathcal{L}[h, \ s, \ \mathbf{a}]y(x) := a_0 y(x-h) + a_1 y(X) + y(x+sh) - hb_2 y'(x+sh), \quad (5.84)$$

which reduces to (5.5) when $s = 1$. We require that $\mathcal{L}[h, \ s, \ \mathbf{a}]y(x)$ is identically vanishing at any x, h and s when $y(x)$ has the same three preset forms as for the generation of the equidistant partition based A0, A1 and A2, see Section 1.1. The new, variable step versions will be abbreviated as VSA0, VSA1 and VSA2, respectively. Their coefficients will depend on the additional parameter s, to reduce to the coefficients of the former versions when $s = 1$. The technique for the derivation of the coefficients is the same as in the Section 1.1. Most of the expressions which follow have been calculated by MATHEMATICA, [65].

VSA0. The three forms are $y(x) = 1$, x, x^2 and the resultant coefficients are

$$a_0(s) = \frac{s^2}{1 + 2s}, \quad a_1(s) = -\frac{(1+s)^2}{1+2s}, \quad b_2(s) = \frac{s(1+s)}{1+2s},$$

With these coefficients the series expansion of $\mathcal{L}[h, \ s, \ \mathbf{a}]y(x)$ reads:

$$\mathcal{L}[h, \ s, \ \mathbf{a}(s)]y(x) = -\frac{s^2(1+s)^2}{6(1+2s)}h^3 y^{(3)}(X) \quad (5.85)$$

$$-\frac{s^2(1+s)^2(-1+2s)}{24(1+2s)}h^4 y^{(4)}(X) + \mathcal{O}(h^5).$$

The first term in the r.h.s. of it represents the *lte*; the second term has been added for comparisons later on.

VSA1. Here the three functions are $y(x) = 1$, x and $\exp(\mu x)$. With $z = \mu h$ the coefficients are:

$$a_0(z, \ s) = \frac{\exp(z)(1 - \exp(sz) + \exp(sz)sz)}{1 - \exp(z) + z\exp[(1+s)z]},$$

$$a_1(z, \ s) = -(1 + a_0(z, \ s)),$$

$$b_2(z, \ s) = \frac{\exp(z)(-1 + \exp(sz)) + s(1 - \exp(z))}{1 - \exp(z) + z\exp[(1+s)z]}.$$

Series expansion: With

$$t_1 = \frac{s^2(1+s)^2}{3(1+2s)^2}, \quad t_2 = \frac{s^2(1+s)^2}{36(1+2s)^3},$$

$$t_3 = -\frac{s^2(1+s)^2(1 + 12s + 24s^2 + 24s^3 + 12s^4)}{540(1+2s)^4},$$

$$t_4 = \frac{s^2 (1+s)^2 \left(-1 - 6s + 6s^2 + 24s^3 + 12s^4\right)}{6480 (1+2s)^5},$$

$$t_5 = \frac{s^2 (1+s)^2 \tau}{136080 (1+2s)^6},$$

where

$$\tau = 5 + 72s + 405s^2 + 1098s^3 + 1917s^4 + 2448s^5 + 2160s^6 + 1152s^7 + 288s^8,$$

and $t = t_1 z + t_2 z^2 + t_3 z^3 + t_4 z^4 + t_5 z^5$, we have

$$a_0(z, s) = \frac{s^2}{1+2s} + t, \quad b_2(z, s) = \frac{s(1+s)}{1+2s} - t,$$

with an accuracy of $\mathcal{O}(z^6)$. For the computation in double precision arithmetics the analytic formulae should be used for $|z| > 10^{-2}$ and the truncated series otherwise. As for the characterization of the error, we have

$$\mathcal{L}[h, s, \mathbf{a}(z, s)]y(x) = T_0^*(z, s, \mathbf{a}(z, s))h^3(y^{(3)}(X) - \mu y^{(2)}(X))$$

$$+ T_1^*(z, s, \mathbf{a}(z, s))h^4(y^{(4)}(X) - \mu y^{(3)}(X)) + \mathcal{O}(h^5), \qquad (5.86)$$

where

$$T_0^*(z, s, \mathbf{a}(z, s)) = \frac{\theta + \rho}{2z[1 - \exp(z) + z \exp((1+s)z)]},$$

with

$$\theta = \exp(z)(-1 + \exp(sz)) - s\exp(z)(2 + (z-2)\exp(sz)),$$

$$\rho = -s^2(-1 + \exp(z) + z\exp[(1+s)z])$$

and

$$T_1^*(z, s, \mathbf{a}(z, s)) = \frac{\alpha + \beta}{6z^2(1 - \exp(z) + z\exp[(1+s)z])},$$

with

$$\alpha = \exp(z)(1 - (z-3)\exp(sz)) + 3s^2(1 - (1+z)\exp(z)),$$

$$\beta = s^3 z(2 - 2\exp(z)) - z\exp[(1+s)z]) - s\exp(z)(6 - \exp(sz)(6 - 3z + z^2)).$$

When $z \to 0$, all these tend to the corresponding expressions for VSA0.

VSA2. The set is $y(x) = 1$, $\exp(\pm\mu x)$. With $Z = \mu^2 h^2$ the coefficients are:

$$a_0(Z,\ s) = \frac{1 - \eta_{-1}(s^2 Z)}{D(Z,\ s)},$$

$$a_1(Z,\ s) = -(1 + a_0(Z,\ s)),$$

$$b_2(Z,\ s) = \frac{\eta_0(Z) + s\eta_0(s^2 Z) + (1+s)\eta_0((1+s)^2 Z)}{D(Z,\ s)},$$

where

$$D(Z,\ s) = \eta_{-1}(s^2 Z) - \eta_{-1}((1+s)^2 Z).$$

Series expansions:

$$a_0(Z,\ s) = \frac{s^2}{1+2s} + \frac{s^2 (1+s)^2 Z}{12 + 24s} + \frac{s^2 (1+s)^2 (1 + 2s + 2s^2) Z^2}{240 (1+2s)}$$

$$+ \frac{s^2 (1+s)^2 (10 + 40s + 91 s^2 + 102 s^3 + 51 s^4) Z^3}{60480 (1+2s)}$$

$$+ \frac{s^2 (1+s)^2 (21 + 126 s + 449 s^2 + 956 s^3 + 1253 s^4 + 930 s^5 + 310 s^6) Z^4}{3628800 (1+2s)},$$

$$b_2(Z,\ s) = \frac{s (1+s)}{1+2s} + \frac{s^2 (1+s)^2 Z}{12 + 24s} + \frac{s^2 (1+s)^2 (1 + 6s + 6 s^2) Z^2}{720 (1+2s)}$$

$$+ \frac{s^2 (1+s)^2 (2 + 20s + 71 s^2 + 102 s^3 + 51 s^4) Z^3}{60480 (1+2s)}$$

$$+ \frac{s^2 (1+s)^2 (3 + 42s + 239 s^2 + 704 s^3 + 1127 s^4 + 930 s^5 + 310 s^6) Z^4}{3628800 (1+2s)},$$

with an accuracy of $\mathcal{O}(Z^5)$. Analytic formulae should be used if $|Z| > 10^{-2}$ and these truncated series expansions otherwise. Also,

$$\mathcal{L}[h,\ s,\ \mathbf{a}(Z,\ s)]y(x) = Q_0^*(Z,\ s, \mathbf{a}(Z,\ s))h^3 (y^{(3)}(X) - \mu^2 y'(X))$$

$$+ Q_1^*(Z,\ s, \mathbf{a}(Z,\ s))h^4 (y^{(4)}(X) - \mu^2 y^{(2)}(X)) + \mathcal{O}(h^5), \qquad (5.87)$$

where

$$Q_0^*(Z,\ s, \mathbf{a}(Z,\ s)) = \frac{1}{ZD(Z,\ s)}[1 + \eta_0(Z) - (1+s)\eta_{-1}(s^2 Z)$$

$$+ s\eta_{-1}((1+s)^2 Z) + s\eta_0(s^2 Z) - (1+s)\eta_0((1+s)^2 Z)],$$

$$Q_1^*(Z, s, \mathbf{a}(Z, s)) = -\frac{1}{2ZD(Z, s)}[1 - 2s\eta_0(Z) - (1 - s^2)\eta_{-1}(s^2Z)$$

$$-s^2\eta_{-1}((1 + s)^2Z) - 2s^2\eta_0(s^2Z) + 2s(1 + s)\eta_0((1 + s)^2Z)].$$

Again, when $Z \to 0$, we re-obtain VSA0.

The formulae (5.61) and (5.62) give the optimal values for μ and μ^2 in VSA1 and VSA2, respectively, in the current two-step interval. As for the numerical values of the involved derivatives, the following variable step generalization of (5.73) must be used:

$$y^{(m)}(X + s_2h) \approx h^{-m}[c_0y(X - h) + c_1y(X) + c_2y(X + s_1h) + c_3y(X + s_2h)],$$
$$(5.88)$$

($m = 1, 2, 3$) with

$$lte = Kh^{4-m}y^{(4)}(X).$$

The coefficients are:

$m = 1$:

$$c_0 = \frac{(s_1 - s_2)\, s_2}{(1 + s_1)\,(1 + s_2)}, \quad c_1 = -\frac{(s_1 - s_2)\,(1 + s_2)}{s_1\, s_2}, \quad c_2 = \frac{s_2 + s_2{}^2}{(s_1 + s_1{}^2)\,(s_1 - s_2)},$$

$$c_3 = \frac{s_1 - 2\, s_2 + 2\, s_1\, s_2 - 3\, s_2{}^2}{(s_1 - s_2)\,(s_2 + s_2{}^2)}, \quad K = -\frac{(s_1 - s_2)\, s_2\,(1 + s_2)}{24}.$$

$m = 2$:

$$c_0 = \frac{2\,(s_1 - 2\, s_2)}{1 + s_1 + s_2 + s_1\, s_2}, \quad c_1 = \frac{2 - 2\, s_1 + 4\, s_2}{s_1\, s_2}, \quad c_2 = \frac{2\,(1 + 2\, s_2)}{(s_1 + s_1{}^2)\,(s_1 - s_2)},$$

$$c_3 = \frac{2\,(-1 + s_1 - 3\, s_2)}{(s_1 - s_2)\, s_2\,(1 + s_2)}, \quad K = -\frac{(s_1 + 2\, s_1\, s_2 - s_2\,(2 + 3\, s_2))}{12}.$$

$m = 3$:

$$c_0 = \frac{-6}{(1 + s_1)\,(1 + s_2)}, \quad c_1 = \frac{6}{s_1\, s_2}, \quad c_2 = \frac{6}{(s_1 + s_1{}^2)\,(s_1 - s_2)},$$

$$c_3 = \frac{-6}{(s_1 - s_2)\,(s_2 + s_2{}^2)}, \quad K = -\frac{(-1 + s_1 - 3\, s_2)}{4}.$$

As for the predictor in the Newton–Raphson iteration process, the approximation

$$y(X + s_2h) \approx d_0y(X - h) + d_1y(X) + d_2y(X + s_1h) + h\, d_3y'(X + s_1h),$$
$$(5.89)$$

where

$$d_0 = -\frac{(s_1 - s_2)^2\, s_2}{(1 + s_1)^2}, \quad d_1 = \frac{(s_1 - s_2)^2\,(1 + s_2)}{s_1{}^2},$$

$$d_2 = -\frac{s_2\,(1+s_2)\,(-3\,s_1{}^2 + 2\,s_1\,(-1+s_2) + s_2)}{s_1{}^2\,(1+s_1)^2},$$

$$d_3 = -\frac{(s_1-s_2)\,s_2\,(1+s_2)}{s_1\,(1+s_1)},$$

and whose *lte* is

$$lte = \frac{(s_1-s_2)^2\,s_2\,(1+s_2)}{24}h^4 y^{(4)}(X),$$

should be used instead of (5.78).

1.5.1 Choosing the stepsize

Let *tol* be the required accuracy. Since *lte* $\sim h^3$ for VSA0 while *lte* $\sim h^4$ for VSA1 or VSA2 (if the optimal frequencies are used), we introduce the local error bounds $\epsilon = tol^{3/2}$ and $\epsilon = tol^{4/3}$, respectively. The determination of the suited stepsize means the determination of s such that the local error in y_{n+2} is kept within this bound. This first implies the computation of an approximation of $y(x_{n+1} + sh)$ which is sufficiently accurate to be accepted as 'exact' for the local problem. The variable step Milne–Simpson formula is sufficient for this purpose. This is

$$y(X+sh) \approx e_0 y(X-h) + e_1 y(X) + h[e_3 y'(X-h) + e_4 y'(X) + e_5 y'(X+sh)].$$
$$(5.90)$$

with

$$e_0 = \frac{s^3\,(2+s)}{1+2\,s}, \quad e_1 = -\frac{(-1+s)\,(1+s)^3}{1+2\,s},$$

$$e_2 = \frac{s^3\,(1+s)}{2+4\,s}, \quad e_4 = \frac{s\,(1+s)^3}{2+4\,s}, \quad e_5 = \frac{s\,(1+s)}{2+4\,s}.$$

Its *lte*, viz:

$$lte = -\frac{s^3\,(1+s)^3}{120(1+2\,s)}h^5 y^{(5)}(X),$$

behaves like h^5 and then this approximation is convenient, indeed.

The procedure for advancing the solution of the system (5.75) one step is as follows.

We assume that y_{n-2}, y_{n-1}, y_n and y_{n+1} are known. We choose some guess value $\bar{s}$ for s (taking $\bar{s} = 1$ is a good choice). By applying (5.90) on each component we determine the 'exact' solution at $x_{n+2} = x_{n+1} + \bar{s}h$, to be denoted y_{n+2}^*.

The same input data are also used for the computation by (5.88) of the derivatives at x_{n+1} for each component of the solution. These allow the determination of the optimal μ and μ^2 for each equation and, on this basis, of the convenient selection of the version for each k. The selected set of versions is then used to advance the solution one step and let the output be denoted $y_{n+2}(\bar{s})$. We compute $\delta = |y_{n+2}^* - y_{n+2}(\bar{s})|/(|y_{n+2}^*| + 1)$ on each component and evaluate the root mean square norm error

$$\Delta = \sqrt{\frac{1}{N} \sum_{k=1}^{N} \delta_k^2} \; .$$

Two situations may appear. If VSA0 appears at least once in the set of the selected versions it is accepted that the error is well described by the first term in (5.85) and then the desired value of s should satisfy the equation

$$\frac{\epsilon}{\Delta} = \frac{s^2(1+s)^2(1+2\bar{s})}{\bar{s}^2(1+\bar{s})^2(1+2s)}. \tag{5.91}$$

This nonlinear equation for s is solved by the Newton–Raphson procedure with $s = \bar{s}$ for starting. If it is obtained that $s \geq \bar{s}$ the current result is accepted and the computation advances to the next interval for which $h_{new} = \bar{s}h$ is taken and the new guess $\bar{s}_{new} = 0.9s/\bar{s}$ is considered. The factor 0.9 is a safety factor which reduces considerably the number of rejected steps. If $s < \bar{s}$ the result on the current step is rejected and the whole computation is repeated over the same two-step interval with s for $\bar{s}$.

If VSA0 does not appear in the set of selected versions then, as shown in [36], s should be calculated by solving

$$\frac{\epsilon}{\Delta} = \frac{s^2(1+s)^2(1+2\bar{s})(f-1+2s)}{\bar{s}^2(1+\bar{s})^2(1+2s)(f-1+2\bar{s})}. \tag{5.92}$$

where $f = \frac{1}{h}(3x_{n+1} - x_n - x_{n-1} - x_{n-2})$. This formula effectively accounts for the influence of the distorsion induced in the determination of μ or of μ^2 by the approximate formulae for the derivatives.

An upper bound $s_{max} = 1.4$ is imposed for s in both (5.91) and (5.92). In fact, the variable step methods are not covered by the usual theory of multistep methods and then they need a separate theoretical treatment. In particular, the stability theory of such methods is discussed in [18] and in [19]. The mentioned upper bound is consistent with one of the results of [19] which says that the classical VSA0 is A-stable if $s \leq \sqrt{2}$, although we are aware that this value is perhaps a too pessimistic estimate of the actual bound, see also [20].

1.5.2   Numerical illustrations

We compare the classical VSA0 and the variable step versions of AEF and of B0 (for the latter see [36]) on three stiff test problems:

1. Lambert equation (5.81) on $x \in [x_0 = 0, x_{max} = 10]$, for $\beta = -1000$ and with initial conditions $^1y(0) = 2$, $^2y(0) = 3$. The exact solution is (5.83).

2. Robertson equation. This is a nonlinear system of three equations,

$$\begin{cases} {}^1y' = -r_2 \cdot {}^1y + r_1 \cdot {}^2y \cdot {}^3y, \\ {}^2y' = r_2 \cdot {}^1y - r_1 \cdot {}^2y \cdot {}^3y - r_3 \cdot {}^2y^2, \\ {}^3y' = r_3 \cdot {}^2y^2, \quad x > 0, \end{cases} \tag{5.93}$$

$r_1 = 10^4$, $r_2 = 0.04$, $r_3 = 3 \times 10^7$, with the initial conditions ${}^1y(0) = 1$, ${}^2y(0) = {}^3y(0) = 0$. There is no analytic solution of it. The system is discussed in [52] to illustrate the stiffness, which increases with x. We solve this system for $x \in [x_0 = 0, x_{max} = 4 \times 10^5]$. The reference values at x_{max} are ${}^1y = 4.93829 \times 10^{-3}$, ${}^2y = 1.98500 \times 10^{-8}$ and ${}^3y = 0.9950617$, as resulting from a separate run with AEF at $tol = 10^{-7}$.

Table 5.3. Number of steps $nstep$ and resultant scd for three variable step algorithms. For AEF the number of steps in which the classical VSA0 had to be activated (ncl) is added.

tol	VSA0		AEF			B0	
	nstep	scd	nstep	ncl	scd	nstep	scd
Lambert[a]							
10^{-2}	78	2.1	49	0	2.7	65	2.6
10^{-3}	211	3.1	83	0	3.3	106	4.1
10^{-4}	634	4.1	158	0	4.5	200	5.3
10^{-5}	1999	5.1	330	0	5.7	420	6.2
10^{-6}	6545	6.1	684	0	6.7	855	7.0
10^{-7}	21349	7.1	1449	0	7.7	1832	7.9
Robertson[b]							
10^{-2}	100	1.0	115	25	1.2	131	1.3
10^{-3}	205	1.8	166	9	1.6	179	2.0
10^{-4}	545	2.7	281	7	2.5	289	2.9
10^{-5}	1641	3.7	506	4	3.5	536	3.8
10^{-6}	5098	4.7	988	2	4.5	1075	4.8
HIRES[c]							
10^{-2}	72	-0.2	73	20	0.4	98	0.1
10^{-3}	153	1.2	111	8	0.7	133	1.3
10^{-4}	411	2.2	191	7	1.8	214	2.4
10^{-5}	1234	3.2	364	7	2.8	394	3.4
10^{-6}	3846	4.2	754	7	3.8	790	4.4

CPU time/step (in ms) for VSA0, AEF and B0:
[a] $2.4\,10^{-2}$, $3.3\,10^{-2}$, $2.8\,10^{-2}$.
[b] $2.7\,10^{-2}$, $3.7\,10^{-2}$, $3.1\,10^{-2}$.
[c] $1.3\,10^{-1}$, $1.6\,10^{-1}$, $1.5\,10^{-1}$.

3. HIRES equation. HIRES is the short for 'High Irradiance RESponse' and it is a system of eight equations taken from a list of difficult problems

[42]. The domain is $[x_0 = 0, x_{max} = 321.8122]$ and the initial conditions are ${}^1y(0) = 1$, ${}^2y(0) = \cdots = {}^7y(0) = 0$, ${}^8y(0) = 0.0057$. For the error estimation at x_{max} we use the highly accurate values listed in the same reference.

All computations were carried out in double precision on a laptop with a Pentium II processor. A rather large range of tolerances was used, between 10^{-2} and 10^{-7} for the Lambert equation and between 10^{-2} and 10^{-6} for the other two. The big tolerances were mainly considered to illustrate that the stiffness is treated correctly by all these algorithms. The error is measured in terms of the scd factor

$$scd = -\log_{10} \max[\Delta_1, \Delta_2, \ldots, \Delta_n],$$

where

$$\Delta_i = \frac{|{}^iy^{ex}(x_{max}) - {}^iy^{comp}(x_{max})|}{|{}^iy^{ex}(x_{xmax})|},$$

which indicates the minimum number of significant correct digits in the numerical solution at x_{max}.

In Table 5.3 we collect the data for the number of accepted steps ($nstep$) and for the scd factor. As expected, the variable step versions of AEF and B0 are substantially more efficient than VSA0. The first two show comparable efficiencies; compare also the CPU times. The number of rejected steps, not included in the table, is very comparable as well. For AEF we also mention the number of intervals at which the classical A0 had to be activated at least once. This number is in general small, just zero for Lambert equation. For the other two it is typically larger at bigger tolerances, a fact that is easily understood by noticing that the conditions for the selection of the version are dependent on h.

1.6   Function fitting form of the two-step bdf algorithm

The components of the ef-based AEF algorithm depend locally on one real parameter (z in A1 and Z in A2) which, when properly selected, allows increasing the order by one unit. It makes sense to ask whether a version depending on a bigger number of local parameters can be constructed such that, if these parameters are selected in some advantageous way, the order of the two-step bdf algorithm is increased even more.

The answer is positive but such a construction requires some extension of exponential fitting approach. To see the things we consider the generalization of the ef-based version A1. The coefficients (5.11) of the latter have been constructed on the condition that the algorithm has to be exact for the solutions of the RDE $y^{(3)} - \mu y'' = 0$ where μ is a constant. Now we want to obtain the coefficients by taking the equation $y^{(3)} - \alpha(x)y'' = 0$ for reference, where $\alpha(x)$ is a given function. The novelty is that now the reference differential equation is no longer an equation with constant coefficients, as it is in all ef-based applications, but with x dependent coefficients and for this reason such an approach will be called a function fitting approach.

Let us concentrate our attention on the current two-step interval $[x_{n-1}, x_{n+1}]$, denote $X = x_n$ (therefore $x_{n\pm1} = X \pm h$), and take $\alpha(x)$ as a parabola,

$$\alpha(x) = p_0 + p_1(x - X) + p_2(x - X)^2, \ x \in [X - h, X + h], \quad (5.94)$$

where the coeffficients p_0, p_1 and p_2 are open for fit in the same way as it was for z in A1. The new algorithm, call it F1, will depend on three dimensionless parameters $q_0 = p_0 h$, $q_1 = p_1 h^2$ and $q_2 = p_2 h^3$, to be collected in vector $\mathbf{q} = [q_0, q_1, q_2]$; A1 will be the particular case of F1 when $q_0 = z$ and $q_1 = q_2 = 0$. The equation

$$y^{(3)} - [p_0 + p_1(x - X) + p_2(x - X)^2]y'' = 0 \quad (5.95)$$

has three linearly independent solutions. Two of them are $y(x) = 1$, x while for the third we take the solution which satisfies the initial conditions $y(X) = y'(X) = 0$, $y''(X) = 2$. This solution, denoted $\Phi(x)$, is generated by the series expansion

$$\Phi(x) = \sum_{k=2} \phi_k(x - X)^k. \quad (5.96)$$

If this is introduced in (5.95) the following recurrence results for the coefficients:

$$\begin{aligned} \phi_{k+3} &= \frac{1}{(k + 3)(k + 2)(k + 1)}[(k + 2)(k + 1)p_0\phi_{k+2} \quad (5.97) \\ &+ (k + 1)kp_1\phi_{k+1} + k(k - 1)p_2\phi_k], \ k = 0, \ 1, \ 2, \ldots \end{aligned}$$

with $\phi_2 = 1$ for start. The expressions of the first coefficients are

$$\phi_3 = p_0/3, \ \phi_4 = (p_0^2 + p_1)/12, \ \phi_5 = (p_0^3 + 3p_0p_1 + 2p_2)/60.$$

We use the $\mathcal{L}$ operator defined by (5.5). Conditions that (5.5) identically vanishes for $y(x) = 1$, x lead to $L_0^*(\mathbf{a}) = L_1^*(\mathbf{a}) = 0$, that is,

$$a_1(\mathbf{q}) = -1 - a_0(\mathbf{q}), \ b_2(\mathbf{q}) = 1 - a_0(\mathbf{q}).$$

To determine $a_0(\mathbf{q})$ we ask that $\mathcal{L}[h, \mathbf{a}]\Phi(x)|_{x=X} = 0$ to obtain

$$a_0(\mathbf{q}) = \frac{h\Phi'(X + h) - \Phi(X + h)}{h\Phi'(X + h) + \Phi(X - h)} = \frac{1 + 2\phi_3 h + 3\phi_4 h^2 + 4\phi_5 h^3 + \mathcal{O}(h^4)}{T}, \quad (5.98)$$

where

$$T = 3 + 2\phi_3 h + 5\phi_4 h^2 + 4\phi_5 h^3 + \mathcal{O}(h^4).$$

Equation (2.76) for $x = X$ reads

$$\mathcal{L}[h, \mathbf{a}]y(X) = \sum_{m=0}^{\infty} \frac{h^m}{m!} L_m^*(\mathbf{a})y^{(m)}(X), \quad (5.99)$$

with the moments (5.8). With the coefficients just derived these moments are found to be

$$L_m^*(\mathbf{a}) = \frac{2}{T}[(2-m)+2\phi_3 h+(4-m)\phi_4 h^2+4\phi_5 h^3+\mathcal{O}(h^4)] \text{ for even } m \geq 2,$$

(5.100)

and

$$L_m^*(\mathbf{a}) = \frac{2(1-m)}{T}[1+\phi_4 h^2+\mathcal{O}(h^4)] \text{ for odd } m \geq 1,$$ (5.101)

and then F1 is of the second order.

Up to now p_0, p_1 and p_2 were arbitrary. How should they be chosen in order to obtain an increased accuracy ?

To answer this it is convenient to split the sum in (5.99) into two parts, main and residual,

$$\mathcal{L}[h,\mathbf{a}]y(X) = \mathcal{L}^{main}y(X) + \mathcal{L}^{res}y(X),$$ (5.102)

with

$$\mathcal{L}^{main}y(X) = \sum_{m=2}^{5} \frac{h^m}{m!} L_m^*(\mathbf{a})y^{(m)}(X)$$

and

$$\mathcal{L}^{res}y(X) = \sum_{m=6}^{\infty} \frac{h^m}{m!} L_m^*(\mathbf{a})y^{(m)}(X).$$

The forms (5.100) and (5.101) show that $\mathcal{L}^{res}y(X) \sim h^6$ irrespective of the values of the parameters while the behaviour of $\mathcal{L}^{main}y(X)$ depends on how they are chosen.

With the function

$$s(x) := \frac{y^{(3)}(x)}{y''(x)},$$ (5.103)

where $y(x)$ is the exact solution of equation (5.1), we have

$$y^{(3)}(x) = s(x)y''(x), \ y^{(4)}(x) = (s^2(x)+s'(x))y''(x),$$
$$y^{(5)}(x) = (s^3(x)+3s(x)s'(x)+s''(x))y''(x), \ \dots$$

and then

$$\mathcal{L}^{main}y(X) = \frac{1}{2}\sum_{m=2}^{5} h^m w_m L_m^*(\mathbf{a})y''(X),$$

where $w_2 = 1$, $w_3 = s(X)/3$, $w_4 = (s^2(X)+s'(X))/12$ and $w_5 = (s^3(X)+3s(X)s'(X)+s''(X))/60$. On using the expressions of the involved moments we obtain

$$\mathcal{L}^{main}y(X) = \frac{2}{T}\{(\phi_3-w_3)h^3 + (\phi_4-w_4)h^4$$
$$+ [2(\phi_5-w_5)-\phi_4 w_3+\phi_3 w_4]h^5 + \mathcal{O}(h^6)\}y''(X),$$

a formula which suggests that the determination of the parameters in terms of $s(x)$ represents a convenient way of fixing them: it is required that the parabola $\alpha(x)$ is interpolating $s(x)$ at the three knots X and $X \pm h$, that is,

$$p_0 = s_0, \quad p_1 = \frac{1}{2h}(s_1 - s_{-1}), \quad p_2 = \frac{1}{2h^2}(s_1 - 2s_0 + s_{-1}), \qquad (5.104)$$

where $s_j = s(X + jh)$. Indeed, since p_1 and p_2 obtained in this way are the central difference representations of $s'(X)$ and of $s''(X)/2$, respectively, we have $p_1 = s'(X) + \mathcal{O}(h^2)$ and $2p_2 = s''(X) + \mathcal{O}(h^2)$ and therefore $\phi_3 = w_3$, $\phi_4 = w_4 + \mathcal{O}(h^2)$ and $\phi_5 = w_5 + \mathcal{O}(h^2)$, such that, finally, $\mathcal{L}^{main}y(X) \sim h^6$.

In short, we have obtained the following result: *The version F1 of the two-step bdf algorithm is in general of the second order. However, if the function fitting coefficients p_0, p_1 and p_2 are fixed by the parabolic interpolation of function $s(x)$ defined by equation (5.103), the order becomes five.*

The method F1 is then comparable for accuracy with the classical five-step bdf alorithm and therefore it is interesting to also investigate how these two methods compare for stability.

A direct detailed investigation of the stability properties of F1 is difficult because the shape of the boundary $\partial \mathcal{R}_A$ of the region of absolute stability depends on three parameters. However, relevant information can be still obtained on the basis of a number of simpler tests. One such set of tests consists in investigating the dependence of the angle α (see Definition 5.5) with respect to two parameters while the third is kept fixed.

In Tables 5.4, 5.5 and 5.6 we give α for a set of values for q_0 and q_1 when $q_2 = 0$, for various q_0 and q_2 when $q_1 = 0$, and for various q_1 and q_2 when $q_0 = 0$, respectively. It is seen that the data in the lower triangle of each table are close or just equal to $90°$ ($\alpha = 90°$ means that the method is A-stable) while in the upper triangle the deviations from $90°$ tend to increase with the distance from the first diagonal.

The worst values of α in the first two tables are $79.10°$ (for $q_0 = q_1 = 1$ and $q_2 = 0$) and $80.21°$ (for $q_0 = q_2 = 1$ and $q_1 = 0$), respectively, but even so they are much better than $\alpha = 73.35°$ and $\alpha = 51.84°$, which correspond to the classical four-step and five-step bdf methods, respectively; the coefficients of the latter are:

$$a_0 = -\frac{12}{137}, \ a_1 = \frac{75}{137}, \ a_2 = -\frac{200}{137}, \ a_3 = \frac{300}{137}, \ a_4 = -\frac{300}{137}, \ a_5 = 1,$$

$$b_0 = b_1 = b_2 = b_3 = b_4 = 0, \ b_5 = \frac{60}{137}.$$

The boundaries of the region of absolute stability for the classical five-step and the two-step F1 algorithms (with the two mentioned sets of parameters in the

Table 5.4. Angle α (in degrees) for the function fitting version F1 of the two-step bdf algorithm at various values of q_0 and q_1 for fixed $q_2 = 0$.

q_0	q_1								
	-1.00	-0.75	-0.50	-0.25	0.00	0.25	0.50	0.75	1.00
1.00	85.70	85.00	84.27	83.50	82.68	81.84	80.96	80.04	79.10
0.75	87.60	87.03	86.41	85.75	85.05	84.30	83.52	82.69	81.84
0.50	89.15	88.74	88.28	87.76	87.20	86.58	85.92	85.22	84.47
0.25	90.00	89.90	89.66	89.35	88.97	88.53	88.04	87.49	86.89
0.00	90.00	90.00	90.00	90.00	90.00	89.87	89.62	89.29	88.89
-0.25	90.00	90.00	90.00	90.00	90.00	90.00	90.00	90.00	90.00
-0.50	90.00	90.00	90.00	90.00	90.00	90.00	90.00	90.00	90.00
-0.75	90.00	90.00	90.00	90.00	90.00	90.00	90.00	90.00	90.00
-1.00	90.00	90.00	90.00	90.00	90.00	90.00	90.00	90.00	90.00

Table 5.5. The same as in Table 5.4 for fixed $q_1 = 0$ and variable q_0 and q_2.

q_0	q_2								
	-1.00	-0.75	-0.50	-0.25	0.00	0.25	0.50	0.75	1.00
1.00	84.85	84.34	83.81	83.26	82.68	82.09	81.48	80.85	80.21
0.75	86.95	86.51	86.04	85.56	85.05	84.52	83.96	83.39	82.79
0.50	88.72	88.38	88.02	87.62	87.20	86.75	86.28	85.78	85.25
0.25	89.91	89.74	89.53	89.27	88.97	88.64	88.28	87.88	87.45
0.00	90.00	90.00	90.00	90.00	90.00	89.90	89.73	89.50	89.22
-0.25	90.00	90.00	90.00	90.00	90.00	90.00	90.00	90.00	90.00
-0.50	90.00	90.00	90.00	90.00	90.00	90.00	90.00	90.00	90.00
-0.75	90.00	90.00	90.00	90.00	90.00	90.00	90.00	90.00	90.00
-1.00	90.00	90.00	90.00	90.00	90.00	90.00	90.00	90.00	90.00

latter) are presented in Figure 5.9. The comparison of these is clearly in the favour of F1; it is perhaps even more impressive than only the comparison of the angles α might have suggested.

All these data enable putting forwards an optimistic message. While quite often it is believed that the multistep methods did already reach their limits and that, for really difficult problems, the one-step methods deserve a definitely better credit, the function fitting appears as providing a promising way towards the enhancement of the quality of the multistep algorithms at a level which may make these algorithms highly competitive again.

We note finally that the function $s(x)$, introduced for the construction of the optimal values of the parameters, was assumed as being defined and admitting a number of derivatives on $[X - h, X + h]$. This assumption is clearly violated

Table 5.6. The same as in Table 5.4 for fixed $q_0 = 0$ and variable q_1 and q_2.

q_1	q_2								
	−1.00	−0.75	−0.50	−0.25	0.00	0.25	0.50	0.75	1.00
1.00	89.93	89.76	89.52	89.23	88.89	88.51	88.10	87.65	87.16
0.75	90.00	89.95	89.79	89.56	89.29	88.96	88.60	88.20	87.76
0.50	90.00	90.00	89.97	89.83	89.62	89.35	89.04	88.69	88.31
0.25	90.00	90.00	90.00	89.99	89.87	89.67	89.42	89.13	88.79
0.00	90.00	90.00	90.00	90.00	90.00	89.90	89.73	89.50	89.22
−0.25	90.00	90.00	90.00	90.00	90.00	90.00	89.94	89.78	89.57
−0.50	90.00	90.00	90.00	90.00	90.00	90.00	90.00	89.97	89.84
−0.75	90.00	90.00	90.00	90.00	90.00	90.00	90.00	90.00	90.00
−1.00	90.00	90.00	90.00	90.00	90.00	90.00	90.00	90.00	90.00

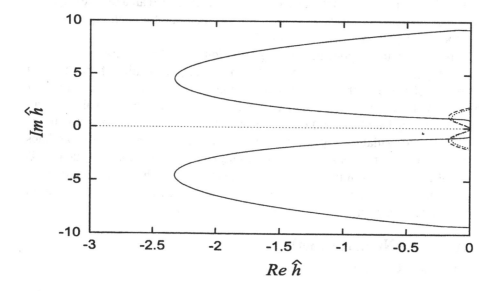

Figure 5.9. The boundary $\partial \mathcal{R}_A$ of the region of absolute stability in the left-half $\hat{h}$ plane for the classical five-step bdf algorithm (solid), and for the version F1 of the two-step bdf algorithm with $q_0 = q_1 = 1$ and $q_2 = 0$ (broken) and with $q_0 = q_2 = 1$ and $q_1 = 0$ (dashed). The region of absolute stability lies on the left-hand side the boundary.

when $y''(x)$ has a zero in the quoted interval and then F1 suffers of the same restriction as A1. A way to overcome this difficulty consists in replacing it by a version which has A2 as its ef counterpart, in the same way as in AEF. The generation of such a version along the described lines is possible but this is beyond our purpose here.

2.   Second order equations

Here we consider the numerical solution of the initial value problem for second order differential equations of special form. The problem reads

$$y'' = f(x, y), \ x \in [a, b], \ y(a) = y_0, \ y'(a) = y_0', \qquad (5.105)$$

where $f(x, y)$, $f : \Re \times \Re \to \Re$, the ends a and b of the interval and the initial values y_0 and y_0' are given. The important feature is that f depends on x and y but not on y'.

Of course, since (5.105) can be written as an initial value problem for a system of two equations of the first order, this problem can be solved by algorithms for first order equations. However, this will be less efficient than if methods specially devised for the given problem would be used. In fact, problems of form (5.105) appear so often in the usual practice that the construction of solvers specialized for them is currently seen as a well-established area of investigation. For the theory of (classical) linear multistep methods for such equations see Chapter 6 of [22] and Chapter III of [20].

The use of the exponential fitting technique for the construction of new multistep methods (or, better saying, for improving the quality of the classical versions of these methods) is specially suited when some specific behaviour is expected for the solution. The case when it is known that the solution will behave piecewise like (3.38) or like (3.39), where some good approximations for ω or λ are available in advance, received special attention. There are perhaps almost 200 such papers in the literature but, since the mentioned behaviour is typical for the Schrödinger equation (see Section 1 of Chapter 3), there is not surprising that a vast majority of these papers were concentrated on the solution of the latter equation.

2.1   The Numerov method

This is of the form

$$y_{n+1} + a_1 y_n + y_{n-1} = h^2[b_0(f_{n+1} + f_{n-1}) + b_1 f_n], \qquad (5.106)$$

where $x_{n\pm 1} = x_n \pm h$, y_n is an approximation to $y(x_n)$ and $f_n = f(x_n, y_n)$. The coefficients of the classical version are

$$a_1 = -2, \ b_0 = \frac{1}{12}, \ b_1 = \frac{5}{6}. \qquad (5.107)$$

This algorithm has been prompted mainly due to Numerov, [44, 45], though it was known earlier; it appears as one of the Cowell methods, see [12]. Hereinafter this scheme will be denoted S_0.

In the case of the Schrödinger equation function $f(x, y)$ has the particular form $f(x, y) = (V(x) - E)y$, where $V(x)$ is the potential and E is the energy, see Section 1 of Chapter 3. Then, in the regions on the x axis where $V(x) < E$, the so called classically allowed regions, the solution is well described by functions of type (3.38) while in the regions where $V(x) > E$ (classically forbidden regions) the solution is described by functions of form (3.39). The generation of a tuned formula to cover both situations is thus strongly motivated.

We follow the $\mathcal{L}$ operator procedure. With $\mathbf{a} := [a_1, b_0, b_1]$ we define $\mathcal{L}[h, \mathbf{a}]$ by

$$\begin{aligned} \mathcal{L}[h, \mathbf{a}]y(x) \quad := \quad & y(x+h) + a_1 y(x) + y(x-h) \\ & - \quad h^2[b_0(y''(x+h) + y''(x-h)) + b_1 y''(x)]. \end{aligned} \qquad (5.108)$$

$\mathcal{L}$ is dimensionless, that is $l = 0$.

Let $z = \mu h$. We construct $E_0^*(z, \mathbf{a})$ by applying $\mathcal{L}$ on $y(x) = \exp(\mu x)$ (see (2.61) and (2.69)) to obtain

$$E_0^*(z, \mathbf{a}) = \exp(z) + \exp(-z) + a_1 - z^2[b_0(\exp(z) + \exp(-z)) + b_1],$$

and then

$$G^+(Z, \mathbf{a}) = 2\eta_{-1}(Z) + a_1 - Z[2b_0\eta_{-1}(Z) + b_1], \ G^-(Z, \mathbf{a}) = 0,$$

where $Z = z^2$. It follows that

$$G^{+(1)}(Z, \mathbf{a}) = \eta_0(Z) - (2\eta_{-1}(Z) + Z\eta_0(Z))b_0 - b_1,$$

$$G^{+(m)}(Z, \mathbf{a}) = 2^{-m+1}[\eta_{m-1}(Z) - (3\eta_{m-2}(Z) + \eta_{m-3}(Z))b_0], \ m = 2, 3, \ldots$$

$$G^{-(m)}(Z, \mathbf{a}) = 0, \ m = 1, 2, \ldots$$

The expressions of the starred classical moments are (use (3.49)):

$$L_0^*(\mathbf{a}) = 2 + a_1, \ L_2^*(\mathbf{a}) = 2(1 - 2b_0 - b_1),$$

$$L_{2k}^*(\mathbf{a}) = 2 - 4k(2k - 1)b_0, \ k = 2, 3, \ldots$$

$$L_{2k+1}^*(\mathbf{a}) = 0, \ k = 0, 1, \ldots$$

System $L_k^*(\mathbf{a}) = 0$, $0 \leq k \leq 5$ is compatible and it has solution (5.107). Since $L_6^*(\mathbf{a}) = -3$, the *lte* is

$$lte_{class} = -\frac{h^6}{240}y^{(6)}(x_n) \qquad (5.109)$$

and also $M = 6$.

Three options for tuning are then available: (i) $P = 0$, $K = 3$, (ii) $P =$

1, $K = 1$, and (iii) $P = 2$, $K = -1$ and these lead to the schemes S_1, S_2 and S_3, respectively; see Ixaru and Rizea [28].

S_1. The six functions to be integrated exactly by the algorithm are 1, x, x^2, x^3, and the pair $\exp(\pm\mu x)$ and therefore the system to be solved is $L_k^*(\mathbf{a}) = 0$, $0 \leq k \leq 3$ and $G^{\pm}(Z, \mathbf{a}) = 0$. The system is compatible and its solution is

$$a_1(Z) = -2, \quad b_0(Z) = \frac{1}{Z} - \frac{1}{2(\eta_{-1}(Z) - 1)}, \quad b_1(Z) = 1 - 2b_0(Z). \quad (5.110)$$

The expression of the *lte* is

$$lte_{ef} = -h^6 \frac{1 - 12b_0(Z)}{12Z}(-\mu^2 y^{(4)}(x_n) + y^{(6)}(x_n)). \quad (5.111)$$

Critical values appear in the trigonometric case $\mu = i\omega$ (that is for negative Z). With $\theta = \omega h$ these are $\theta = 2k\pi$, $k = 1, 2, \ldots$.

The trigonometric case of this scheme, that is with $b_0(Z)$ expressed in terms of θ,

$$b_0(\theta) = \frac{\theta^2 - 2(1 - \cos(\theta))}{2\theta^2(1 - \cos(\theta))}, \quad (5.112)$$

has been first derived by Stiefel and Bettis, [60], for orbit problems of the form (5.105), based on the idea of integrating products of ordinary and Fourier polynomials exactly. The same scheme has been re-derived, for any real Z, by Raptis and Allison, [49], on the basis of the work of Lyche [43]. These authors also gave the series expansion to be used for the computation of $b_0(Z)$ when Z is close to zero. As a matter of fact, by using (3.35), $b_0(Z)$ can be re-written in the form

$$b_0(Z) = \frac{(\eta_0(Z/4) + 1)(\eta_0^2(Z/16) - 2\eta_1(Z/4))}{8\eta_0^2(Z/4)},$$

which is free of computational difficulties irrespective of the value of Z, see [31]; in particular, no separate series expansion form is needed. Vanden Berghe et al., [62], also re-derived the scheme S_1 for the trigonometric case but they were the first to express the *lte* in a form consistent with (5.111).

S_2. The functions to be integrated exactly are now 1, x, and the pairs $\exp(\pm\mu x)$ and $x\exp(\pm\mu x)$. Since all classical moments with odd indices and all G functions with a minus sign for the upper index are identically vanishing, the system to be solved is simply $L_0^*(\mathbf{a}) = G^+(Z, \mathbf{a}) = G^{+(1)}(Z, \mathbf{a}) = 0$, with the solution

$$a_1(Z) = -2, \quad b_0(Z) = \frac{1}{Z} - \frac{2(\eta_{-1}(Z) - 1)}{Z^2\eta_0(Z)} = \frac{\eta_1(Z/4)}{4\eta_{-1}(Z/4)},$$

$$b_1(Z) = 2[\frac{\eta_{-1}(Z) - 1}{Z} - b_0(Z)\eta_{-1}(Z)] = \eta_0^2(Z/4) - 2b_0(Z)\eta_{-1}(Z).$$

$$(5.113)$$

The last expressions in $b_0(Z)$ and in $b_1(Z)$ are obtained from the previous ones via (3.35). They are free of computational difficulties when Z is in the vicinity of zero.

The leading term of the error is:

$$lte_{ef} = h^6 \frac{Z^2 \eta_0(Z) - 4(\eta_{-1}(Z) - 1)^2}{Z^4 \eta_0(Z)} [\mu^4 y''(x_n) - 2\mu^2 y^{(4)}(x_n) + y^{(6)}(x_n)]$$

(5.114)

and the critical values are $\theta = (2k+1)\pi$, $k = 0, 1, \ldots$ This scheme has been derived by Ixaru and Rizea in [25]. In the trigonometric case the coefficients read:

$$a_1(\theta) = -2, \quad b_0(\theta) = \frac{2\tan(\theta/2) - \theta}{\theta^3}, \quad b_1(\theta) = \frac{2[\theta - 2\tan(\theta/2)\cos(\theta)]}{\theta^3}.$$

(5.115)

S_3. The reference set of six functions is $\exp(\pm\mu x)$, $x\exp(\pm\mu x)$, $x^2\exp(\pm\mu x)$ and thus the system to be solved is

$$G^+(Z, \mathbf{a}) = G^{+(1)}(Z, \mathbf{a}) = G^{+(2)}(Z, \mathbf{a}) = 0,$$

with the solution

$$a_1(Z) = -(6\eta_{-1}(Z)\eta_0(Z) - 2\eta_{-1}^2(Z) + 4)/D(Z),$$

(5.116)

$$b_0(Z) = \eta_1(Z)/D(Z), \quad b_1(Z) = (4\eta_0^2(Z) - 2\eta_1(Z)\eta_{-1}(Z))/D(Z),$$

where $D(Z) = 3\eta_0(Z) + \eta_{-1}(Z)$. This set of coefficients were derived in [47] and in [28] based on the Lyche's theory. A special issue in the latter reference is the examination of the convergence of the scheme. This is needed because, due to the fact that a_1 is not a constant, the convergence of this scheme is not automatically covered by that theory. The construction of the summed form of the algorithm (the summed form enables the reduction of the influence of the round-off errors) and rules for the application of this version for systems of coupled Schrödinger equations were also considered in that reference.

The leading term of the error is

$$lte_{ef} = -h^6 \frac{N(Z)}{F(Z)} [-\mu^6 y(x_n) + 3\mu^4 y^{(2)}(x_n) - 3\mu^2 y^{(4)}(x_n) + y^{(6)}(x_n)]$$

(5.117)

where $N(Z) = 6\eta_0(Z) + 2\eta_{-1}(Z) - 6\eta_{-1}(Z)\eta_0(Z) + 2\eta_{-1}^2(Z) - 4$ and $F(Z) = Z^3 D(Z)$, see [30].

The expressions of the coefficients in the trigonometric case are

$$a_1(\theta) = -\frac{2\{2\theta + \cos(\theta)[3\sin(\theta) - \theta\cos(\theta)]\}}{d(\theta)},$$

$$b_0(\theta) = \frac{\sin(\theta) - \theta\cos(\theta)]}{\theta^2 d(\theta)}, \qquad\qquad (5.118)$$

$$b_1(\theta) = \frac{2\{2\theta - \cos(\theta)[\sin(\theta) + \theta\cos(\theta)]\}}{\theta^2 d(\theta)},$$

where $d(\theta) = 3\sin(\theta) + \theta\cos(\theta)$. The critical values are given by the roots of $d(\theta)$. The first four of them are $\theta = 0.782\pi$, 1.666π, 2.612π and 3.583π, see [11].

The relative merits of each of the four versions can be evaluated by comparing the expressions of the *lte*. Thus, it is seen that the *lte* has one and the same h^6 dependence and then the order of each version is four; it is recalled that for multistep methods for second order equations, if the *lte* behaves like h^M then the order is $p = M - 2$. Also the factors in the middle are close to $-1/240$ when Z is around zero and then the difference for accuracy is contained in the third factor. The situation is very similar to the one met for the versions A1 and A2 of the two-step bdf algorithm and, as there, the structure of the third factor can be exploited for the determination of the optimal frequencies which would normally lead to an increase of the order.

However, we shall not follow this direction. Instead, we consider the case of the Schrödinger equation where, as explained in Section 1 of Chapter 3 on the basis of [25], the knowledge of the potential function $V(x)$ and of the energy E is sufficient to get reasonable approximations for frequencies: the domain $[a, b]$ is divided in subintervals and on each of them the function $V(x)$ is approximated by a constant $\bar{V}$. On all steps in such a subinterval one and the same μ^2 is used, $\mu^2 = \bar{V} - E$.

If this is done the order of each version remains four but the errors will be very different when big values of the energy are involved. To see this let us denote $\Delta V(x) = V(x) - \bar{V}$, express the higher order derivatives of y in terms of y, y', μ^2, $\Delta V(x)$ and the derivatives of $V(x)$, for example $y''(x) = (V(x) - E)y(x) = (\mu^2 + \Delta V(x))y(x)$, $y^{(3)}(x) = V'(x)y(x) + (\mu^2 + \Delta V(x))y'(x)$ etc., and finally introduce them in the expressions of the last factors in the *lte*, which will be denoted $\Delta_i(x_n)$, $i = 0, 1, 2, 3$. The expression of each $\Delta_i(x_n)$ resulting from such a treatment will consist in a sum of y and y' with coefficients which depend on μ^2, $\Delta V(x)$ and on the derivatives of $V(x)$.

If $E >> \bar{V}$, $|\mu^2|$ has a big value and then the μ^2 dependence of $\Delta_i(x_n)$ will become dominating; the approximation $\mu^2 \approx -E$ will hold as well. To compare the errors it is then sufficient to organize the coefficients of y and of y'

as polynomials in E and to retain only the terms with the highest power. This gives:

$$\Delta_0(x_n) \approx -E^3 y - 6EV'y',$$

$$\Delta_1(x_n) \approx E^2 \Delta V y - 4EV'y',$$

$$\Delta_2(x_n) \approx -E[5V^{(2)} + (\Delta V)^2]y - 2EV'y',$$

$$\Delta_3(x_n) \approx -4EV^{(2)}y + (4V^{(3)} + 6V'\Delta V)y'.$$

Since in the discussed range of energies the solution is of oscillatory type with almost constant coefficients, the amplitude of the first derivative is bigger by a factor $E^{1/2}$ than that of the solution itself and then the error from the four schemes increases with E as E^3, E^2, $E^{3/2}$ and E, respectively.

For illustration we take the sum of the Woods–Saxon potential (3.5) and its first derivative, that is

$$V(x) = v_0/(1 + t) + v_1 t/(1 + t)^2, \ t = \exp[(x - x_0)/a],$$

where $v_0 = -50$, $x_0 = 7$, $a = 0.6$ and $v_1 = -v_0/a$. Its shape is such that only two values for $\bar{V}$ are sufficient: $\bar{V} = -50$ for $0 \le x \le 6.5$ and $\bar{V} = 0$ for $x \ge 6.5$.

We solve the resonance problem which consists in the determination of the positive eigenvalues corresponding to the boundary conditions

$$y(0) = 0, \ y(x) = \cos(E^{1/2}x) \text{ for big } x.$$

The physical interval $x \ge 0$ is cut at $b = 20$ and the eigenvalues are obtained by shooting at $x_c = 6.5$. For any given E the solution is propagated forwards with the starting values $y(0) = 0$, $y(h) = h$ up to $x_c + h$, and backwards with the starting conditions $y(b) = \cos(E^{1/2}b)$, $y(b - h) = \cos(E^{1/2}(b - h))$ up to x_c. If E is an eigenvalue the forwards and backwards solutions are proportional and then the numerical values of quantities as the logarithmic derivatives $y^{f'}(x_c)/y^f(x_c)$ and $y^{b'}(x_c)/y^b(x_c)$, or as the products $y^f(x_c + h)y^b(x_c)$ and $y^b(x_c + h)y^f(x_c)$, must coincide. However, since the Numerov method does not compute the derivatives of the solution, the latter has to be adopted. If so, the resonance eigenenergies searched for are the roots of the mismatch function

$$\Delta(E) = y^f(x_c + h)y^b(x_c) - y^b(x_c + h)y^f(x_c),$$

which can be evaluated directly on the basis of the available data. The error in the eigenvalues will then reflect directly the quality of the solvers for the initial value problem used for the determination of the solution $y(x)$.

Table 5.7. Absolute errors $E_{exact} - E_{comput}$ in 10^{-6} units from the four schemes of the Numerov method for the resonance eigenenergy problem of the Schrödinger equation in the Woods–Saxon potential. The empty areas indicate that the corresponding errors are bigger than the format adopted in the table.

	h	S_0	S_1	S_2	S_3
$E_{exact} = 53.588852$					
	1/16	−259175	6178	−1472	587
	1/32	−15872	367	−84	35
	1/64	−989	22	−5	1
	1/128	−62	1	0	0
$E_{exact} = 163.215298$					
	1/16		79579	−9093	721
	1/32	−595230	4734	−525	46
	1/64	−36661	292	−32	2
	1/128	−2287	18	−1	0
$E_{exact} = 341.495796$					
	1/16		661454	−40122	1600
	1/32		36703	−2116	126
	1/64	−560909	2215	−126	7
	1/128	−34813	136	−8	0
$E_{exact} = 989.701700$					
	1/16			−466720	−1417
	1/32		806507	−17541	421
	1/64		46269	−975	28
	1/128		2812	−59	1

In Table 5.7 we list the absolute errors in four such eigenvalues for all four schemes of the Numerov method; reference values, which are exact in the written figures, have been generated in a separate run with the method CPM(2) from [23] at $h = 1/16$. It is seen that, as expected, all these versions are of order four but the way in which the error increases with the energy differs from one version to another.

2.1.1   Subroutine EFNUM

This computes the coefficients a_1, b_0 and b_1 for any of the four versions S_i ($0 \leq i \leq 3$) of the Numerov method. It uses the expressions written before in

terms of functions $\eta_s(Z)$.
Calling this subroutine is

$$\text{CALL EFNUM}(I, A1, B0, B1, Z)$$

where:

- I - the label i of the version (input).

- Z - double precision value of Z (input).

- A1, B0, B1, - resultant values of the coefficients (double precision).

The following driving program computes these coefficients for all $0 \le i \le 3$ and for three values of Z: -1 (trigonometric case), 0 (classical case), and 1 (hyperbolic case).

```
      implicit double precision(a-h,o-z)
      do j=1,3
        if(j.eq.1)z=-1.d0
        if(j.eq.2)z=0.d0
        if(j.eq.3)z=1.d0
        write(*,51)z
        do i=0,3
          call efnum(i,a1,b0,b1,z)
          write(*,50)i,a1,b0,b1
        enddo
      enddo
50    format(2x,'i=',i2,': a1=',f9.6,' b0=',f9.6,' b1=',f9.6)
51    format(/,2x,'z=',f4.1)
      stop
      end
```

The output is:

```
 z=-1.0
 i= 0: a1=-2.000000 b0= 0.083333 b1= 0.833333
 i= 1: a1=-2.000000 b0= 0.087671 b1= 0.824657
 i= 2: a1=-2.000000 b0= 0.092605 b1= 0.819326
 i= 3: a1=-2.004767 b0= 0.098270 b1= 0.817971

 z= 0.0
 i= 0: a1=-2.000000 b0= 0.083333 b1= 0.833333
 i= 1: a1=-2.000000 b0= 0.083333 b1= 0.833333
```

```
i= 2: a1=-2.000000 b0= 0.083333 b1= 0.833333
i= 3: a1=-2.000000 b0= 0.083333 b1= 0.833333

z= 1.0
i= 0: a1=-2.000000 b0= 0.083333 b1= 0.833333
i= 1: a1=-2.000000 b0= 0.079326 b1= 0.841347
i= 2: a1=-2.000000 b0= 0.075766 b1= 0.852336
i= 3: a1=-1.996255 b0= 0.072579 b1= 0.865916
```

It is seen that, as expected, when $Z = 0$ all versions give the classical set of coefficients.

2.2   Linear stability theory and P-stability

The classical linear J-step method for (5.105) is a method with the algorithm

$$\sum_{j=0}^{J} a_j y_{n+j} = h^2 \sum_{j=0}^{J} b_j f_{n+j}, \; n = 0, \, 1, \, \ldots \qquad (5.119)$$

where y_{n+j} is an approximation to $y(x_{n+j})$ and $f_{n+j} = f(x_{n+j}, y_{n+j})$. The coefficients a_j and b_j are constants which do not depend on n subject to the conditions

$$a_J = 1, \; |a_0| + |b_0| \neq 0.$$

For the convergence theory see [22]. In essence this theory extends the one for the multistep algorithms for first order ODEs: two issues, that is local and propagation problems, are considered separately and concepts like consistency, order of the method and zero-stability are conveniently adapted to finally reach a convergence theorem like 5.2. The adaptation of this theory to the ef-based methods is also similar to the one for first order equations, see Section 1.2, and thus it will not be repeated here. In contrast, the linear stability theory is different enough to deserve a special consideration.

The linear stability theory for algorithm (5.119) with symmetric coefficients ($a_j = a_{J-j}$, $b_j = b_{J-j}$) was considered by Lambert and Watson in [39]. In the same way as for first order equations, a test equation is introduced and conditions are formulated such that, for given $h > 0$, the numerical solution of this equation mimics the behaviour of the exact solution.

The test equation is

$$y'' = -k^2 y, \; x \geq 0, \qquad (5.120)$$

where k is a nonnegative constant. Its solution is

$$y(x) = y_0 \cos(kx) + y_0' \sin(kx)/k \, .$$

This is a periodic function with period $2\pi/k$ for all non-trivial initial conditions y_0 and y_0'. The algorithm (5.119) applied to this equation reads

$$\sum_{j=0}^{J}[a_j + \nu^2 b_j]y_{n+j} = 0, \ n = 0, 1, \ldots \tag{5.121}$$

where $\nu = kh$. The characteristic equation is

$$\sum_{j=0}^{J}[a_j + \nu^2 b_j]d^j = 0, \tag{5.122}$$

and the answer to the problem of whether the numerical solution mimics the behaviour of the exact solution depends on the properties of the roots of this equation. Lambert and Watson introduced the concepts of the periodicity interval and of the P-stability in the following way:

DEFINITION 5.10 *The method (5.119) with symmetric coefficients has an interval of periodicity* $(0, \nu_0^2)$ *if, for all* $\nu^2 \in (0, \nu_0^2)$, *the roots d_j of the characteristic equation satisfy*

$$d_1 = \exp(i\phi(\nu)), \ d_2 = \exp(-i\phi(\nu)), \ |d_j| \leq 1, \ 3 \leq j \leq J, \quad (5.123)$$

where $\phi(\nu)$ is real; it is said to be P-stable if the periodicity interval is $(0, \infty)$.

The theory of Lambert and Watson was re-considered by Coleman and Ixaru, [11], for methods whose coefficients depend on one parameter which is proportional with the steplength h. We follow the main ideas of that paper. As there, a special attention will be given to the class of algorithms which, when applied on the test equation, are of the form

$$D_m(\nu^2; \theta)y_{n+1} - 2N_q(\nu^2; \theta)y_n + D_m(\nu^2; \theta)y_{n-1} = 0, \ n = 1, 2, \ldots \tag{5.124}$$

where N_q and D_m are polynomials in ν^2 of degree q and m, respectively, with coefficients which depend on the parameter θ which specifies the method of concern,

$$N_q(\nu^2; \theta) = \sum_{k=0}^{q} \alpha_k(\theta)\nu^{2k}, \ D_m(\nu^2; \theta) = 1 + \sum_{k=1}^{m} \beta_k(\theta)\nu^{2k}. \tag{5.125}$$

As said, it is assumed that θ is proportional to the steplength, $\theta = \omega h$.
The rational function

$$R_{qm}(\nu^2; \theta) = \frac{N_q(\nu^2; \theta)}{D_m(\nu^2; \theta)} \tag{5.126}$$

is called the stability function. With it equation (5.124) reads simply

$$y_{n+1} - 2R_{qm}(\nu^2; \theta)y_n + y_{n-1} = 0, \ n = 1, 2, \ldots \tag{5.127}$$

Function $R_{qm}(\nu^2; \theta)$ depends on two parameters; one of them, θ, is related to the method while the other one, ν, to the test equation.

A long series of methods belongs to this class. This is clearly the case for the three exponential fitting versions of the Numerov method where the two fitting frequencies are $\pm i\omega$. Indeed, algorithm (5.106) applied on the test equation reads

$$[1 + \nu^2 b_0(\theta)]y_{n+1} + [a_1(\theta) + \nu^2 b_1(\theta)]y_n + [1 + \nu^2 b_0(\theta)]y_{n-1} = 0,$$

with the set of coefficients given by (5.112), (5.115) and (5.118), respectively. The classical form (5.107) also belongs to this class insomuch as its coefficients can be seen as functions of θ which, incidentally, remain unchanged for any θ. The version with $a_1 = -2$, $b_0(\theta) = 1/12 + \theta^2/240$ and $b_1(\theta) = 5/6 - \theta^2/120$, call it S_1', is still another version in this class, although this is not a genuine exponential fitting version; its coefficients result by retaining the first two terms from the representation in powers of θ^2 of the coefficients of S_1.

In all these cases the two polynomials are of the first degree and the stability function is

$$R_{11}(\nu^2; \theta) = -\frac{a_1(\theta) + \nu^2 b_1(\theta)}{2[1 + \nu^2 b_0(\theta)]}. \tag{5.128}$$

Methods with bigger values for q and m also exist. Two such examples follow.

Anantha Krishnaiah, [2], proposed the method

$$y_{n+1} - 2y_n + y_{n-1} = h^2[\lambda f_{n+1} + (1 - 2\lambda)f_n + \lambda f_{n-1}]$$

$$+ h^4 \eta(f_{n+1}'' - 2\cos(\theta)f_n'' + f_{n-1}''),$$

where λ and η are functions of θ. Its stability function is

$$R_{22}(\nu^2; \theta) = \frac{1 + (\lambda - 1/2)\nu^2 - \eta\nu^4 \cos(\theta)}{1 + \lambda\nu^2 - \eta\nu^4}.$$

The hybrid algorithm of Simos, [54],

$$\bar{y}_{n\pm 1/2} = (y_n + y_{n\pm 1})/2 - h^2(34f_n + 19f_{n\pm 1} - 5f_{n\mp 1})/384,$$

$$\bar{y}_n = y_n - a(\theta)h^2[(f_{n+1} - 2f_n + f_{n-1}) - 4(\bar{f}_{n+1/2} - 2f_n + \bar{f}_{n-1/2})],$$

$$y_{n+1} - 2y_n + y_{n-1} = h^2[f_{n+1} + 26\bar{f}_n + f_{n-1} + 16(\bar{f}_{n+1/2} + \bar{f}_{n-1/2})]/60,$$

has the stability function

$$R_{33}(\nu^2; \theta) = \frac{1 - 7\nu^2/20 - (17 + 312a(\theta))\nu^4/720 - 221a(\theta)\nu^6/1440}{1 + 3\nu^2/20 + (7 - 312a(\theta))\nu^4/720 - 91a(\theta)\nu^6/1440}.$$

The expression of $a(\theta)$ is

$$a(\theta) = \frac{1440(1 - \cos(\theta)) - 72(7 + 3\cos(\theta))\theta^2 - 2(17 + 7\cos(\theta))\theta^4}{624(1 - \cos(\theta))\theta^4 - 13(17 + 7\cos(\theta))\theta^6}.$$

There exist also methods for which $q \neq m$. For example, the stability function of the method

$$y_{n+1} + a_1(\theta)y_n + y_{n-1} = h^2 b_1(\theta)f_n,$$

(three versions of this are discussed in [11]) is

$$R_{10}(\nu^2; \theta) = -\frac{a_1(\theta) + \nu^2 b_1(\theta)}{2}.$$

It is also worth mentioning that stability functions of form (5.126) can be written also for some methods of the Runge-Kutta-Nyström type. The mixed collocation method with two or three symmetric collocation points provides such an example, see Coleman and Duxbury [10].

Let us consider the relation

$$y(x_{n+1}) - 2\bar{R}y(x_n) + y(x_{n-1}) = 0, \tag{5.129}$$

and search for $\bar{R}$ such that this is exactly satisfied when $y(x)$ is an arbitrary solution of the test equation. Functions $\exp[\pm ik(x - x_n)]$ are two linearly independent solutions. When each of them is introduced in (5.129) one and the same equation results, viz.:

$$\exp(i\nu) - 2\bar{R} + \exp(-i\nu) = 0,$$

whence

$$\bar{R} = \cos(\nu).$$

The stability function $R_{qm}(\nu^2; \theta)$ can be then regarded as a rational approximation of $\cos(\nu)$. Such a perspective has been adopted by Coleman [9] to provide a general framework for the study of the stability properties of a variety of one- and two-step classical methods and the same is also used in [11].

The characteristic equation of (5.127) is

$$d^2 - 2R_{qm}(\nu^2; \theta)d + 1 = 0, \tag{5.130}$$

and thus its roots depend on $R_{qm}(\nu^2; \theta)$.

If $|R_{qm}(\nu^2; \theta)| < 1$ a real ρ exists such that $R_{qm}(\nu^2; \theta) = \cos(\rho h)$. The two roots are $d_1 = \exp(i\rho h)$ and $d_2 = \exp(-i\rho h)$ and then

$$y_n = C_1 \exp(i\rho n h) + C_2 \exp(-i\rho n h)$$

$$= (C_1 + C_2)\cos(\rho x_n) + i(C_1 - C_2)\sin(\rho x_n), \quad x_n = nh,$$

where the constants C_1 and C_2 can be fixed in terms of the starting values y_0 and y_1. The important fact is that the numerical solution exhibits periodicity with period $2\pi/\rho$. The difference between this ρ and k does not matter at this moment but it becomes important when the accuracy of the method is of concern. In such a case the difference between $\nu = kh$ and $\rho h = \cos^{-1} R_{qm}(\nu^2; \theta)$ is monitored and the following definition, which is a slight adaptation of a definition originally formulated by Van der Houwen and Sommeijer in [63], is pertinent for this purpose:

DEFINITION 5.11 *For any method with the stability function* $R_{qm}(\nu^2; \theta)$ *the quantity*

$$\phi(\nu^2; \theta) = \nu - \cos^{-1} R_{qm}(\nu^2; \theta) \qquad (5.131)$$

is called the dispersion (or phase error or phase lag). If $\phi(\nu^2; \theta) = \mathcal{O}(\nu^{s+1})$ *as* $\nu \to 0$ *the order of dispersion is s.*

Coleman and Duxbury, [10], formulated the following alternative but equivalent definition for the order of dispersion:

DEFINITION 5.12 *The order of dispersion is s if*

$$\cos(\nu) - R_{qm}(\nu^2; r\nu) = \gamma(r)\nu^{s+2} + \mathcal{O}(\nu^{s+4}) \text{ as } \nu \to 0, \qquad (5.132)$$

with $r = \theta/\nu$ *and* $\gamma(r) \neq 0$.

The main difference between these two definitions is that the pair ν, θ used in the first definition is replaced in the second by the pair ν, r. However, the latter choice is better suited for the analysis of the properties of the function $R_{qm}(\nu^2; \theta)$ in the $\nu - \theta$ plane because such an analysis becomes more transparent if rays $\theta = r\nu$ with fixed values of r are followed.

The case $r = 1$ represents the situation when the fitting ω equals the test k. If the equality

$$R_{qm}(\theta^2; \theta) = \cos(\theta) \qquad (5.133)$$

holds for all θ at which the method is defined, then we have $\phi(\theta^2; \theta) = 0$ and the method is said phase-fitted. As a matter of fact, relation (5.133) has been used to obtain the written expression of the parameter $a(\theta)$ in Simos' method.

Note however that not all methods mentioned before are phase-fitted. The ef-based methods are automatically phase-fitted simply because they are exact for the pair $\exp(\pm\omega x)$. Expressed in other words, for these methods there is no phase lag if $\theta = \nu$ and therefore $\gamma(1) = 0$. In contrast, the standard Numerov method and the method S_1' are not phase-fitted.

If $|R_{qm}(\nu^2; \theta)| = 1$ the two roots of (5.130) are equal. When $R_{qm}(\nu^2; \theta) = 1$ they are $d_1 = d_2 = 1$ and then

$$y_n = C_1' + nC_2',$$

while when $R_{qm}(\nu^2; \theta) = -1$ they are $d_1 = d_2 = -1$, such that

$$y_n = (-1)^n [C_1' + nC_2'],$$

where the coefficients C_1' and C_2' are fixed in terms of the starting conditions. It is seen that $|y_n|$ increases with n, thus contrasting with the behaviour of the exact solution.

If $|R_{qm}(\nu^2; \theta)| > 1$ a real number κ exists such that $d_1 = \sigma \exp(\kappa h)$ and $d_2 = \sigma \exp(-\kappa h)$ where σ is the sign of $R_{qm}(\nu^2; \theta)$, and then

$$y_n = \sigma^n [C_1 \exp(\kappa nh) + C_2 \exp(-\kappa nh)].$$

As before, $|y_n|$ increases indefinitely with n.

The numerical solution of the test equation is therefore either periodic or divergent and then the periodicity condition, which holds only if $|R_{qm}(\nu^2; \theta)| < 1$, acts as a stability condition. This explains why $R_{qm}(\nu^2; \theta)$ is called a stability function and also suggests adopting the following definition:

DEFINITION 5.13 *([11]) A region of stability is a region in the $\nu - \theta$ plane, throughout which $|R_{qm}(\nu^2; \theta)| < 1$. Any closed curve defined by $|R_{qm}(\nu^2; \theta)| = 1$ is a stability boundary.*

Equations of the stability boundary curves can be written without difficulty for $R_{11}(\nu^2; \theta)$ of the form (5.128). Relations $R_{11}(\nu^2; \theta) = \pm 1$ hold for

$$\nu = \nu_\pm := \sqrt{-\frac{a_1(\theta) \pm 2}{b_1(\theta) \pm 2b_0(\theta)}} \tag{5.134}$$

and also along the horizontal lines corresponding to any value of θ for which

$$a_1(\theta) \pm 2 = b_1(\theta) \pm 2b_0(\theta) = 0. \tag{5.135}$$

For the classical Numerov method S_0 we have $\nu_+ = 0$ and $\nu_- = \sqrt{6}$ and then the only region of stability is the vertical strip bounded by $\nu = 0$ and $\nu = \sqrt{6}$, see Figure 5.10. This result reflects directly the fact that the coefficients of this version are independent of θ, and it actually represents only an equivalent way of expressing the result which is normally obtained in the frame of the standard approach of Lambert and Watson. In contrast, the stability maps for the genuine ef-based versions deserve some specific comments.

For S_1 the stability boundaries are the vertical axis $\nu = \nu_+ = 0$ and the curves $\nu = \nu_-(\theta)$ which are asymptotic to lines of constant θ corresponding to the zeros of $b_1(\theta) - 2b_0(\theta)$; the first three of these are 4.058, 8.987 and 9.863, to three decimal places.

The existence of the critical values for θ leads to additional boundaries but these are of a special type. The critical values of S_1 are $\theta = 2n\pi$, $n =$

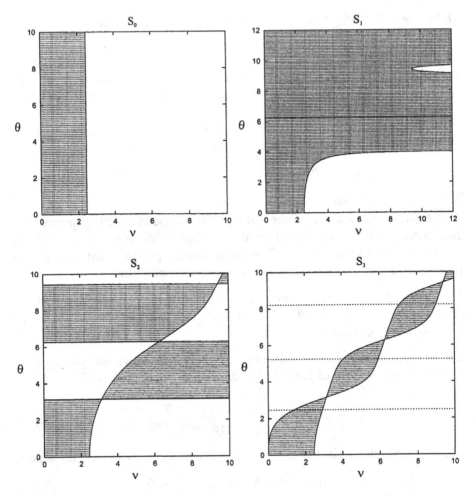

Figure 5.10. The stability maps for versions S_0, S_1, S_2 and S_3 of the Numerov method.

1, 2, ... At them the method coefficients are undetermined but, as shown in [11], $R_{11}(\nu^2; \theta)$ tends to 1 when $\theta \to 2n\pi$. Each horizontal line $\theta = 2n\pi$ is therefore a stability boundary, but, in contrast to the other boundaries, this is common for two neighbouring stability regions.

For S_2 the vertical axis $\nu = \nu_+ = 0$ is one of the boundaries but this is an effective border of a stability region only when $\theta \in (2n\pi, (2n + 1)\pi)$, $n = 0, 1, \ldots$ The other boundary, $\nu = \nu_-(\theta)$, is a continuous curve which oscillates about the line $\theta = \nu$. Horizontal boundaries also appear and these are of two types: the ones with $\theta = 2n\pi$, $n = 1, 2 \ldots$ are of the type (5.135) while those with $\theta = (2n - 1)\pi$ are related to the critical values. Indeed, $R_{11}(\nu^2; \theta)$ tends to -1 when $\theta \to (2n - 1)\pi$.

For S_3 the equations $\nu = \nu_\pm(\theta)$ define two curves which oscillate about the line $\theta = \nu$. The modulus of the stability function equals 1 also when $\theta = 2n\pi$ but these lines do not act as effective borders of any stability region. Interesting enough, the values of θ for which the coefficients of this version are undefined (broken lines on the figure) have no bearing on the stability properties.

The following example shows how these maps can be used for practical purposes. The equation

$$y'' = -k^2 y + (k^2 - \omega^2)\sin(\omega x), \ x \geq 0, \ y(0) = y_0, \ y'(0) = y'_0, \quad (5.136)$$

which is a slight generalization of a test equation used by Simos et al., [59], has the solution

$$y(x) = y_0 \cos(kx) + y'_0 \sin(kx)/k + \sin(\omega x).$$

When $y_0 = y'_0 = 0$, as we shall assume from now on, the solution is simply $y(x) = \sin(\omega x)$. For $\theta = \omega h$ any of the ef-based versions of the Numerov method will solve the problem exactly (except for round-off), provided this θ does not happen to be a critical value and also provided no stability problems are encountered. For fixed ω the stability depends on k and on h, and the method is stable if the point $(\nu = kh, \theta)$ is inside a stability region.

Table 5.8. Stability tests for the scheme S_2 on equation (5.136) with $\omega = 10$ and $y_0 = y'_0 = 0$: absolute error at $x = 80$ for fixed h and variable k.

	k	ν	$\Delta y(x)$
$h = 0.2, \ \theta = 2.0, \ \nu_-(\theta) = 2.708$			
	13.4	2.68	0.113(−13)
	13.5	2.70	0.325(−13)
	13.6	2.72	0.179(+09)
	13.7	2.74	0.248(+23)
$h = 0.5, \ \theta = 5.0, \ \nu_-(\theta) = 4.580$			
	9.0	4.50	0.123(+89)
	9.1	4.55	0.552(+41)
	9.2	4.60	−0.676(−14)
	9.3	4.65	0.210(−13)
$h = 0.8, \ \theta = 8.0, \ \nu_-(\theta) = 8.546$			
	10.5	8.40	0.169(−13)
	10.6	8.48	−0.154(−14)
	10.7	8.56	−0.171(−02)
	10.8	8.64	−0.835(+16)

Table 5.9. Stability tests for the scheme S_2 on equation (5.136) with $\omega = 10$ and $y_0 = y_0' = 0$: absolute error at $x = 200h$ for fixed k and variable h.

h	θ	ν	$\Delta y(x)$	h	θ	ν	$\Delta y(x)$
		$k = 5.0$				$k = 15.0$	
0.30	3.00	1.50	$-0.325(-14)$	0.16	1.60	2.40	$-0.218(-12)$
0.31	3.10	1.55	$0.182(-13)$	0.17	1.70	2.55	$0.904(-13)$
0.32	3.20	1.60	$0.156(+53)$	0.18	1.80	2.70	$0.169(+09)$
0.33	3.30	1.65	$0.740(+111)$	0.19	1.90	2.85	$-0.140(+28)$
0.55	5.50	2.75	$0.251(+01)$	0.30	3.00	4.50	$0.624(+25)$
0.60	6.00	3.00	$0.132(-10)$	0.31	3.10	4.65	$-0.547(+08)$
0.65	6.50	3:25	$-0.143(-12)$	0.32	3.20	4.80	$-0.210(-13)$
0.70	7.00	3.50	$-0.680(-13)$	0.33	3.30	4.95	$-0.137(-13)$
0.93	9.30	4.65	$-0.733(-12)$	0.55	5.50	8.25	$-0.149(-13)$
0.94	9.40	4.70	$-0.291(-12)$	0.60	6.00	9.00	$-0.383(-13)$
0.95	9.50	4.75	$-0.459(+145)$	0.65	6.50	9.75	$-0.139(-08)$
0.96	9.60	4.80	$0.210(+229)$	0.70	7.00	10.50	$0.317(+16)$

We select the scheme S_2, take $\omega = 10$ and consider two situations. Firstly, we keep h fixed and ask for the range in k such that S_2 satisfies the periodicity condition. We have to follow the horizontal line $\theta = \omega h$. This intersects the boundary curve $\nu_-(\theta)$, which for this method is

$$\nu_-(\theta) = \sqrt{\frac{\theta^3}{\theta - \sin(\theta)}},$$

at the point $(\nu_-(\omega h), \omega h)$. For $h = 0.2$ we have $\omega h = 2$ and $\nu_-(2) = 2.708$ to three decimal places. S_2 is then stable for all $k < \nu_-(2)/0.2 \approx 13.54$. It follows that the absolute error will be in the round-off range for values like $k = 13.4$ or 13.5 while significantly affected by instability for $k = 13.6$ or 13.7, a prediction nicely confirmed by the first set of data in Table 5.8. Results for $h = 0.5$ and 0.8 are also given in that table.

Secondly, we keep k fixed and ask for the values of h such that the periodicity condition is satisfied by S_2. We are actually interested in the cases when ω is either significantly smaller or significantly bigger than the chosen k. The ratio $r = \theta/\nu = \omega/k$ plays an important role. We have to follow the line $\theta = r\nu$ and to search for the values of ν at which this line intersects the boundary curves. When $\omega > k$, that is $r > 1$, only the horizontal boundaries $\theta = n\pi$, $n = 1, 2, \ldots$ are intercepted, see Figure 5.10. The method will be then stable only if $\nu \in (0, \pi/r)$ or if $\nu \in (2n\pi/r, (2n+1)\pi/r)$ which, expressed in terms of h, means that $h \in (0, \pi/\omega)$ or $h \in (2n\pi/\omega, (2n+1)\pi/\omega)$. In the first four columns of Table 5.9 we give numerical results for $k = 5$, that is $r = 2$, for values of h around π/ω, $2\pi/\omega$ and $3\pi/\omega$ to confirm the markedly different behaviour of the absolute error on the two sides with respect to each of the three reference steplengths. When $\omega < k$, that is $r < 1$, the first boundary curve intercepted by the line $\theta = r\nu$ is $\nu_-(\theta)$. The abscissa of the intersection point is the root of the equation $r\nu = \nu_-(r\nu)$. For $k = 15$ ($r = 2/3$) this is $\nu = 2.6487$ to four decimal places. A first interval in h where S_2 is stable is $(0, 2.6487/15 \approx 0.176)$, a result confirmed by the first block of data on the right in Table 5.9. Other intervals are delimited by the abscissas, divided by k, of the points where the line $\theta = r\nu$ is intersecting the horizontal boundaries; these intervals are $((2n-1)\pi/\omega, 2n\pi/\omega)$, $n = 1, 2, \ldots$. Experimental data around the ends of the first of them are also presented in that table.

The following definition, taken from [11], is consistent with the second way of looking at the things:

DEFINITION 5.14 *For a method with the stability function* $R_{qm}(\nu^2; \theta)$, *where* $\nu = kh$ *and* $\theta = \omega h$, *and* k *and* ω *are given, the primary interval of periodicity is the largest interval* $(0, h_0)$ *such that* $|R_{qm}(\nu^2; \theta)| < 1$ *for all steplengths* $h \in (0, h_0)$. *If, when* h_0 *is finite,* $|R_{qm}(\nu^2; \theta)| < 1$ *also for* $\gamma < h < \delta$, *where* $\gamma > h_0$, *then the interval* (γ, δ) *is a secondary interval of periodicity.*

P-stability, as defined for classical methods, ensures that the periodicity condition holds for all h, whatever the value of k in the test equation. The following definition extends this concept to the methods with θ-dependent coefficients:

DEFINITION 5.15 *([11]) A method with the stability function* $R_{qm}(\nu^2; \theta)$, *where* $\nu = kh$ *and* $\theta = \omega h$, *is P-stable if, for each value of* ω *the inequality* $|R_{qm}(\nu^2; \theta)| < 1$ *holds for all values of* k *and for all steplengths* h, *except possibly for a discrete set of exceptional values of* h *determined by the chosen value of* ω.

[The latter provision is meant to account for the critical values of θ, where the method is not defined, and also the stability boundaries.] Expressed in more intuitive terms, a method is P-stable if the whole stability map of it is shadowed, but the presence of a set of curves which correspond to critical values or to stability boundaries is also tolerated.

For equation (5.136) a P-stable method will be exact (except for round-off) without any serious restriction on k and on h, in contrast with what happens for the ef-based versions of the Numerov method. However, asking a method for (5.105) be P-stable is quite demanding (perhaps as much demanding as asking a method for (5.1) be A-stable) and only one such ef-based method is known. This is

$$y_{n+1} - 2y_n + y_{n-1} = h^2[\beta_1(\theta)(f_{n+1} + f_{n-1}) - 2(\gamma - \beta_1(\theta))f_n],$$

where γ is an arbitrary nonnegative constant and

$$\beta_1(\theta) = \frac{1 - \cos(\theta) + \gamma\theta^2}{\theta^2(1 + \cos(\theta))}, \tag{5.137}$$

see [11]. Its order is two.

However, some authors concluded that their ef-based methods are also P-stable; see, e.g., Jain et al., [37] and [38], Anantha Krishnaiah, [2], Denk, [16] and Thomas et al., [61]. The main reason is that in their investigations ω and k were assumed equal, and then only the situation along the line $\theta = \nu$ has been examined. However, since any ef-based method is phase-fitted, such a method is exact along the mentioned line and stability is not an issue. In other words, the examination only along the line $\theta = \nu$ tells nothing about what happens to either side of that line, as it would be relevant to draw conclusions on the stability.

All the above considerations referred to the methods with one stability function, $R_{qm}(\nu^2; \theta)$, but not all existing methods are of this type; the four-step methods provide such an example. The extension of this theory to methods with the stepnumber greater than 2 was also considered in [11].

2.3   Symmetric four-step methods

A symmetric four-step method has the form

$$y_{n+2} + a_1 y_{n+1} + a_2 y_n + a_1 y_{n-1} + y_{n-2}$$
$$= h^2(b_0 f_{n+2} + b_1 f_{n+1} + b_2 f_n + b_1 f_{n-1} + b_0 f_{n-2}). \tag{5.138}$$

The corresponding operator $\mathcal{L}$ is defined through

$$\mathcal{L}[h, \mathbf{a}]y(x) := y(x + 2h) + a_1 y(x + h) + a_2 y(x) + a_1 y(x - h) + y(x - 2h)$$

$$-h^2[b_0 y''(x + 2h) + b_1 y''(x + h) + b_2 y''(x) + b_1 y''(x - h) + b_0 y''(x - 2h)],$$

where $\mathbf{a} = [a_1, a_2, b_0, b_1, b_2]$ is a set with five components.
For the calculation of the coefficients we need the expressions of the starred classical moments, of the functions $G^{\pm}(Z, \mathbf{a})$ and of their derivatives with respect to $Z = \mu^2 h^2$, see the six-point flow chart in Chapter 3. These expressions are:

$$L_{2k}^*(\mathbf{a}) = 2^{2k+1} + 2a_1 + a_2\delta_{k0} - 2k(2k-1)(2^{2k-1}b_0 + 2b_1 + b_2\delta_{k1}),$$

$$L_{2k+1}^*(\mathbf{a}) = 0, \ k = 0, 1, 2, \ldots,$$

$$G^+(Z, \mathbf{a}) = 2\eta_{-1}(4Z) + 2a_1\eta_{-1}(Z) + a_2 - Z[2b_0\eta_{-1}(4Z) + 2b_1\eta_{-1}(Z) + b_2],$$

$$G^{+(1)}(Z, \mathbf{a}) = 4\eta_0(4Z) + a_1\eta_0(Z) - 2b_0[\eta_{-1}(4Z) + 2Z\eta_0(4Z)]$$

$$-b_1[2\eta_{-1}(Z) + Z\eta_0(Z)] - b_2,$$

$$G^{+(p)}(Z, \mathbf{a}) = 2^{p+1}\eta_{p-1}(4Z) + 2^{1-p}a_1\eta_{p-1}(Z) - 2^{p-1}b_0[3\eta_{p-2}(4Z) + \eta_{p-3}(4Z)]$$

$$-2^{1-p}b_1[3\eta_{p-2}(Z) + \eta_{p-3}(Z)], \ p = 2, 3, 4, \ldots,$$

$$G^-(Z, \mathbf{a}) = 0, \ G^{-(p)}(Z, \mathbf{a}) = 0, \ p = 0, 1, 2, \ldots$$

Also following the flow chart, the classical coefficients result by solving the system $L_{2k}^*(\mathbf{a}) = 0$ after taking as many possible successive k from 0 onwards such that the system is still compatible. In our case this would mean to take $0 \le k \le 4$. This linear system of five equations for five unknowns is compatible indeed, but there is a problem of a different nature: the method with the coefficients fixed in this way is not zero-stable and then it is not convergent. Although such a method cannot be used for the propagation of the solution the knowledge of its coefficients is still useful on some occasions. For example, the value of y_{n+2} produced by it is accurate enough to be taken for reference in the estimation of the local error, if this is needed; this formula is $S(q)$ for $q = 159/31$, see below.

A simple technique for the evaluation of the coefficients of the convergent versions consists in taking one of these as a free parameter and then calculating the other coefficients in terms of it; the values or the range of values of this parameter for which the method is zero-stable are subsequently selected in terms of the roots of the characteristic equation.

We introduce the parameter q and define $a_2 = -2q$ (this is the notation used in [29]). For the classical methods the other four coefficients result by solving the system $L_{2k}^*(\mathbf{a}) = 0$, $0 \le k \le 3$, and let the version with these coefficients be denoted $S(q)$. The coefficients of $S(q)$ are:

$$a_1(q) = q - 1, \ a_2(q) = -2q, \ b_0(q) = \frac{17 - q}{240},$$

$$b_1(q) = \frac{29 + 3q}{30}, \ b_2(q) = \frac{111 + 97q}{120},$$

and the *lte* is

$$lte = \frac{31q - 159}{60480}h^8 y^{(8)}(x_n).$$

$S(q)$ is zero-stable only if $-3 < q \le 1$, see [22]. The versions $S(0)$ and $S(1)$ are given in [22] with a different notation, see formulae (6-76) and (6-22) in that book.

With the notations introduced in the six-step flow chart $S(q)$ has $M = 8$ and then four exponential fitting extensions of it, depending on one parameter Z, exist: (*i*) $P = 0$, $K = 5$, (*ii*) $P = 1$, $K = 3$, (*iii*) $P = 2$, $K = 1$, and (*iv*) $P = 3$, $K = -1$. Any such extension will be abbreviated $S(q, Z)$. Raptis constructed the version $S(1, Z)$ for the choice (*i*) in [48], and for the choice (*iv*) in [46]. Some properties of these, which are pertinent for the solution of the Schrödinger equation, are discussed in [26]. The expressions of the coefficients of $S(0, Z)$ for all four choices, and of $S(1, Z)$ for choice (iv) are given by Simos in [55] and in [53], respectively. In [33], Ixaru et al. considered an exponential fitting extension with four parameters Z_i; its coefficients are given by solving the system

$$G^+(Z_i, \mathbf{a}) = 0, \; i = 1, 2, 3, 4. \tag{5.139}$$

The main intention of that paper was to use this version, denoted $S(q, Z)$ where $\mathbf{Z} = [Z_1, Z_2, Z_3, Z_4]$, for the solution of nonlinear systems of second-order equations and a particular attention has been devoted to the stability properties when this is used in the predictor-corrector mode. The paper also contains a proposal for the automatic choice of the frequencies at each step. A computer code based on this version, called EXPFIT4, is published in [34].

2.3.1   Subroutine EF4STEP

The expressions of the coefficients of $S(q, Z)$ for each of the choices (i)–(iv), and of $S(q, \mathbf{Z})$ are rather long but the computation of these coefficients becomes easy if the subroutine REGSOLV2 is used with versions of subroutines INFF and INFG adapted to the system (5.139).

All these adaptations are gathered in the subroutine EF4STEP which, for given values of q and of the four components of vector $\mathbf{Z}$, returns the values of a_1, a_2, b_0, b_1 and b_2. To compute the classical coefficients the components of $\mathbf{Z}$ are simply $Z_i = 0$, $(1 \leq i \leq 4)$ but for $S(q, \mathbf{Z})$ these components depend on the value of P for the considered choice, viz.: $Z_i = Z$ $(1 \leq i \leq P + 1)$ and $Z_i = 0$ $(P + 2 \leq i \leq 4)$. For example, the choice (ii) has $P = 1$ and then $Z_1 = Z_2 = Z$ and $Z_3 = Z_4 = 0$.

Calling this subroutine is

$$\text{CALL   EF4STEP}(Q, Z, A1, A2, B0, B1, B2)$$

The arguments are:
Input:

- Q - double precision value of q.

- Z - double precision vector with four components for Z_1, Z_2, Z_3 and Z_4.

Output:

- A1, A2, B0, B1, B2 - resultant values of the coefficients (double precision).

The following driving program serves for the computation of the coefficients of $S(q = 1, Z = -1)$ for $P = 1$, and of $S(q = 0, \mathbf{Z})$ with $\mathbf{Z} = [0, -1/2, -1, -1]$. In the program they are identified by j=1 and j=2, respectively.

```
    implicit double precision(a-h,o-z)
    dimension z1(4),z2(4)
    data z1/-1.d0,-1.d0,0.d0,0.d0/
    data z2/0.d0,-.5d0,-1.d0,-1.d0/
    do j=1,2
      if(j.eq.1)then
        q=1
        call ef4step(q,z1,a1,a2,b0,b1,b2)
      else
        q=0
        call ef4step(q,z2,a1,a2,b0,b1,b2)
      endif
      write(*,50)j,a1,a2,b0,b1,b2
    enddo
50  format(2x,'j=',i2,': a1=',f9.6,' a2=',f9.6,/,
   !7x,' b0=',f9.6,' b1=',f9.6,' b2=',f9.6,/)
    stop
    end
```

The output is:

```
j= 1: a1= 0.000000  a2=-2.000000
         b0= 0.071140   b1= 1.050926  b2= 1.755868

j= 2: a1=-1.000000  a2= 0.000000
         b0= 0.078156   b1= 0.943159  b2= 0.958765
```

2.4   Other methods

There exist multistep methods which consist in a set of formulae and many such hybrid algorithms have been investigated in the spirit of the exponential fitting. Some examples follow.

Cash et al., [5], consider a method with three formulae. This is :

$$y_{n+1/2} + y_{n-1/2} = A_1(y_{n+1} + y_{n-1}) + A_2 y_n + h^2[A_3(y''_{n+1} + y''_{n-1}) + A_4 y''_n],$$

$$y_{n-1/2} = B_1(y_{n+1} + y_{n-2}) + B_2(y_n + y_{n-1})$$

$$+ h^2[B_3(y''_{n+1} + y''_{n-2}) + B_4(y''_n + y''_{n-1})],$$

$$y_{n+1} + a y_n + y_{n-1} = h^2[\beta_0(y''_{n+1} + y''_{n-1}) + \beta_1(y''_{n+1/2} + y''_{n-1/2}) + \gamma y''_n],$$

where for practical application y''_{n+s} must be replaced by $f(x_{n+s}, y_{n+s})$ for all s involved. The method allows the computation of y_{n+1} in terms of y_{n-2}, y_{n-1} and y_n by using the ad-hoc approximations $y_{n\pm1/2}$ as intermediary data.

The coefficients are determined by imposing the condition that each of the three formulae is exact for the set

$$1, \; x, \; x^2, \; x^3, \; \exp(\pm\mu x), \; x\exp(\pm\mu x). \tag{5.140}$$

Their expressions are written as functions of the product μh and the expansions of these in powers of $\mu^2 h^2$ are also given. The latter are needed for the evaluation of the coefficients when $|\mu h|$ is small. The method is of order six.

The same method is also considered by Simos, [56], for the reference set

$$1, \; x, \; \exp(\pm\mu x), \; x\exp(\pm\mu x), x^2\exp(\pm\mu x)$$

instead of (5.140).

In [57] Simos considers the following two-step hybrid algorithm:

$$\bar{y}_n = y_n - ah^2(f_{n+1} - 2f_n + f_{n-1}),$$

$$y_{n+1} - 2y_n + y_{n-1} = h^2(f_{n+1} + 10\bar{f}_n + f_{n-1})/12.$$

To determine a the phase-fitting technique is used. Specifically, when applied on the test equation $y'' = -k^2 y$ this becomes of the form (5.127) with

$$R_{22}(\nu^2; \theta) = \frac{1 - 5\nu^2/12 + 5a\nu^4/6}{1 + \nu^2/12 + 5a\nu^4/6},$$

and then the phase-fitting condition $R_{22}(\theta^2; \theta) = \cos(\theta)$ gives

$$a(\theta) = \frac{-\theta^2[5 + \cos(\theta)] + 12[1 - \cos(\theta)]}{-10\theta^4[1 - \cos(\theta)]}.$$

Hybrid four-step methods have been also investigated in several papers, see, e.g., [51] and [58].

The stability of the hybrid methods received attention in a series of papers and some methods, in particular the ones in which the phase-fitting was involved, were reported as P-stable although only the situation along the line $\theta = \nu$ has been considered for this purpose; see also the comment after equation (5.137).

In contrast, Ixaru and Paternoster [32] have analysed the exponential fitting version of a hybrid method with full awareness of the definition 5.13. They considered Chawla's algorithm with two formulae:

$$\bar{y}_{n\pm s} = a_0 y_n + a_1 y_{n\pm s} + a_{-1} y_{n\mp s} + h^2(b_0 f_n + b_1 f_{n\pm 1} + b_{-1} f_{n\mp s}),$$

$$y_{n+1} - 2y_n + y_{n-1} = h^2[2A_0 f_n + A_1(f_{n+1} + f_{n-1}) + A_s(\bar{f}_{n+s} + \bar{f}_{n-s})],$$

where four coefficients, that is, $s \in (0, 1)$, a_0, b_0 and A_0, are kept free and the other ones are given by

$$a_{\pm 1} = \frac{1 - a_0 \pm s}{12}, \quad b_{\pm 1} = \frac{-1 + a_0 - 2b_0 + s^2 \pm s(s^2 - 1)/3}{4},$$

$$A_1 = \frac{6s^2 - 1 - 12s^2 A_0}{12(s^2 - 1)}, \quad A_s = \frac{-5 + 12A_0}{12(s^2 - 1)}.$$

see [6]. The algorithm has the order four and for some values of the free parameters it is P-stable in the sense of definition 5.10. In particular, for $a_0 = b_0 = A_0 = 0$, it is P-stable irrespective of $s \in (0, 1)$. By examining the exponential fitting version of this algorithm authors of [32] have found that the stability region is a strip in the (ν, θ)-plane bounded by the horizontal lines $\theta = 0$ and $\theta = \theta_{max} > 0$, and called this behaviour as a conditional P-stability. For $s = 1/2$, $a_0 = b_0 = 0$ and $A_0(\theta) = -0.06134514(\theta - \pi/2)^2 + 0.40525$ they obtained $\theta_{max} \approx 3.4$.

There exist also multistep methods which compute both the solution and its first derivative. Examples are the methods of de Vogelaere, [17], and of Coleman, [7]. The latter has the algorithm

$$y_{2n+1} = y_{2n} + hy'_{2n} + h^2(7f_{2n} + 6f_{2n+1} - f_{2n+2})/24,$$

$$y_{2n+2} = y_{2n} + 2hy'_{2n} + 2h^2(f_{2n} + 2f_{2n+1})/3,$$

$$y'_{2n+2} = y'_{2n} + h(f_{2n} + 4f_{2n+1} + f_{2n+2})/3,$$

such that it allows computing y_{2n+2} and y'_{2n+2} in terms of y_{2n} and y'_{2n}. The method is implicit but it becomes easy to apply if $f(x, y)$ is linear in y, as for the Schrödinger equation. This is because y_{2n+1} and y_{2n+2} can be readily explicited from the first two equations. For the exponential fitting version of this method see [27].

2.5   Function fitting methods

Here we describe two methods for the radial Schrödinger equation $y'' = (V(x) - E)y$ which exploit the specific form of the potential function $V(x)$ for x close to the origin, and in the asymptotic region, respectively.

In many physical problems $V(x)$ is of the form

$$V(x) = l(l+1)/x^2 + g_1(x)/x + g_2(x),$$

where l (the orbital quantum number) is a nonnegative integer, and $g_1(x)$ and $g_2(x)$ are well-behaved functions; a possible physical significance of the term $g_1(x)/x$ is that it represents a screened Coulomb potential.

For the first method let us concentrate on the eigenvalue problem in a shooting process with backwards integration by the Numerov method at each trial value of E. Let the domain be $[a = 0, b = Nh]$ with given N and h. For given E and suitable values of y_N and y_{N-1} the Numerov method

$$y_{n+1} + a_1 y_n + y_{n-1} = h^2[b_0(f_{n+1} + f_{n-1}) + b_1 f_n]$$

(see Section 2.1), where $f_n = (V(nh) - E)y_n$, is applied successively at $n = N-1, N-2, \ldots$, and the calculation of an eigenvalue means determining that E for which $y(x)$ is regular at the origin, that is $y(0) = 0$.

The problem is that such an algorithm cannot be used at $n = 1$ because this involves $V(0)$, which is infinite. However, since the values of $y_1, y_2, \ldots$ are available, it makes sense to search for an expression based on these values, which can be used as a convenient substitute for $y(0)$ in the shooting procedure.

This problem is discussed by Ixaru in [24]. The approach relies on the fact that the regular solution of the radial Schrödinger equation with $V(x)$ of the indicated form can be written as a series,

$$y(x) = \sum_{i=0}^{\infty} c_i x^{l+i+1}.$$

The form of the expression searched for is

$$P(y_1, y_2, y_3) := [1 - h^2 \beta_1 (V(h) - E)]y_1$$
$$+ [\alpha_2 - h^2 \beta_2 (V(2h) - E)]y_2 + [\alpha_3 - h^2 \beta_3 (V(3h) - E)]y_3,$$

where the five coefficients α_i and β_i have to be determined. To this aim the operator $\mathcal{L}$ is introduced by

$$\mathcal{L}[h, a]y(x) := y(h) + \alpha_2 y(2h) + \alpha_3 y(3h) - h^2 [\beta_1 y''(h) + \beta_2 y''(2h) + \beta_3 y''(3h)],$$

where $a = [\alpha_2, \alpha_3, \beta_1, \beta_2, \beta_3]$, and it is required that this vanishes for the functions

$$y(x) = x^m, \quad m = l+1, l+2, \ldots, l+5.$$

Notice that the first function in this set is x^{l+1}. The set is then lacunary and for this reason the six-step flow chart cannot be applied; see note 1 in Section 3 of Chapter 3. However, it can be easily found that $\mathcal{L}[h, a]x^m = h^m L_m^*(a)$ where

$$L_m^*(a) = 1 + 2^m \alpha_2 + 3^m \alpha_3 - m(m-1)(\beta_1 + 2^{m-2}\beta_3 + 3^{m-2}\beta_3).$$

The stated conditions lead to the linear system $L_m^*(a) = 0$, $l+1 \leq m \leq l+5$, with the solution

$$\alpha_i = A_i/Q \;\; (i = 2, 3), \quad \beta_i = B_i/Q \;\; (i = 1, 2, 3),$$

where

$$
\begin{aligned}
A_2 &= 2^{-(l+1)}(4l^4 + 28l^3 - l^2 - 217l + 240), \\
A_3 &= 3^{-(l+2)}(2l^4 + 41l^3 + 232l^2 + 301l - 540), \\
B_1 &= 2l^2 + 15l - 5, \\
B_2 &= 2^{1-l}(4l^4 + 24l - 25), \\
B_3 &= 3^{-(l+1)}(6l^2 + 9l - 15), \\
Q &= 2l^4 + 5l^3 - 68l^2 + 85l - 60.
\end{aligned}
\tag{5.141}
$$

The expression $P(y_1, y_2, y_3)$ with the l-dependent coefficients just determined represents the optimal three-point formula for checking whether the solution is regular. More exactly, the eigenvalues are those E for which $P(y_1, y_2, y_3) = 0$. Other expressions can also be derived in a similar way. If, for example, the values $l = 0$ and $\alpha_3 = 0$ are taken by default then the system to be solved is $L_m^*(\mathbf{a}) = 0$, $1 \le m \le 4$. It leads to the expression

$$D(y_1, y_2, y_3) = -[2 + 13h^2(V(h) - E)/12]y_1$$

$$+[1 + h^2(V(2h) - E)/6]y_2 - h^2(V(3h) - E)y_3/12,$$

which has been first obtained by Buendia and Guardiola [3] by a different technique. For the error analysis of such expressions see [24].

At big x the physically relevant expressions of $g_1(x)$ and $g_2(x)$ are of forms which tend rapidly to some constant values $\bar{g}_1$ and $\bar{g}_2$, respectively. As a rule one has $\bar{g}_2 = 0$ but the value of $\bar{g}_1$ depends on the physical context. However, in some cases one has $\bar{g}_1 = 0$, as well. If so, only the centrifugal term $l(l+1)/x^2$ survives in $V(x)$ and then the Schrödinger equation reads

$$y'' = [l(l+1)/x^2 - E]y.$$

Let $E > 0$ and denote $k = E^{1/2}$. The two linearly independent solutions of this equation are $kxj_l(kx)$ and $kx\eta_l(kx)$, where j_l and η_l are the spherical Bessel and Neumann functions, respectively; see [1].

In [50] Raptis and Cash consider a two-step method of the form

$$y_{n+1} - 2y_n + y_{n-1} = h^2(\beta_2 f_{n+1} + \beta_1 f_n + \beta_0 f_{n-1}),$$

whose associated operator is defined by

$$\mathcal{L}[h, \mathbf{a}]y(x) := y(x+h) - 2y(x) + y(x-h) - h^2[\beta_2 y''(x+h) + \beta_1 y''(x) + \beta_0 y''(x-h)].$$

They require that this identically vanishes in x and in h for $y(x) = 1$, x and x^2, and also that $\mathcal{L}[h, \mathbf{a}]y(x_n) = 0$ for $y(x) = kxj_l(kx)$ and $kx\eta_l(kx)$. The first two conditions, that is $\mathcal{L}[h, \mathbf{a}]1 = \mathcal{L}[h, \mathbf{a}]x = 0$, are automatically satisfied but the other three lead to a linear system of equations for the unknown coefficients. With

$$F_j = h^2[l(l+1)/x_{n+j-1}^2 - E],$$

$$J_j = kx_{n+j-1}j_l(kx_{n+j-1}),\ Y_j = kx_{n+j-1}\eta_l(kx_{n+j-1})\ (j = 0, 1, 2),$$

$$R_{i,j} = x_{n-i+1}x_{n-j+1}[j_l(kx_{n+i-1})\eta_l(kx_{n+j-1}) - j_l(kx_{n+j-1})\eta_l(kx_{n+i-1})],$$

$$(i, j = 0, 1, 2),$$

$$A = F_1 F_2 R_{2,1} - F_1 F_0 R_{1,0} - F_0 F_2 R_{2,0},$$

the solution is

$$\beta_0 = [(F_1 F_2 - F_1 + 2F_2)R_{2,1} + F_1 R_{1,0} - F_2 R_{2,0}]/A \,,$$

$$\beta_1 = [(-F_0 F_2 + F_0 + F_2)R_{2,0} - 2F_2 R_{2,1} - 2F_0 R_{1,0}]/A \,,$$

$$\beta_2 = [(F_1 F_0 - F_1 + 2F_0)R_{1,0} + F_1 R_{2,1} - F_0 R_{2,0}]/A \,.$$

Notice that these coefficients depend on x_n and thus they must be updated at each n.

When $\bar{g}_1 \neq 0$ the form of the Schrödinger equation in the asymptotic region becomes

$$y'' = [l(l+1)/x^2 + \bar{g}_1/x - E]y \,.$$

Its linearly independent solutions are the Coulomb functions; see, e.g., [1]. Multistep methods suited to this case can be constructed in the same way as above.

References

[1] Abramowitz, M. and Stegun, I. (1964). *Handbook of Mathematical Functions*. Nat. Bur. Stand. Appl. Math. Ser. No. 55, U.S. Govt. Printing Office, Washington, D.C.

[2] Anantha Krishnaiah, U. (1982). Adaptive methods for periodic initial value problems of second order differential equations. *J. Comput. Appl. Math.*, 8: 101–104.

[3] Buendia, E. and Guardiola, R. (1985). The Numerov method and singular potentials. *J. Comput. Phys.*, 60: 561–564.

[4] Butcher, J. C. (2003). *Numerical Methods for Ordinary Differential Equations*. Wiley, New York.

[5] Cash, J. R. , Raptis, A. D. and Simos, T. E. (1990). A sixth-order exponentially fitted numerical solution of the radial Schrödinger equation. *J. Comput. Phys.*, 91: 413–423.

[6] Chawla, M. M. (1981). Two-step P-stable methods for second-order differential equations. *BIT*, 21: 190–193.

[7] Coleman, J. P. (1980). A new fourth-order method for $y'' = g(x)y + r(x)$. *Comput. Phys. Commun.*, 19: 185–195.

[8] Coleman, J. P. (2003). Private communication.

[9] Coleman, J. P. (1989). Numerical methods for $y'' = f(x, y)$ via rational approximations for the cosine. *IMA J. Numer. Anal.*, 9: 145–165.

[10] Coleman, J. P. and Duxbury, S. C. (2000). Mixed collocation methods for $y'' = f(x,y)$. *J. Comp. Appl. Math.*, 126: 47–75.

[11] Coleman, J. P. and Ixaru, L. Gr. (1996). P-stability and exponential-fitting methods for $y'' = f(x, y)$. *IMA J. Numer. Anal.*, 16: 179–199.

[12] Cowell, P. H. and Crommelin, A. C. D. (1910). Investigation of the motion of Harley's comet from 1729 to 1910. In *Appendix to Greenwich Observations for 1909*. Edinburgh.

[13] Cryer, C. W. (1973). A new class of highly stable methods: A_0-stable methods. *BIT*, 13: 153–159.

[14] Dahlquist, G. (1956). Convergence and stability in the numerical integration of ordinary differential equations. *Math. Scand.*, 4: 33–53.

[15] Dahlquist, G. (1963). A special stability problem for linear multistep methods. *BIT*, 3: 27–43.

[16] Denk, G. (1993). A new numerical method for the integration of highly oscillatory second-order ordinary differential equations. *Appl. Numer. Math.*, 13: 57–67.

[17] de Vogelaere, R. (1955). A method for the numerical integration of differential equations of second order without explicit derivatives. *J. Res. Nat. Bur. Standards*, 54: 119–125.

[18] Gear, C. W. and Tu, K. W. (1974). The effect of the variable mesh size on the stability of the multistep methods. *SIAM J. Num. Anal.*, 11: 1025–1043.

[19] Grigorieff, R. (1983). Stability of multistep methods on variable grids. *Numer. Math.*, 42: 359–377.

[20] Hairer, E. , Nørsett, S. P. and Wanner, G. (1987). *Solving Ordinary Differential Equations I, Non-stiff Problems*. Springer, Berlin.

[21] Hairer, E. and Wanner, G. (1991). *Solving Ordinary Differential Equations II, Stiff and Differential–algebraic Problems*. Springer, Berlin.

[22] Henrici, P. (1962). *Discrete Variable Methods in Ordinary Differential Equations*. Wiley, New York.

[23] Ixaru, L. Gr. (1984). *Numerical Methods for Differential Equations and Applications*. Reidel, Dordrecht - Boston - Lancaster.

[24] Ixaru, L. Gr. (1987). The Numerov method and singular potentials. *J. Comput. Phys.*, 72: 270–274.

[25] Ixaru, L. Gr. and Rizea, M. (1980). A Numerov - like scheme for the numerical solution of the Schroedinger equation in the deep continuum spectrum of energies. *Comput. Phys. Commun.*, 19: 23 - 27.

[26] Ixaru, L. Gr. and Rizea, M. (1985). Comparison of some four-step methods for the numerical integration of the Schroedinger equation. *Comput. Phys. Commun.*, 38: 329–337.

[27] Ixaru, L. Gr. and Berceanu, S. (1987). Coleman's method maximally adapted to the Schrödinger equation. *Comput. Phys. Commun.*, 44: 14–20.

[28] Ixaru, L. Gr. and Rizea, M. (1987). Numerov method maximally adapted to the Schrödinger equation. *J. Comput. Phys.*, 73: 306–324.

[29] Ixaru, L. Gr. and Rizea, M. (1997). Four step methods for $y'' = f(x, y)$. *J. Comput. Appl. Math.*, 79: 87–99.

[30] Ixaru, L. Gr. (1997). Operations on oscillatory functions. *Comput. Phys. Commun.*, 105: 1–19.

[31] Ixaru, L. Gr. (2001). Numerical operations on oscillatory functions. *Computers and Chemistry*, 25: 39–53.

[32] Ixaru, L. Gr. and Paternoster, B. (1999). A conditionally P-stable fourth-order exponential-fitting method for $y'' = f(x, y)$. *J. Comput. Appl. Math.*, 106: 87–98.

[33] Ixaru, L. Gr. , Vanden Berghe, G. , De Meyer, H. and Van Daele, M. (1997). Four step exponential fitted methods for nonlinear physical problems. *Comput. Phys. Commun.*, 100: 56–70.

[34] Ixaru, L. Gr. , Vanden Berghe, G. , De Meyer, H. and Van Daele, M. (1977). EXPFIT4 –A FORTRAN program for the numerical solution of systems of nonlinear second order initial value problems. *Comput. Phys. Commun.*, 100: 71–80.

[35] Ixaru, L. Gr. , Vanden Berghe, G. and De Meyer, H. (2002). Frequency evaluation in exponential mutistep algorithms for ODEs. *J. Comput. Appl. Math.*, 140: 423–434.

[36] Ixaru, L. Gr. , Vanden Berghe, G. and De Meyer, H. (2003). Exponentially fitted variable two-step BDF algorithm for first order ODE. *Comput. Phys. Commun.*, 150: 116–128.

[37] Jain, M. K. , Jain, R. K. and Anantha Krishnaiah, U. (1979). P-stable methods for periodic initial-value problems of second order differential equations. *BIT*, 19: 347–355.

[38] Jain, M. K. , Jain, R. K. and Anantha Krishnaiah, U. (1979). P-stable single step methods for periodic initial-value problems involving second-order differential equations. *J. Eng. Math.*, 13: 317–326.

[39] Lambert, J. D. and Watson, I. A. (1976). Symmetric multistep methods for periodic initial-value problems. *J. Inst. Math. Applic.*, 18: 189–202.

[40] Lambert, J. D. (1991). *Numerical Methods for Ordinary Differential Equations*. Wiley, New York.

[41] Liniger, W. , Willoughby, R. A. (1970). Efficient integration methods for stiff systems of ordinary differential equations. *SIAM J. Numer. Anal.*, 7: 47–66.

[42] Lioen, W. M. and de Swart, J. J. B. (1999). Test set for initial value solvers. Available on internet at http://hilbert.dm.uniba.it/~testset/

[43] Lyche, T. (1974). Chebyshevian multistep methods for ordinary differential equations. *Numer. Math.*, 19: 65–75.

[44] Noumerov, B. V. (1924). A method of extrapolation of perturbations. *Roy. Ast. Soc. Monthly Notices*, 84: 592–601.

[45] Numerov,, B. (1927). Note on the numerical integration of $d^2x/dt^2 = f(x, t)$. *Astron. Nachrichten*, 230: 359–364.

[46] Raptis, A. D. (1981). On the numerical solution of the Schrödinger equation. *Comput. Phys. Commun.*, 24: 1–4.

[47] Raptis, A. D. (1982). Two-step methods for the numerical solution of the Schrödinger equation. *Computing*, 28: 373–378.

[48] Raptis, A. D. (1983). Exponentially-fitted solutions of the eigenvalue Schrödinger equation with automatic error control. *Comput. Phys. Commun.*, 28: 427–431.

[49] Raptis, A. D. and Allison, A. C. (1978). Exponential-fitting methods for the numerical solution of the Schrödinger equation. *Comput. Phys. Commun.*, 44: 95–103.

[50] Raptis, A. D. and Cash, J. R. (1987). Exponential and Bessel fitting methods for the numerical solution of the Schrödinger equation. *Comput. Phys. Commun.*, 44: 95–103.

[51] Raptis, A. D. and Simos, T. E. (1991). A four-step phase-fitted method for the numerical integration of second-order initial-value problems. *BIT*, 31: 160–168.

[52] Shampine, L. F. (1994). *Numerical Solution of Ordinary Differential Equations*. Chapman and Hall, New York.

[53] Simos, T. E. (1990). A four-step method for the numerical solution of the Schrödinger equation. *J. Comput. Appl. Math.*, 30: 251–255.

[54] Simos, T. E. (1991). A two-step method with phase-lag of order infinity for the numerical integration of second order periodic initial-value problems. *Int. J. Comput. Math.*, 39: 135–140.

[55] Simos, T. E. (1991). Some new four-step exponential-fitting methods for the numerical solution of the radial Schrödinger equation. *IMA J. Numer. Anal.*, 11: 347–356.

[56] Simos, T. E. (1992). Exponential fitted methods for the numerical integration of the Schrödinger equation. *Comput. Phys. Commun.*, 71: 32–38.

[57] Simos, T. E. (1992). Two-step almost P-stable complete in phase methods for the numerical integration of second order periodic initial-value problems. *Int. J. Comput. Math.*, 46: 77–85.

[58] Simos, T. E. (1994). An explicit four-step phase-fitted method for the numerical integration of second-order initial-value problems. *J. Comput. Appl. Math.*, 55: 125–133.

[59] Simos, T. E. , Dimas, E. and Sideridis, A. B. (1994). A Runge-Kutta-Nyström method for the numerical integration of special second-order periodic initial-value problems. *J. Comput. Appl. Math.*, 51: 317–326.

[60] Stiefel, E. and Bettis, D. G. (1969). Stabilization of Cowell's method. *Numer. Math.*, 13: 154–175.

[61] Thomas, R. M. , Simos, T. E. and Mitsou, G. V. (1996). A family of Numerov-type exponentially fitted predictor-corrector methods for the numerical integration of the radial Schrödinger equation. *J. Comput. Appl. Math.*, 67: 255–270.

[62] Vanden Berghe, G. , De Meyer, H. and Vanthournout, J. (1990). A modified Numerov integration method for second order periodic initial-value problems. *Int. J. Comput. Math.*, 32: 233–242.

[63] Van der Houwen, P. J. and Sommeijer, B. P. (1987). Explicit Runge-Kutta(-Nyström) methods with reduced phase errors for computing oscillating solutions. *SIAM J. Numer. Anal.*, 24: 595–517.

[64] Widlund, O. B. (1967). A note on unconditionally stable linear multistep methods. *BIT*, 7: 65–70.

[65] Wolfram, S. (1992). *MATHEMATICA   A system for doing mathematics by computer.* Addison–Wesley, Reading.

Chapter 6

RUNGE-KUTTA SOLVERS FOR
ORDINARY DIFFERENTIAL EQUATIONS

Since the original papers of Runge [24] and Kutta [17] a great number of papers and books have been devoted to the properties of Runge-Kutta methods. Reviews of this material can be found in [4], [5], [12], [18]. Kutta [17] formulated the general scheme of what is now called a Runge-Kutta method.

1. Formalism and construction scheme for classical Runge-Kutta methods

1.1 General formulation of explicit Runge-Kutta methods

DEFINITION 6.1 *Let s be an integer (the number of stages) and a_{21}, a_{31}, a_{32}, ..., a_{s1}, a_{s2}, ..., $a_{s,s-1}$, $b_1, b_2, \ldots, b_s, c_2, \ldots, c_s$ real coefficients. Then the method ($n = 0, 1, \ldots$)*

$$
\begin{aligned}
k_1 &= f(x_n, y_n)\,, \\
k_2 &= f(x_n + c_2 h, y_n + h a_{21} k_1)\,, \\
k_3 &= f(x_n + c_3 h, y_n + h(a_{31} k_1 + a_{32} k_2))\,, \qquad (6.1) \\
k_4 &= f(x_n + c_4 h, y_n + h(a_{41} k_1 + a_{42} k_2 + a_{43} k_3))\,, \\
&\quad \cdots \\
k_s &= f(x_n + c_s h, y_n + h(a_{s1} k_1 + \ldots + a_{s\,s-1} k_{s-1}))\,, \\
y_{n+1} &= y_n + h(b_1 k_1 + \ldots + b_s k_s)\,,
\end{aligned}
$$

is called an s-stage explicit Runge-Kutta method (ERK) for the initial value problem

$$
y' = f(x, y), \quad y(x_0) = y_0\,. \qquad (6.2)
$$

Usually, the c_i satisfy the *row-sum condition*

$$c_i = \sum_j a_{ij} . \tag{6.3}$$

DEFINITION 6.2 *A Runge-Kutta method (6.1) for sufficiently smooth problems (6.2) has (algebraic) order p if the Taylor series for the exact solution $y(x_n + h)$ and for the numerical solution y_{n+1} coincide up to (and including) the term h^p, taking into account the localizing assumption that $y_n = y(x_n)$. If this happens then a K exists such that for small h*

$$\|y(x_n + h) - y_{n+1}\| \leq Kh^{p+1} . \tag{6.4}$$

With the paper of Butcher [3] it became customary to symbolize method (6.1) by the tableau (for explicit methods $c_1 = 0$):

$$
\begin{array}{c|cccccc}
c_1 & & & & & & \\
c_2 & a_{21} & & & & & \\
c_3 & a_{31} & a_{32} & & & & \\
c_4 & a_{41} & a_{42} & a_{43} & & & \\
\vdots & \vdots & \vdots & \vdots & \vdots & & \\
c_s & a_{s1} & a_{s2} & a_{s3} & \cdots & a_{s\,s-1} & \\
\hline
 & b_1 & b_2 & b_3 & \cdots & b_{s-1} & b_s
\end{array}
\tag{6.5}
$$

which can be written in a short-hand notation as

$$
\begin{array}{c|c}
c & A \\
\hline
 & b^T
\end{array}
$$

Since Runge-Kutta methods of order up to 4 for scalar problems have the same order for systems [12], we shall restrict the discussion of the derivation of the method to the case of scalar problems. The original technique, used by Runge and Kutta, for deriving the order conditions consists of matching the expansion of the solution generated by one step of the Runge-Kutta method with the Taylor expansion of the exact solution, the terms in the expansions being calculated essentially by brute force. In this section and for subsequent comparison we derive in this way explicit Runge-Kutta methods up to four stages. We follow the presentation of [12] and [18].

The 3-stage explicit Runge-Kutta method is given by (6.1) with $s = 3$. We assume that $f(x, y)$ is sufficiently smooth and introduce the shortened notation

$$f := f(x, y), \quad f_x := \frac{\partial f(x, y)}{\partial x},$$

$$f_{xx} := \frac{\partial^2 f(x,y)}{\partial^2 x}, \quad f_{xy}(\equiv f_{yx}) := \frac{\partial^2 f(x,y)}{\partial x \partial y},$$

etc. all evaluated at the point $(x_n, y(x_n))$. Then, the Taylor series expansion of $y(x_{n+1})$ around x_n results in:

$$y(x_{n+1}) = y(x_n) + h y^{(1)}(x_n) + \frac{1}{2} h^2 y^{(2)}(x_n) + \frac{1}{6} h^3 y^{(3)}(x_n) + \mathcal{O}(h^4).$$

Now

$$
\begin{aligned}
y^{(1)}(x_n) &= f, \\
y^{(2)}(x_n) &= f_x + f f_y, \\
y^{(3)}(x_n) &= f_{xx} + 2 f f_{xy} + f^2 f_{yy} + f_y (f_x + f f_y).
\end{aligned}
$$

Let us shorten the notation by defining

$$F := f_x + f f_y, \quad G := f_{xx} + 2 f f_{xy} + f^2 f_{yy}, \tag{6.6}$$

so that we can write the expansion for $y(x_{n+1})$ as

$$y(x_{n+1}) = y(x_n) + h f + \frac{1}{2} h^2 F + \frac{1}{6} h^3 (F f_y + G) + \mathcal{O}(h^4). \tag{6.7}$$

For the numerical solution y_{n+1} we have a similar expansion. Expanding the k_i given by (6.1) and taking into account (6.3) we have $k_1 = f$ and

$$k_2 = f + c_2 h (f_x + k_1 f_y) + \frac{1}{2} c_2^2 h^2 (f_{xx} + 2 k_1 f_{xy} + k_1^2 f_{yy}) + \mathcal{O}(h^3).$$

On substituting f for k_1, and using the notation (6.6), we get

$$k_2 = f + c_2 h F + \frac{1}{2} c_2^2 h^2 G + \mathcal{O}(h^3). \tag{6.8}$$

We can treat k_3 similarly to obtain:

$$
\begin{aligned}
k_3 &= f + h \{ c_3 f_x + [(c_3 - a_{32}) k_1 + a_{32} k_2] f_y \} \\
&\quad + \frac{1}{2} h^2 \{ c_3^2 f_{xx} + 2 c_3 [(c_3 - a_{32}) k_1 + a_{32} k_2] f_{xy} \\
&\quad + [(c_3 - a_{32}) k_1 + a_{32} k_2]^2 f_{yy} \} + \mathcal{O}(h^3).
\end{aligned}
$$

Now write f for k_1, substitute the form (6.8) for k_2 and retain terms up to $\mathcal{O}(h^2)$ to get

$$k_3 = f + c_3 h F + h^2 (c_2 a_{32} F f_y + \frac{1}{2} c_3^2 G) + \mathcal{O}(h^3). \tag{6.9}$$

If the forms (6.8) and (6.9) for k_2 and k_3, respectively, are introduced into (6.1) with $s = 3$ we obtain the following expansion for y_{n+1}:

$$
\begin{aligned}
y_{n+1} &= y(x_n) + h\,(b_1 + b_2 + b_3)\,f + h^2\,(b_2\,c_2 + b_3\,c_3)\,F \\
&\quad + \frac{1}{2}\,h^3\,[2\,b_3\,c_2\,a_{32}\,F\,f_y + (b_2\,c_2^2 + b_3\,c_3^2)\,G] + \mathcal{O}(h^4)\,.(6.10)
\end{aligned}
$$

It is now a question of trying to match the expansions (6.7) and (6.10). Let us in first instance see what can be reached with one, two and three stages.

- one-stage : the method (6.1) becomes one-stage if we set $b_i = 0, i \geq 2$. Then (6.10) reduces to

$$ y_{n+1} = y(x_n) + h\,b_1\,f + \mathcal{O}(h^4)\,. $$

  If one subtracts this from (6.7) the first non-vanishing term behaves as h^2, if b_1 is set to be 1. Thus there exists only one explicit one-stage Runge-Kutta method of order 1, namely Euler's rule.

- two-stage : The method becomes two-stage if we set $b_i = 0, i \geq 3$, when (6.1) becomes

$$ y_{n+1} = y(x_n) + h\,(b_1 + b_2)\,f + h^2\,b_2\,c_2\,F + \frac{1}{2}\,h^3\,b_2\,c_2^2\,G + \mathcal{O}(h^4)\,. $$

  On comparing with (6.7) we see that the order 2 can be achieved by choosing

$$ b_1 + b_2 = 1, \quad b_2\,c_2 = \frac{1}{2}\,. \tag{6.11} $$

  This is a pair of equations in three unknowns, so there exists a singly infinite λ dependent family of explicit two-stage Runge-Kutta methods of order 2,

$$ c_2 = \lambda \neq 0, \quad b_1 = 1 - \frac{1}{2\,\lambda}, \quad b_2 = \frac{1}{2\,\lambda}\,. $$

  Two particular values of λ yield well-known methods:

  (i) The *midpoint*-method: by choosing $\lambda = \frac{1}{2}$ we get $b_1 = 0$, $b_2 = 1$, $c_2 = \frac{1}{2}$. Its corresponding Butcher array is Table 6.1:

  (ii) the *modified Euler*-method: for $\lambda = 1$ we get $b_1 = b_2 = \frac{1}{2}$, $c_2 = 1$, with Butcher array given in Table 6.2.

- three-stage : we can achieve order 3 if we can satisfy the following conditions:

$$ b_1 + b_2 + b_3 = 1\,, $$

Table 6.1. Two-stage midpoint-method.

$$
\begin{array}{c|cc}
0 \\
\frac{1}{2} & \frac{1}{2} \\
\hline
& 0 & 1
\end{array}
$$

Table 6.2. Two-stage modified Euler-method.

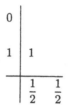

There are now four equations in six unknowns and there exists a doubly infinite family of solutions; consideration of the h^4 terms, which we ignored in the derivation, confirms that none of these solutions leads to a method of order greater than three. Two particular solutions lead to well-known methods:

(i) *Heun's third-order formula* with Butcher array given in Table 6.3.

(ii) *Kutta's third-order formula* with Butcher array given in Table 6.4.

By a similar approach it is possible to show that there exists a doubly infinite family of explicit four-stage Runge-Kutta methods of order 4, none of which has order greater than 4. In theory, this is a completely trivial task with the known rules of differential calculus; by the use of (6.3), the following conditions result:

$$
\sum_i b_i = b_1 + b_2 + b_3 + b_4 = 1 , \tag{6.12a}
$$

Table 6.3. Three-stage Heun's method.

$$
\begin{array}{c|ccc}
0 & & & \\
\dfrac{1}{3} & \dfrac{1}{3} & & \\
\dfrac{2}{3} & 0 & \dfrac{2}{3} & \\
\hline
& \dfrac{1}{4} & 0 & \dfrac{3}{4}
\end{array}
$$

Table 6.4. Three-stage Kutta's method.

$$
\begin{array}{c|ccc}
0 & & & \\
\dfrac{1}{2} & \dfrac{1}{2} & & \\
1 & -1 & 2 & \\
\hline
& \dfrac{1}{6} & \dfrac{2}{3} & \dfrac{1}{6}
\end{array}
$$

$$\sum_i b_i c_i = b_2 c_2 + b_3 c_3 + b_4 c_4 = \frac{1}{2} \,, \tag{6.12b}$$

$$\sum_i b_i c_i^2 = b_2 c_2^2 + b_3 c_3^2 + b_4 c_4^2 = \frac{1}{3} \,, \tag{6.12c}$$

$$\sum_{i,j} b_i a_{ij} c_j = b_3 a_{32} c_2 + b_4 (a_{42} c_2 + a_{43} c_3) = \frac{1}{6} \,, \tag{6.12d}$$

$$\sum_i b_i c_i^3 = b_2 c_2^3 + b_3 c_3^3 + b_4 c_4^3 = \frac{1}{4} \,, \tag{6.12e}$$

$$\sum_{i,j} b_i c_i a_{ij} c_j = b_3 c_3 a_{32} c_2 + b_4 c_4 (a_{42} c_2 + a_{43} c_3) = \frac{1}{8} \,, \tag{6.12f}$$

$$\sum_{i,j} b_i a_{ij} c_j^2 = b_3 a_{32} c_2^2 + b_4 (a_{42} c_2^2 + a_{43} c_3^2) = \frac{1}{12} \,, \tag{6.12g}$$

$$\sum_{i,j,k} b_i a_{ij} a_{jk} c_k = b_4 a_{43} a_{32} c_2 = \frac{1}{24} \,. \tag{6.12h}$$

However, these computations are very tedious and they grow enormously with higher orders. Due to Butcher [4] and by using an appropriate notation they can become very elegant. For the explicit methods the above approach is still usefull. Kutta gave the general solution of the nonlinear algebraic system (6.12) without comment. A detailed derivation of the solution is given in Runge and König [23]. In more recent books [4],[12] use has been made during the derivation of the solution of the so-called *simplifying assumptions (or conditions)*, which are also applied to higher order cases. We just summarize them in the form of lemmas for futher reference (an interested reader can find proofs in [12, 18]).

LEMMA 6.1 *If*

$$\sum_i b_i a_{ij} = b_j(1 - c_j), \quad j = 1, \ldots, 4, \tag{6.13}$$

then the equations d, g and h in (6.12) follow from the others.

Equation (6.13) is the simplest of the so-called *column-simplifying conditions*, which are defined as followed:

DEFINITION 6.3 *The column-simplifying assumptions are linear conditions on the columns of the Butcher matrix A*

$$D(\xi): \quad \sum_i b_i c_i^{k-1} a_{ij} = \frac{1}{k} b_j(1 - c_j^k), \quad j = 1, \ldots, s, \quad k = 1, \ldots, \xi.$$
$$\tag{6.14}$$

REMARK 6.1 *For explicit methods it is not possible to satisfy (6.14) with $k = 1$ as well as $k = 2$ and obtain an order greater than unity (see [4]).*

Also *row-simplifying conditions* have been introduced:

DEFINITION 6.4 *The conditions*

$$C(\eta): \quad \sum_j a_{ij} c_j^{k-1} = \frac{1}{k} c_i^k, \tag{6.15}$$
$$\text{with } i = 1, \ldots, s, \quad k = 1, \ldots, \eta$$

are called row-simplifying conditions.

REMARK 6.2 *These conditions are also used to define the stage-order, which is the highest integer value of k for which (6.15) holds. Sometimes, the stage vector is used; the ith entry of this s-vector is q_i iff q_i is the highest values* *for which (6.15) holds for the i-th stage. Clearly $\eta = \min_{1 \le i \le s} q_i$. The sum-row conditions (6.3) are the simplest row-simplifying conditions. For explicit methods it follows from $q_2 \le 1$ that $\eta = 1$ and the maximum stage-order for such methods is 1.*

LEMMA 6.2 *For $s = 4$, the equations (6.12) and (6.3) imply (6.13).*

THEOREM 6.1 *Under the assumption (6.3) the equations (6.12) are equivalent to*

$$b_1 + b_2 + b_3 + b_4 = 1 , \tag{6.16a}$$

$$b_2 c_2 + b_3 c_3 + b_4 c_4 = \frac{1}{2} , \tag{6.16b}$$

$$b_2 c_2^2 + b_3 c_3^2 + b_4 c_4^2 = \frac{1}{3} , \tag{6.16c}$$

$$b_2 c_2^3 + b_3 c_3^3 + b_4 c_4^3 = \frac{1}{4} , \tag{6.16e}$$

$$b_3 c_3 a_{32} c_2 + b_4 c_4 (a_{42} c_2 + a_{43} c_3) = \frac{1}{8} , \tag{6.16f}$$

$$b_3 a_{32} + b_4 a_{42} = b_2 (1 - c_2) , \tag{6.16i}$$

$$b_4 a_{43} = b_3 (1 - c_3) , \tag{6.16j}$$

$$0 = b_4 (1 - c_4) . \tag{6.16k}$$

In particular (6.16k) implies $c_4 = 1$. Note that equations (6.16, a–e, k) just state that b_i and c_i are the coefficients of a fourth order quadrature formula with $c_1 = 0$ and $c_4 = 1$.

DEFINITION 6.5 *A method has a quadrature order q if the conditions*

$$\sum_{i=1}^{s} b_i c_i^{k-1} = \frac{1}{k} \tag{6.17}$$

hold for all $k \le q$.

Two particular choices of Kutta [17] have become particularly popular (see Table (6.5) and Table (6.6)):

Table 6.5. "The" fourth-order Runge-Kutta method.

$$
\begin{array}{c|cccc}
0 \\
\dfrac{1}{2} & \dfrac{1}{2} \\
\dfrac{1}{2} & 0 & \dfrac{1}{2} \\
1 & 0 & 0 & 1 \\
\hline
 & \dfrac{1}{6} & \dfrac{2}{6} & \dfrac{2}{6} & \dfrac{1}{6}
\end{array}
$$

Table 6.6. 3/8 Runge-Kutta method.

$$
\begin{array}{c|cccc}
0 \\
\dfrac{1}{3} & \dfrac{1}{3} \\
\dfrac{2}{3} & -\dfrac{1}{3} & 1 \\
1 & 1 & -1 & 1 \\
\hline
 & \dfrac{1}{8} & \dfrac{3}{8} & \dfrac{3}{8} & \dfrac{1}{8}
\end{array}
$$

2. Exponential-fitted explicit methods
2.1 Introduction

In the last decades particular tuned methods for (6.2) have been proposed when the solution exhibits a pronounced oscillatory character. Behaviour of pendulum-like systems, vibrations, resonances or wave propagation are all phenomena of this type in classical mechanics, while the same is true for the typical behaviour of quantum particles. In the class of Runge-Kutta(-Nyström) methods, a few methods with a reduced or null phase error have been considered [25, 33, 34, 35]. Recently two authors [22, 26] have constructed Runge-Kutta(-Nyström) methods for which they claim that they exactly integrate trigonometric functions with known periodicity. Paternoster [22] used the linear stage representation of a Runge-Kutta method given in Albrecht's approach and derived some examples of implicit Runge-Kutta-(Nyström) methods of low algebraic order (for the definition of this property see [22] and Definition 6.2). On the other hand, Simos [26] constructed an explicit Runge-Kutta method of algebraic order 4, which integrates certain first-order initial value problems with periodic or exponential solutions. In a separate section we shall discuss the merits of this approach and its region of applicability. We furthermore shall introduce other ways of constructing explicit Runge-Kutta methods up to algebraic order 4, which integrate exactly first-order systems with solutions which can be expressed as linear combinations of $\exp(\mu x)$ and $\exp(-\mu x)$ ($\mu \in \mathbf{R}$ or $i\mathbf{R}$) (see also [28, 29]). We conclude this chapter with some numerical experiments to test the methods.

2.2   The explicit exponential-fitted methods

A general s-stage Runge-Kutta method can integrate exactly at most a set of s linearly independent functions. The number of these functions depends in the first instance on the stage-order and not on the quadrature order of the method. This property is a consequence of the following theorem (see also theorem 367A of [4] for an analogous approach and [28]):

THEOREM 6.2 *Consider an s-stage explicit Runge-Kutta method with stage-order $r = 1$ and quadrature order q ($q \geq 1$). Consider also a problem $y' = f(x, y)$ whereby f satisfies the Lipschitz condition with Lipschitz constant $L > 0$ and whose solution is sufficiently differentiable on a closed interval I which contains x_n, x_{n+1} and each of $x_n + hc_i$ for $i = 1, 2, \ldots, s$. Then the local truncation error is bounded by an expression of the form*

$$\sum_{i=1}^{s} m_i h^{q_i+2} \|y^{(q_i+1)}\| + m h^{q+1} \|y^{(q+1)}\|, \qquad (6.18)$$

where m_i and m are positive constants and $\|y^{(k)}\| = \sup_{x \in I} |y^{(k)}(x)|$. Note that q_i is defined in Remark 6.2.

Proof   Let $Y_1, Y_2, \ldots, Y_s, y_{n+1}$ denote, as above, stage values and the result at the end of the step starting from an initial value $y_n = y(x_n)$. Let $Z_1, Z_2, \ldots, Z_s, z_{n+1}$ denote the corresponding exact quantities $Z_i = y(X_i)$, ($i = 1, 2, \ldots, s$), $z_{n+1} = y(x_{n+1})$ where $X_i = x_n + c_i h, (i = 1, 2, \ldots, s)$. For explicit Runge-Kutta methods we have

$$Z_i = y_n + h \sum_{j=1}^{i-1} a_{ij} f(X_j, Z_j) + \alpha_i h^{q_i+1} y^{(q_i+1)}(\xi_i), \quad i = 1, 2, \ldots, s, ,$$

with $\xi_i \in I$ and $\alpha_1 = 0$ and

$$z_{n+1} = y_n + h \sum_{i=1}^{s} b_i f(X_i, Z_i) + \alpha h^{q+1} y^{(q+1)}(\xi), \quad \xi \in I .$$

The local truncation error is then given by

$$|y_{n+1} - z_{n+1}| \leq h \sum_{i=1}^{s} |b_i| |f(X_i, Y_i) - f(X_i, Z_i)| + |\alpha| h^{q+1} \|y^{(q+1)}\|,$$

which by using the Lipschitz condition on f becomes

$$|y_{n+1} - z_{n+1}| \leq h L \sum_{i=1}^{s} |b_i| |Y_i - Z_i| + |\alpha| h^{q+1} \|y^{(q+1)}\|. \tag{6.19}$$

Herein, one can rely on

$$|Y_i - Z_i| \leq h \sum_{j=1}^{i-1} |a_{ij}| |f(X_j, Y_j) - f(X_j, Z_j)| + |\alpha_i| h^{q_i+1} |y^{(q_i+1)}(\xi_i)|$$

$$\leq h L \sum_{j=1}^{i-1} |a_{ij}| |Y_j - Z_j| + |\alpha_i| h^{q_i+1} \|y^{(q_i+1)}\|.$$

We can conclude that each of the differences $|Y_i - Z_i|$, $i = 2, 3, \ldots, s$ can be bounded by linear combinations of the quantities $\|y^{(q_j+1)}\|$, $(j = 2, \ldots, i-1)$. Substituting this in (6.19), the theorem follows. Q.E.D.

The bound for the local truncation error in Theorem 6.2 is thus a linear combination of total derivatives of the solution of (possibly) different orders. One of the orders of the derivatives which appear in this expression is $q_2 + 1$ and since $q_2 = 1$, there is a term in $\|y^{(2)}\|$. Theorem 6.2 thus shows that an explicit Runge-Kutta method applied to (6.2) automatically integrates exactly the functions 1 and x. If however the function f is independent of y, only the last term in the right-hand side of (6.18) remains and the Runge-Kutta methods (which then reduces to a quadrature formula of order q) exactly integrates the set $1, x, \ldots, x^q$. In what follows we only consider the general case, where y is present.

Theorem 6.2 is related to the following well-known property which is valid for all Runge-Kutta methods : if a Runge-Kutta method has order p and stage-order q ($q \geq r$), then the local truncation error can be re-expressed in terms of products of partial derivatives of f and total derivatives $y^{(k)}$ where $k > r$. Note again that for the considered explicit methods $r = 1$.

Since the above arguments clearly indicate that an explicit Runge-Kutta method exactly integrates a set of two linearly independent functions one can choose the set $\exp(\mu x), \exp(-\mu x)$ instead of the set $1, x$. This means that every stage equation and the final step equation has to integrate exactly the introduced set of functions. In order to realize this goal we have defined an EFRK method as:

$$y_{n+1} = y_n + h \sum_{i=1}^{s} b_i f(x_n + c_i h, Y_i), \tag{6.20}$$

with

$$Y_i = \gamma_i y_n + \sum_{j=1}^{s} a_{ij} f(x_n + c_j h, Y_j), \qquad (6.21)$$

or in tableau form

c_1	γ_1	a_{11}	a_{12}	$\cdots$	a_{1s}
c_2	γ_2	a_{21}	a_{22}	$\cdots$	a_{2s}
			$\cdots$		
c_s	γ_s	a_{s1}	a_{s2}	$\cdots$	a_{ss}
		b_1	b_2	$\cdots$	b_s

This formulation is unusual and it can be seen as an extension of the classical definition of a Butcher tableau (6.5). The addition of γ_i factors in the stage definition is absolutely necessary in the second stage. If γ_2 has the classical value 1, this second stage would integrate exactly 1 and any other function of x, but never the combination $\exp(\mu x), \exp(-\mu x)$. It is obvious that in the limit of $\mu \to 0$ the γ_i should tend to 1.

Following Albrecht's approach [1, 2, 18] we observe that each of the s internal stages (6.21) and the final stage (6.20) of a RK-method are linear, in the sense a linear multistep method is linear. Nonlinearity only arises when one substitutes from one stage into another. We can regard each of the s stages and the final stage as being a generalized linear multistep method on a nonequidistant grid and associate with it a linear operator in exactly the same way as has been done in the previous chapters for other problems, i.e.

$$\mathcal{L}_i[h, \mathbf{a}]y(x) := y(x + c_i h) - \gamma_i\, y(x) - h \sum_{j=1}^{i-1} a_{ij}\, y'(x + c_j h),$$
$$i = 1, 2, \ldots, s \qquad (6.22)$$

and

$$\mathcal{L}[h, \mathbf{b}]y(x) := y(x + h) - y(x) - h \sum_{i=1}^{s} b_i y'(x + c_i h). \qquad (6.23)$$

For classical explicit Runge-Kutta methods one can take $y = 1, x, x^2, \ldots, x^m$, where m cannot be more than 1 for $\mathcal{L}_i$ and is limited by s for $\mathcal{L}$. Under these circumstances the operators can be rewritten as in (2.76) as:

$$\mathcal{L}_i[h, \mathbf{a}]y(x) = L_{i0}^*(\mathbf{a})y + \frac{h}{1!}L_{i1}^*(\mathbf{a})y' + \ldots + \frac{h^m}{m!}L_{im}^*(\mathbf{a})y^{(m)}, \qquad (6.24)$$

$$\mathcal{L}[h, \mathbf{b}]y(x) = L_i^*(\mathbf{b})y + \frac{h}{1!}L_1^*(\mathbf{b})y' + \ldots + \frac{h^m}{m!}L_m^*(\mathbf{b})y^{(m)}. \qquad (6.25)$$

where the symbols $L^*_{ij}(\mathbf{a})$ and $L^*_i(\mathbf{b})$ are the dimensionless moments. They are related to the expressions for $\mathcal{L}_i[h, \mathbf{a}]x^j$ and $\mathcal{L}[h, \mathbf{b}]x^i$ at $x = 0$, respectively. Since only 1 and x can be integrated exactly by an explicit Runge-Kutta method only $L^*_{i0}(\mathbf{a})$ and $L^*_{i1}(\mathbf{a})$ can be made equal to zero at the same time; this gives rise to the equations:

$$L^*_{i0}(\mathbf{a}) \;=\; 1 - \gamma_i = 0 \,,$$

$$L^*_{i1}(\mathbf{a}) \;=\; c_i - \sum_{j=1}^{i-1} a_{ij} = 0 \,.$$

This clearly shows that for classical explicit Runge-Kutta methods all γ_i's are equal to 1 and that the row-sum condition (6.3) has to be fulfilled. (Note that Oliver [20] has proved that one can derive new formulae of second and third order, by introducing an additional parameter into the first stage of an explicit Runge-Kutta method. These methods offer improved error bounds in the second-order case. However then the row-sum condition is not fulfilled anymore).

The end-step relation can integrate exactly $1, x, x^2, \ldots, x^s$; this means that $L^*_0(\mathbf{b}), L^*_1(\mathbf{b}), \ldots, L^*_s(\mathbf{b})$ must vanish simultaneously, resulting in the equations

$$1 - i \sum_{j=1}^{s} b_j c_j^{i-1} = 0, \quad i = 1, 2, \ldots, s \,, \tag{6.26}$$

showing that such method can have a quadrature order s.

When the exponential fitting set $\exp(\pm \mu x)$ is used for y in the expressions (6.22) and (6.23) they get the following form:

$$\mathcal{L}_i[h, \mathbf{a}] \exp(\mu x) \;=\; \exp(\pm \mu x)(\exp(\pm c_i z) - \gamma_i \mp v \sum_{j=1}^{i-1} a_{ij} \exp(\pm c_j z))$$

$$=\; \exp(\pm \mu x) E^*_{0,i}(z, \mathbf{a}), \quad i = 1, \ldots, s \,,$$

$$\mathcal{L}[h, \mathbf{b}] \exp(\mu x) \;=\; \exp(\pm \mu x)(\exp(\pm z) - 1 \mp v \sum_{j=1}^{s} b_j \exp(\pm c_j z))$$

$$=\; \exp(\pm \mu x) E^*_0(z, \mathbf{b}) \,,$$

where $z = \mu h$. One can construct as usual the formal expressions of

$$G^+_i(Z, \mathbf{a}) = \frac{1}{2} \left[E^*_{0,i}(z, \mathbf{a}) + E^*_{0,i}(-z, \mathbf{a}) \right],$$

$$G^-_i(Z, \mathbf{a}) = \frac{1}{2z} \left[E^*_{0,i}(z, \mathbf{a}) - E^*_{0,i}(-z, \mathbf{a}) \right], \qquad i = 1, 2, \ldots, s$$

$$G^+(Z, \mathbf{b}) = \frac{1}{2} \left[E^*_0(z, \mathbf{b}) + E^*_0(-z, \mathbf{b}) \right],$$

$$G^-(Z, \mathbf{b}) = \frac{1}{2z} \left[E_0^*(z, \mathbf{b}) - E_0^*(-z, \mathbf{b}) \right] ,$$

with $Z = z^2$, which can be expressed in terms of η_{-1} and η_0 functions, i.e.

$$G^+(Z, \mathbf{b}) = \eta_{-1}(Z) - 1 - Z \sum_{i=1}^{s} b_i c_i \eta_0(c_i^2 Z) , \tag{6.27}$$

$$G^-(Z, \mathbf{b}) = \eta_0(Z) - \sum_{i-1}^{s} b_i \eta_{-1}(c_i^2 Z) , \tag{6.28}$$

$$G_i^+(Z, \mathbf{a}) = \eta_{-1}(c_i^2 Z) - \gamma_i - Z \sum_{j=1}^{i-1} a_{ij} c_j \eta_0(c_j^2 Z) , \tag{6.29}$$

$$G_i^-(Z, \mathbf{a}) = c_i \eta_0(c_i^2 Z) - \sum_{j=1}^{i-1} a_{ij} \eta_{-1}(c_j^2 Z) , \tag{6.30}$$

for $i = 1, \ldots, s$. For the occurring $c_i, (i = 1, \ldots, s)$ we preserve the components of c corresponding to the classical case. Let us consider some results for two, three and four stages.

- two-stage: Equations (6.27–6.30) fully define $b_1, b_2, \gamma_1, \gamma_2$ and a_{21}.

  (i) the exponential-fitted *midpoint*-method, that is with $c_2 = \frac{1}{2}$. Its corresponding Butcher array is Table 6.7. For small values of Z the following series expansions for the occurring arguments of the Butcher array are obtained:

$$\gamma_2 = 1 + \frac{1}{8}Z + \frac{1}{384}Z^2 + \frac{1}{46080}Z^3 + \frac{1}{10321920}Z^4 + \mathcal{O}(Z^5) ,$$

$$a_{21} = \frac{1}{2} + \frac{1}{48}Z + \frac{1}{3840}Z^2 + \frac{1}{645120}Z^3 + \frac{1}{185794560}Z^4 + \mathcal{O}(Z^5) ,$$

$$b_2 = 1 + \frac{1}{24}Z + \frac{1}{1920}Z^2 + \frac{1}{322560}Z^3 + \frac{1}{92897280}Z^4 + \mathcal{O}(Z^5) ,$$

  showing that in the limit $Z \to 0$ this exponential-fitted method reduces to the classical scheme (see Table 6.1). As a matter of fact none of these series formulae should be programmed in practice if care is taken to re-formulate the analytical expressions in terms of (3.35). For example b_2 can be written as:

$$b_2 = \eta_0(Z/4) .$$

  (ii) the exponential fitted *modified Euler*-method, i.e. $c_2 = 1$, with Butcher array given in Table 6.8. For small values of Z the following series ex-

Table 6.7. Two-stage exponential-fitted midpoint-method.

$$
\begin{array}{c|cc}
0 & & \\
\frac{1}{2} & \eta_{-1}(Z/4) & \eta_0(Z/4)/2 \\
\hline
& 0 & \dfrac{2(\eta_{-1}(Z)-1)}{Z\,\eta_0(Z/4)}
\end{array}
$$

pansions for the occurring arguments of the Butcher array are obtained:

$$
\gamma_2 = 1 + \frac{1}{2}Z + \frac{1}{24}Z^2 + \frac{1}{720}Z^3 + \frac{1}{40320}Z^4 + \mathcal{O}(Z^5)\,,
$$

$$
a_{21} = 1 + \frac{1}{6}Z + \frac{1}{120}Z^2 + \frac{1}{5040}Z^3 + \frac{1}{362880}Z^4 + \mathcal{O}(Z^5)\,,
$$

$$
b_1 = b_2 = \frac{1}{2} - \frac{1}{24}Z + \frac{1}{240}Z^2 - \frac{17}{40320}Z^3 + \frac{31}{725760}Z^4 + \mathcal{O}(Z^5)\,,
$$

showing that in the limit $Z \to 0$ this exponential-fitted method reduces to the classical scheme (see Table 6.2). As before the expression for b_1 or b_2 can re-formulated by (3.35) as

$$
b_1 = b_2 = \frac{\eta_0^2(Z/4)}{2\eta_0(Z)}\,.
$$

Table 6.8. Two-stage exponential fitted modified Euler-method.

$$
\begin{array}{c|cc}
0 & 1 & \\
1 & \eta_{-1}(Z) & \eta_0(Z) \\
\hline
& \dfrac{\eta_{-1}(Z)-1}{Z\eta_0(Z)} & \dfrac{\eta_{-1}(Z)-1}{Z\eta_0(Z)}
\end{array}
$$

- three-stage: equations (6.27-6.28) leave one degree of freedom. In order to fully define all components of b we add the following supplementary equation (see (6.26), $i = 1$):

$$
L_0^*(\mathbf{b}) = 1 - b_1 - b_2 - b_3 = 0\,.
$$

Equations (6.29-6.30) leave one degree of freedom as well; as an example we consider the exponential fitting version of the three-stage Kutta's method (Table 6.4), where we have chosen $a_{31} = -1$ as in the classical case. The array given in Table 6.9 emerges.

Table 6.9. Three-stage exponential-fitted Kutta's method.

0		
$\frac{1}{2}$	$\eta_{-1}(Z/4)$	$\eta_0(Z/4)/2$
1	$\dfrac{\eta_{-1}(Z/4) - Z/2\,\eta_0(Z/4)}{\eta_{-1}(Z/4)}$ $\qquad -1$	$\dfrac{\eta_0(Z)+1}{\eta_{-1}(Z/4)}$
	b_1	b_2 $\qquad b_3$

where

$$b_1 = b_3 = \frac{\eta_0(Z/4) - 1}{2(\eta_{-1}(Z/4) - 1)} = \frac{\eta_0^2(Z/16) - 2\eta_1(Z/4)}{2\eta_0^2(Z/16)},$$

$$b_2 = \frac{\eta_{-1}(Z/4) - \eta_0(Z/4)}{\eta_{-1}(Z/4) - 1} = \frac{2\eta_1(Z/4)}{\eta_0^2(Z/16)}.$$

- four-stage: equations (6.27-6.28) leave two degrees of freedom. In order to fully define all components of b we add the following supplementary equations (see (6.26), $i = 1, 2$):

$$L_0^*(\mathbf{b}) = 1 - \sum_{i=1}^{4'} b_i = 0, \tag{6.31}$$

and

$$L_1^*(\mathbf{b}) = 1/2 - \sum_{i=1}^{4} b_i c_i = 0. \tag{6.32}$$

Equations (6.29-6.30) leave three degrees of freedom; as an example we consider the exponential fitting version of "the" fourth-order Runge-Kutta method (Table 6.5), where we have chosen $a_{31} = 0$, $a_{42} = 0$ as in the classical case, and additionally, also $\gamma_4 = 1$. In the exponential fitting frame the Table 6.10 emerges.

For small values of Z it is preferable to use series expansions for the values of the b-components, the non-zero elements of A and the γ_i, $(i = 1, 2)$ or

Table 6.10. "The" fourth-order exponential-fitted Runge-Kutta method.

0	1				
$\frac{1}{2}$	$\eta_{-1}(Z/4)$	$\eta_0(Z/4)/2$			
$\frac{1}{2}$	$\dfrac{1}{\eta_{-1}(Z/4)}$	0	$\dfrac{\eta_0(Z/4)}{2\eta_{-1}(Z/4)}$		
1	1	0	0	$\eta_0(Z/4)$	
		b_1	b_2	b_3	b_4

with

$$b_1 = b_4 = \frac{\eta_0(Z/4) - 1}{2(\eta_{-1}(Z/4) - 1)} = \frac{\eta_0^2(Z/16) - 2\eta_1(Z/4)}{2\eta_0^2(Z/16)} ,$$

$$b_2 = b_3 = \frac{\eta_{-1}(Z/4) - \eta_0(Z/4)}{2(\eta_{-1}(Z/4) - 1)} = \frac{\eta_1(Z/4)}{\eta_0^2(Z/16)} .$$

the r.h.s. expressions given above.

$$b_1 = b_4 = \frac{1}{6} - \frac{1}{720}Z + \frac{1}{80640}Z^2 - \frac{1}{9676800}Z^3 + \frac{1}{1226244096}Z^4 + \cdots ,$$

$$b_2 = b_3 = \frac{1}{2} - b_1 ,$$

$$a_{21} = \frac{1}{2} + \frac{1}{48}Z + \frac{1}{3840}Z^2 + \frac{1}{645120}Z^3 + \frac{1}{185794560}Z^4 + \cdots ,$$

$$a_{32} = \frac{1}{2} - \frac{1}{24}Z + \frac{1}{240}Z^2 - \frac{17}{40320}Z^3 + \frac{31}{725760}Z^4 + \cdots ,$$

$$a_{43} = 2a_{21} ,$$

$$\gamma_2 = 1 + \frac{1}{8}Z + \frac{1}{384}Z^2 + \frac{1}{46080}Z^3 + \frac{1}{10321920}Z^4 + \cdots ,$$

$$\gamma_3 = 1 + \frac{1}{8}Z + \frac{5}{384}Z^2 + \frac{61}{46080}Z^3 + \frac{277}{2064384}Z^4 + \cdots$$

It is clear that when $Z \to 0$ the well-known classical fourth-order method is reproduced.

As for the algebraic order conditions we have (compare with equations under (6.12)):

$$\sum_{i=1}^{4} b_i = 1 ,$$

$$\sum_{i=1}^{4} b_i c_i = \frac{1}{2},$$

$$\sum_{i=1}^{4}\sum_{j=1}^{4} b_i a_{ij} = \frac{1}{2} + \frac{17}{11520} Z^2 + \cdots,$$

$$\sum_{i=1}^{4} b_i c_i^2 = \frac{1}{3} - \frac{1}{1440} Z + \frac{1}{161280} Z^2 - \cdots,$$

$$\sum_{i=1}^{4}\sum_{j=1}^{4} b_i a_{ij} c_j = \frac{1}{6} - \frac{11}{2880} Z + \frac{661}{967680} Z^2 - \cdots,$$

$$\sum_{i=1}^{4} b_i c_i^3 = \frac{1}{4} - \frac{1}{960} Z + \frac{1}{107520} Z^2 - \cdots,$$

$$\sum_{i=1}^{4}\sum_{i=1}^{4} b_i c_i a_{ij} c_j = \frac{1}{8} - \frac{1}{1920} Z + \frac{227}{645120} Z^2 - \cdots,$$

$$\sum_{i=1}^{4}\sum_{j=1}^{4}\sum_{k=1}^{4} b_i a_{ij} a_{jk} c_k = \frac{1}{24} - \frac{1}{480} Z + \frac{13}{53760} Z^2 - \cdots,$$

$$\sum_{i=1}^{4}\sum_{j=1}^{4} b_i a_{ij} c_j^2 = \frac{1}{12} - \frac{11}{5760} Z + \frac{661}{1935360} Z^2 - \cdots$$

From this it follows that the exponential-fitted method is algebraically of fourth-order, in the sense that (see also (6.4))

$$|y(x_n + h) - y_{n+1}| = \mathcal{O}(h^5).$$

2.3   Error analysis

In order to get an idea about the form of the local truncation error (LTE) we have calculated the difference $y(x_{n+1}) - y_{n+1}$, where $y(x_{n+1})$ is the exact solution at x_{n+1}, for the exponential-fitted modified Euler method, the three-stage Kutta's method and "the" fourth-order method. Note that the well-known theory of Butcher based on rooted trees is developed for the case of autonomous systems, that means $y' = f(y)$. In contrast to the development by Butcher (see [5, 18]) the row-sum rule is here not fulfilled. We present the LTE for a non-autonomous scalar equation. All occurring partial derivatives of $f(x, y(x))$ with respect to x and y are collected in order to express these LTEs in terms of total derivatives of $y(x)$. The following results were obtained:

- For the two-stage method (see Table 6.8)

$$LTE = -\frac{h^3}{12}\left[y^{(3)} - 3f_y y^{(2)} - \mu^2(y' - 3f_y y)\right] + \mathcal{O}(h^4) \,. \tag{6.33}$$

- For the three-stage method (see Table 6.9)

$$LTE = \frac{h^4}{24}\left[-f_y y^{(3)} + 2f_y^2 y^{(2)} - \mu^2(-f_y y' + 2f_y^2 y)\right]$$
$$+\mathcal{O}(h^5) \,. \tag{6.34}$$

- For the four-stage method (see Table 6.10)

$$LTE = -\frac{h^5}{2880}\left[y^{(5)} - 5f_y y^{(4)} - 10f_{xy} y^{(3)} - 10f_{yy} f_y y^{(3)} + 10f_y^2 y^{(3)}\right.$$
$$+15f_{yy}(y^{(2)})^2 + 30f_{xy} f_{yy} y^{(2)} + 30f_{yy} f f_{yy} y^{(2)} - 30f_y^3 y^{(2)}$$
$$-\mu^2(y^{(3)} - 5f_y y^{(2)} - 10f_{xy} y' - 10f_{yy} f y'$$
$$+10f_y^2 y' + 15f_{yy} y y^{(2)} + 30f_{xy} f_{yy} y + 30f_{yy} f f_{yy} y - 30f_y^3 y)$$
$$\left.-15\mu^2((f_y + f_{yy}y)y^{(2)} - \mu^2(f_y + f_{yy}y)y)\right] + \mathcal{O}(h^6) \,. \tag{6.35}$$

All functions are evaluated at $x = x_n$ and $y = y_n$. Note that the leading order terms in the given LTEs become zero for $y = \exp(\pm\mu x)$. Moreover in each of the LTE-expressions the derivative of lowest order occurring for $\mu = 0$ is $y^{(2)}$, showing that classically only the set $1, x$ is integrated exactly.

2.4 An approach of Simos

The methods described by Simos [26, 27] are based on the Butcher tableau (6.5) and the Definition 6.1 of explicit Runge-Kutta methods. He introduces some new general definitions, which we repeat here.

DEFINITION 6.6 *The method (6.1) is called exponential of order p if it integrates exactly any linear combination of the linearly independent functions*

$$\exp(\mu_0 x), \exp(\mu_1 x), \dots, \exp(\mu_p x) \,,$$

where $\mu_i, i = 0, 1, \dots, p$ are real or complex numbers.

REMARK 6.3 *If* $\mu_i = \mu$ *for* $i = 0, 1, \ldots, n$, $n \leq p$ *then the method (6.1) integrates exactly any linear combination of*

$$\exp(\mu x), x \exp(\mu x), \ldots, x^n \exp(\mu x), \exp(\mu_{n+1} x), \ldots, \exp(\mu_p x) \ ,$$

(see [19]*).*

REMARK 6.4 *Every exponential method corresponds in a unique way to a (classical) algebraic method by setting* $\mu_i = 0$ *for all* i, *(see* [19]*).*

In two different papers [26, 27] Simos develops some fourth-order Runge-Kutta methods with exponential character based on different assumptions:

- In [26] he considers the special initial value problem

$$y' = f(x, y) = \mu \, y \ . \tag{6.36}$$

Applied to the differential equation (6.36), an explicit four-stage Runge-Kutta method reads as follows:

$$y_{n+1} = y_n + h \sum_{i=1}^{4} b_i k_i \ ,$$

with

$$
\begin{aligned}
k_1 &= \mu y_n \ , \\
k_2 &= \mu y_n + h\mu^2 a_{21} y_n \ , \\
k_3 &= \mu y_n + h\mu^2 (a_{31} + a_{32}) y_n + h^2 \mu^3 a_{32} a_{21} y_n \ , \\
k_4 &= \mu y_n + h\mu^2 (a_{41} + a_{42} + a_{43}) y_n \\
&\quad + h^2 \mu^3 (a_{42} a_{21} + a_{43}(a_{31} + a_{32})) y_n + h^3 \mu^4 a_{43} a_{32} a_{21} y_n \ ,
\end{aligned}
$$

which results in

$$
\begin{aligned}
y_{n+1} &= y_n + h\mu \sum_{i=1}^{4} b_i \, y_n + h^2 \mu^2 \sum_{i=2}^{4} b_i \sum_{j=1}^{i-1} a_{ij} \, y_n \\
&\quad + h^3 \mu^3 (b_3 \, a_{32} \, a_{21} + b_4(a_{42} \, a_{21} + a_{43}(a_{31} + a_{32}))) y_n \quad (6.37) \\
&\quad + h^4 \mu^4 b_4 \, a_{43} \, a_{32} \, a_{21} y_n \ .
\end{aligned}
$$

Requiring that $\exp(\mu x)$ is integrated exactly by (6.37) leads to the equation

$$\exp(z) = 1 + \sum_{q=1}^{4} z^q \, s_q \ , \tag{6.38}$$

where

$$z = \mu\, h, \qquad s_1 = \sum_{i=1}^{4} b_i\,, \qquad s_2 = \sum_{i=2}^{4} b_i \sum_{j=1}^{i-1} a_{ij}\,,$$

$$s_3 = b_3\, a_{32}\, a_{21} + b_4(a_{42}\, a_{21} + a_{43}(a_{31} + a_{32}))\,, \qquad (6.39)$$

$$s_4 = b_4\, a_{43}\, a_{32}\, a_{21}\,.$$

Besides the above assumptions Simos also requires that the method should have a quadrature order three at least. Taking into account the row-conditions (6.3) and the column-symplifying conditions (6.13) the above introduced quantities $s_i, i = 1, 2, 3$ take the values

$$s_1 = 1, \qquad s_2 = \frac{1}{2}, \qquad s_3 = \frac{1}{6}\,. \qquad (6.40)$$

By introducing these values in (6.38) one obtains for s_4:

$$s_4 = -\frac{6 + z^3 - 6\exp(z) + 3z^2 + 6z}{6z^4}\,. \qquad (6.41)$$

Equations (6.38–6.41) supply four conditions for ten unknowns. Therefore Simos ([26]) re-uses the c_i and a_{ij} values of "the" fourth-order Runge-Kutta method (see Table 6.5) and solves the above equations for the b_i. Altogether, the coefficients of Simos' version are:

$$c_1 = a_{21} = \frac{1}{2}, \qquad c_2 = \frac{1}{2}, \qquad c_3 = 1\,,$$

$$a_{31} = a_{41} = a_{42} = 0, \qquad a_{32} = \frac{1}{2}, \qquad a_{43} = 1\,,$$

$$b_1 = b_4 = -\frac{2(6 + z^3 - 6\exp(z) + 3z^2 + 6z)}{3z^4}\,,$$

$$b_2 = \frac{1}{3}, \qquad b_3 = \frac{2(12 + 2z^3 - 12\exp(z) + 6z^2 + 12z + z^4)}{3z^4}\,.$$

It can be checked that this method has an algebraic order 4.

REMARK 6.5 *In Simos' approach the emphasis is put on deriving the coefficients such that the equation $y' = \mu\, y$ is solved exactly but in the spirit of the exponential fitting the condition is that the function $\exp(\mu x)$ has to be integrated exactly. These are not equivalent. Specifically, for the equation $y' = \mu y$ Simos' method is good, but for the innocent looking $y' = \mu\exp(\mu x), y(0) = 1$ it is not, although $\exp(\mu x)$ is its solution, as well!*

- In the same paper [26] Simos also introduces a technique by which $\exp(\mu\, x)$ as well as $\exp(-\mu\, x)$ are solutions of an equation of the type (6.37), i.e. the following system of equations has to be solved:

$$\begin{cases} \exp(z) & = 1 + \sum_{q=1}^{4} z^q\, s_q \\ \exp(-z) & = 1 + \sum_{q=1}^{4} (-z)^q\, s_q \;. \end{cases} \tag{6.42}$$

Following the same ideas as those developed in Section 2.2 we consider the equivalent system:

$$\begin{cases} \dfrac{\exp(z) + \exp(-z)}{2} & = 1 + z^2\, s_2 + z^4\, s_4 \\ \dfrac{\exp(z) - \exp(-z)}{2z} & = s_1 + z^2\, s_3 \end{cases}$$

or

$$\begin{cases} \eta_{-1}(Z) & = 1 + Z\, s_2 + Z^2\, s_4 \\ \eta_0(Z) & = s_1 + Z\, s_3 \;. \end{cases} \tag{6.43}$$

These two equations together with the conditions $s_1 = 1$ and $s_2 = 1/2$ give us the following values for s_3 and s_4:

$$s_3 = \frac{\eta_0(Z) - 1}{Z} = \frac{1}{2}(\eta_0^2(Z/4) - 2\eta_1(Z)) \;,$$

$$s_4 = \frac{2\eta_{-1}(Z) - 2 - Z}{2Z^2} = \frac{1}{16}(\eta_0^2(Z/16) - 2\eta_1(Z/4))(\eta_0(Z/4) + 1) \;.$$

Using again the c_i- and a_{ij}-values of Table 6.5 the following values for the b_i-values are obtained:

$$b_1 = b_4 = 2\frac{2\eta_{-1}(Z) - 2 - Z}{Z^2} = \frac{1}{4}(\eta_0^2(Z/16) - 2\eta_1(Z/4))(\eta_0(Z/4) + 1) \;,$$

$$b_2 = \frac{Z - 4\eta_0(Z) + 4}{Z} = 1 - 2(\eta_0^2(Z/4) - 2\eta_1(Z)) \;, \tag{6.44}$$

$$b_3 = -4\frac{-Z\eta_0(Z) + 2\eta_{-1}(Z) - 2}{Z^2}$$

$$= 2(\eta_0^2(Z/4) - 2\eta_1(Z)) - \frac{1}{2}(\eta_0^2(Z/16) - 2\eta_1(Z/4))(\eta_0(Z/4) + 1) \;.$$

Again it is easy to verify that this method is of algebraic order 4.

REMARK 6.6 *This method has no counterpart in the exponential-fitting approach of this book, since it does not integrate exactly the set*

$$\exp(\mu\, x), \exp(-\mu\, x) \;.$$

It is completely different from "the" exponential-fitted fourth-order Runge-Kutta method (see Table 6.10). The last one integrates exactly the above mentioned set, while Simos' method only integrates in an exact way differential equations with right-hand sides of the form μy.

For more work in this direction we can refer to Simos [27] .

2.5   Some simple examples

In this section we compare the results of two (algebraic) fourth-order exponential-fitted approaches obtained for some simple initial value systems, having as solution a combination of sine and cosine functions. The following problems have been considered:

- **Example 1**

$$u'' = -u(u^2 + v^2)^{-3/2}, \quad v'' = -v(u^2 + v^2)^{3/2} , \tag{6.45}$$

with $u(0) = 1, u'(0) = 0, v(0) = 0, v'(0) = 1$; its exact solution is

$$u(x) = \cos(x), \quad v(x) = \sin(x) .$$

Equation (6.45) has been solved in the interval $0 \le x \le 7$ with several stepsizes.

- **Example 2**

$$y'' = -30 \sin(30x), \quad y(0) = 0, \quad y'(0) = 1 , \tag{6.46}$$

with $y(0) = 0, y'(0) = 1$; its exact solution is $y(x) = \sin(30x)/30$. Equation (6.46) has been solved in the interval $0 \le x \le 10$ with several stepsizes.

- **Example 3**

$$y'' + y + y^3 = B \cos(\Omega x), \quad y(0) = 0.20042672806900, \quad y'(0) = 0 , \tag{6.47}$$

with $B = 0.002, \Omega = 1.01$; its approximate very accurate solution is

$$\begin{aligned} y(x) = \quad & 0.200179477536 \cos(\Omega x) + 0.246946143 \ 10^{-3} \cos(3\Omega x) \\ & + 0.304016 \ 10^{-6} \cos(5\Omega x) + 0.374 \ 10^{-9} \cos(7\Omega x) . \end{aligned}$$

Equation (6.47) has been solved in the interval $0 \le x \le 300$ with several stepsizes.

- **Example 4**

$$u'' + u = 0.001 \cos(x), \quad u(0) = 1, \quad u'(0) = 0, \tag{6.48}$$

with exact solution

$$u(x) = \cos(x) + 0.0005x \sin(x) \ . \tag{6.49}$$

Equation (6.48) has been solved in the interval $0 \le x \le 1000$ with several stepsizes.

After having rewritten the above systems as equivalent first-order systems, we have solved them with Simos' method (with the c_i- and $a_{ij}-$values of Table 6.5 and the b_1-values given in (6.44)), "the" exponential-fitted fourth-order method (see Table 6.10) and the classical fourth-order method (see Table 6.5). For each example we have calculated the Euclidean norm of the error vector at the endpoint of the integration interval; this is defined as

$$\Delta := \sqrt{\sum_{t=1}^{n} ({}^t y(x_{end}) - {}^t y_{end})^2} \ . \tag{6.50}$$

These data are collected in Table 6.11 together with the value of μ used in each example. All computations were carried out in double precision arithmetic (16 significant digits of accuracy).

From the results obtained for examples 1 and 2 it is clear that the method of Table 6.10 integrates the considered systems within machine accuracy. The accuracy of the results obtained with Simos' method and the classical method is poor. For examples 3 and 4 our results and those of Simos are comparable and much better than the classical ones. From all these results it is obvious that for the examples considered the method in Table 6.10 reproduces quite accurately the exact data in all cases, while Simos' method fails whenever the right-hand side function in (6.2) is not proportional in the dominant term with $\pm\mu y$.

2.6   Error and steplength control

2.6.1   Local error estimation and a good choice for μ

In this section we study a method for the determination of μ in an optimal way for the method of Table 6.10.

There exists no mathematical theory to determine μ in an exact way. The only goal we can put forward is to make the LTE as small as possible by calculating μ by means of a heuristically chosen algorithm. Since for a scalar equation the term $y^{(2)} - \mu^2 y$ is present in each of the expressions for the LTEs (6.33),(6.34) and (6.35) we propose to calculate μ in each integration interval $[x_n, x_{n+1}]$ in the following way:

$$\mu = \sqrt{\frac{y^{(2)}(x_n)}{y(x_n)}}, \quad n = 0, \ldots \text{ if } y(x_n) \ne 0 \text{ and } \mu = 0 \text{ otherwise } .$$

Table 6.11. Comparison of Euclidean norms of the end-point global errors in the approximations obtained by one of Simos' method, the method of Table 6.10 and the classical method.

Example 1	$\omega = 1$		
h	Method of Table 6.10	Simos' method (6.44)	Method of Table 6.5
1	$4.38\ 10^{-15}$	37.49	10.97
0.5	$1.65\ 10^{-14}$	0.07	0.11
0.25	$1.26\ 10^{-14}$	$8.00\ 10^{-3}$	$1.10\ 10^{-2}$
0.125	$8.55\ 10^{-15}$	$6.02\ 10^{-4}$	$8.29\ 10^{-4}$
0.0625	$2.03\ 10^{-14}$	$4.05\ 10^{-5}$	$5.53\ 10^{-5}$

Example 2	$\omega = 30$		
1	$5.50\ 10^{-16}$	193.67	82.26
0.5	$7.78\ 10^{-15}$	70.65	53.98
0.25	$2.56\ 10^{-15}$	66.96	36.32
0.125	$1.65\ 10^{-15}$	4.23	1.13
0.0625	$1.70\ 10^{-14}$	0.22	$4.76\ 10^{-2}$

Example 3	$\omega = 1.01$		
1	$1.10\ 10^{-3}$	$1.70\ 10^{-3}$	0.31
0.5	$5.42\ 10^{-5}$	$1.88\ 10^{-4}$	$1.32\ 10^{-2}$
0.25	$1.86\ 10^{-6}$	$1.37\ 10^{-5}$	$4.61\ 10^{-4}$
0.125	$6.19\ 10^{-8}$	$8.70\ 10^{-7}$	$2.67\ 10^{-5}$
0.0625	$2.40\ 10^{-9}$	$5.41\ 10^{-8}$	$1.82\ 10^{-6}$

Example 4	$\omega = 1$		
1	$1.40\ 10^{-2}$	$1.20\ 10^{-3}$	1.14
0.5	$8.52\ 10^{-4}$	$7.54\ 10^{-5}$	0.49
0.25	$5.30\ 10^{-5}$	$4.74\ 10^{-6}$	$3.34\ 10^{-2}$
0.125	$3.31\ 10^{-6}$	$2.96\ 10^{-7}$	$2.09\ 10^{-3}$
0.0625	$2.07\ 10^{-7}$	$1.86\ 10^{-8}$	$1.31\ 10^{-4}$

Note that if $y(x)$ is a linear combination of $\sin(\omega x)$ and $\cos(\omega x)$, and $y(x_n) \neq 0$, we obtain from this recipe $\mu = i\omega$ and $y(x)$ will be integrated exactly (if infinite precision arithmetic is assumed). For a system of n first-order equations we propose to make the Euclidean norm of all ($^t y^{(2)} - \mu^2\ ^t y, t = 1, \ldots n$) present as small as possible. This results in

$$\mu = \sqrt{\frac{\sum_{t=1}^{n} {}^t y(x_n) \, {}^t y^{(2)}(x_n)}{\sum_{t=1}^{n} {}^t y(x_n)^2}} . \tag{6.51}$$

The expressions for the occurring second-order derivatives can be obtained analytically from the given ODEs or calculated numerically using previously derived $y(x_{n-j})$ values. The μ-values used are then in each integration interval taken as the positive square root of the numerically obtained μ^2. If negative μ^2-values are obtained, μ is replaced in the corresponding formulae by $i\omega$. In fact in this case the trigonometric functions instead of the exponential ones are integrated.

Since the μ-values are calculated numerically, they are never exact due to finite precision arithmetics. As a consequence the leading term of the LTE does not vanish. This means that for the chosen μ we can try to estimate the LTE numerically. A technique which can be used is Richardson extrapolation. First we fix μ by (6.51). We consider a Runge-Kutta method of order $p \in \{2, 3, 4\}$ to obtain the solution y_{n+1} at x_{n+1}. Under the usual localizing assumption that $y_n = y(x_n)$ it follows from (6.33), (6.34) and (6.35) that the LTE can be written in the form $T_{n+1} = y(x_{n+1}) - y_{n+1} = C(y, f)h^{p+1} + \mathcal{O}(h^{p+2})$, where $C(y, f)$ is some function of y, its derivatives, $f(x, y)$ and its partial derivatives with respect to x and y, all evaluated at the point (x_n, y_n). Let us now compute a second numerical solution at x_{n+1} by applying the same method twice with steplength $h/2$ and also starting in x_n; denote the solution so obtained by z_{n+1}. By starting in both calculations at the same point x_n one can work during these two processes with the same value of μ. The error is now

$$T_{n+1} = y(x_{n+1}) - z_{n+1} = 2C(y, f)(h/2)^{p+1} + \mathcal{O}(h^{p+2}) .$$

From these two estimates for the LTE one can derive that the error in the second calculation is given by: $error \approx 2C(y, f)(h/2)^{p+1} \approx (z_{n+1} - y_{n+1})/(2^p - 1)$. If the user asks for a given tolerance tol, he can control the steplength and the error in the following way:

- if $|error| <= tol$ accept the step and progress with the z_{n+1} value,

- if $|error| > tol$ reject the step.

The step is then adapted in the following way:

$$h_{new} = h_{old} \, \min(facmax \, , \max(facmin \, , fac \, (tol/error)^{1/(1+p)})) ,$$

with *facmax* and *facmin* representing the maximum and minimum acceptable increasing or decreasing factors, respectively. The symbol *fac* represents a safety factor in order to have an acceptable error in the following step. The method for estimating the LTE and the notation used to define h_{new} is given in Hairer *et al.* [12]. In the developed code the following values for these factors were taken : $facmax = 2$, $facmin = 0.5$, $fac = 0.9$.

2.6.2   Some numerical experiments

In this section we solve some initial value problems having a combination of sine or cosine functions as solution. The following cases have been considered.

- **Example 1**

  Equation (6.46) has been solved in the interval $0 \leq x \leq 10$ with $tol = 10^{-10}$.

- **Example 2**

  Equation (6.47) has been solved in the interval $0 \leq x \leq 300$ with $tol = 10^{-8}$.

- **Example 3**

  Equation (6.48) has been solved in the interval $0 \leq x \leq 1000$ with $tol = 10^{-8}$.

- **Example 4**

$$y'' + 0.2y' + y = 0, \quad y(0) = 1, \quad y'(0) = 0, \qquad (6.52)$$

  with exact solution:

$$y(x) = \exp(-0.1x)(\sin(\sqrt{0.99}x)/\sqrt{99} + \cos(\sqrt{0.99}x)) .$$

  Equation (6.52) has been solved in the interval $0 \leq x \leq 2\pi$ with $tol = 10^{-8}$.

After having re-written the above equations as equivalent first-order systems, we have solved them with the exponential-fitted fourth-order EFRK method (Table 6.10) and the equivalent classical fourth-order RK method (Table 6.5); in both cases we applied the above described Richardson extrapolation technique. The Euclidean norms of the error vectors (6.50) are collected in Table 6.12 together with the number of accepted and rejected steps.

It is clear from Table 6.12 that for equations with a purely trigonometric solution (example 1) the new method is superior to the classical one. In this example (6.51) initially gives a quite accurate value for μ, such that there is a small error and a huge increase in the stepsize in each step. However these errors accumulate and at a certain moment such a large stepsize causes a rejected step. After decreasing the stepsize, this process is repeated. In cases where the solution is of a mixed trigonometric form, although one of the terms dominates the solution (examples 2 and 3) the new method is still more efficient than the classical one. In example 4 where the solution is a mixture of exponential and trigonometric functions both methods considered are practically equivalent.

Table 6.12. Comparison of the Euclidean norms of the end-point global errors obtained by using the fourth-order EFRK and classical RK-methods with steplength and error control based on Richardson extrapolation techniques.

		accepted steps	rejected steps	error
Example 1	Exponentially fitted	33	16	$4.01E - 9$
	Classical	3495	5	$1.01E - 8$
Example 2	Exponentially fitted	1191	0	$1.01E - 7$
	Classical	2167	0	$2.79E - 5$
Example 3	Exponentially fitted	1946	1	$1.14E - 5$
	Classical	9826	0	$5.75E - 5$
Example 4	Exponentially fitted	46	3	$1.96E - 7$
	Classical	57	0	$2.70E - 7$

2.6.3 An embedded pair of exponential-fitted explicit Runge-Kutta methods

Derivation of the embedded pair. There exist other techniques than Richardson's extrapolation to determine the LTE in every step. Franco [10] has derived a five-stage embedded pair of exponential-fitted Runge-Kutta methods which is based on the four-stage explicit method given in Table 6.10. The methods, which form the pair have algebraic order 4 and 3, and this pair corresponds in a unique way to an algebraic pair: the Zonneveld 4(3) pair [12]. This scheme implies that at least five stages are required and therefore Table 6.13 has been considered.

The fifth stage and the weights $\bar{b}_i$ are introduced in order to obtain another explicit EFRK method so that $LTE = |y_{n+1} - \bar{y}_{n+1}|$ becomes a local error estimation at each step with a computational cost smaller than the technique based on Richardson's extrapolation of Section 2.6.1. For the determination of the unknowns a_{5j}, $j = 1, 2, 3, 4$ and the $\bar{b}_i$, $i = 1, 2, 3, 4$ the relations (6.29–6.30) and (6.27–6.28), complimented with (6.26) for the case $i = 1$ are used with $s = 5$. This leaves a lot of degrees of freedom; inspired by the classical Zonneveld 4(3) pair (see [12]) following parameter values are chosen by default:

$$\bar{b}_2 = \bar{b}_3, \qquad \bar{b}_5 = -\frac{16}{5}, \qquad a_{51} = \frac{5}{32}, \qquad a_{52} = \frac{7}{32}.$$

The remaining coefficients follow as solutions of the above cited equations:

$$\bar{b}_1 = \frac{3 - 3\eta_{-1}(Z) - 8\, Z\, \eta_0(Z/16) + 9.5\, Z\, \eta_0(Z/4)}{3\, Z\eta_0(Z/4) - 3\, Z\, \eta_0(Z)},$$

$$\bar{b}_3 = \frac{-16\, \eta_{-1}(Z/16) + 19\, \eta_{-1}(Z/4) - 3\, \eta_0(Z/4)}{3/4\, Z\, \eta_0^2(Z/16)},$$

Table 6.13.   "The" 4(3) exponential-fitted Runge-Kutta Zonneveld method.

c						
0	1					
$\dfrac{1}{2}$	$\eta_{-1}(Z/4)$	$\eta_0(Z/4)/2$				
$\dfrac{1}{2}$	$\dfrac{1}{\eta_{-1}(Z/4)}$	0	$\dfrac{\eta_0(Z/4)}{2\eta_{-1}(Z/4)}$			
$\dfrac{1}{3}$	1	0	0	$\eta_0(Z/4)$	0	
$\dfrac{3}{4}$	1	a_{51}	a_{52}	a_{53}	a_{54}	0
		b_1	b_2	b_3	b_4	0
		$\bar{b}_1$	$\bar{b}_2$	$\bar{b}_3$	$\bar{b}_4$	$\bar{b}_5$

with

$$b_1 = b_4 = \frac{\eta_0(Z/4) - 1}{2(\eta_{-1}(Z/4) - 1)},$$

$$b_2 = b_3 = \frac{\eta_{-1}(Z/4) - \eta_0(Z/4)}{2(\eta_{-1}(Z/4) - 1)}.$$

$$\bar{b}_4 = \frac{3 - 3\eta_{-1}(Z) + 4\,Z\,\eta_0(Z/16) + 9.5\,Z\,\eta_0(Z/4) - 12\,Z\,\eta_0(9Z/16)}{3\,Z\,\eta_0(Z/4) - 3\,Z\,\eta_0(Z)},$$

$$a_{53} = -\frac{32\eta_{-1}(Z/16) - 32\eta_{-1}(Z) + 3.5\,Z\,\eta_0(Z/4) + 5\,Z\,\eta_0(Z)}{16\,Z\eta_0(Z/4)},$$

$$a_{54} = \frac{5 - 64\,\eta_{-1}(Z/4)/(Z\,\eta_0(Z/4)) + 64/(Z\,\eta_0(Z/16))}{32}.$$

For $Z \to 0$ the EFRK pair reduces to the classical Zonneveld 4(3) pair. These formulae can also be re-formulated for small Z-values in terms of (3.35)

Numerical experiments. In order to evaluate the effectiveness of the EFRK pair several model problems with clear periodic solutions have been studied. The frequency determination is here of no importance since the choice of μ follows straighforwardly from the given problems. The criterion used in the numerical comparisons is the maximum global error over the whole integration interval. In Figures 6.1–6.3 the efficiency curves for the tested codes are depicted. These figures show the decimal logarithm of the maximal global error $(sd(e))$ against the computational effort measured by the number of function evaluations required by each code. The codes used in the comparisons have been denoted by:

- **EFRK4(3):** The embedded pair of Table 6.13 implemented in a variable step code following the way given in [12].

- **VBExtrapo:** The variable step code proposed in Chapter 2.6.1.

Following problems have been considered:

- **Problem 1**
  This is a linear problem with variable coefficients:

$$y'' + 4t^2 y = (4t^2 - \omega^2)\sin(\omega\, t) - 2\sin(t^2), \quad t \in [0, t_{end}]\,,$$
$$y(0) = 1, \qquad y'(0) = \omega\,,$$

whose analytic solution is given by

$$y(t) = \sin(\omega\, t) + \cos(t^2)$$

This solution represents a periodic motion that involves a constant frequency and a variable frequency. In the test we take $\omega = 10$ and $t_{end} = 10$. The numerical results plotted in Figure 6.1 have been computed with error tolerances $Tol = 10^i$, $i \geq 2$ and $\mu = i\omega = i\,10$.

- **Problem 2**
  The periodically forced nonlinear problem (undamped Duffin's equation). This is

$$y'' + y + y^3 = (\cos(t) + \epsilon\sin(10t))^3 - 99\epsilon\sin(10t), \qquad t \in [0, t_{end}]\,,$$
$$y(0) = 1, \qquad y'(0) = 10\epsilon\,,$$

with $\epsilon = 10^{-3}$. The analytic solution is given by

$$y(t) = \cos(t) + \epsilon\,\sin(10\, t)\,,$$

and it represents a periodic motion of low frequency with a small perturbation of high frequency. In the test we use $t_{end} = 100$ and the numerical results shown in Figure 6.2 have been computed with error tolerances $Tol = 10^i$, $i \geq 3$ and $\mu = i\omega = i$.

- **Problem 3**
  The nonlinear system

$$y_1'' = -4t^2\, y_1 - \frac{2y_2}{\sqrt{y_1^2 + y_2^2}}\,, y_1(0) = 1, \quad y_1'(0) = 0\,, \quad t \in [0, t_{end}]\,,$$

$$y_2'' = -4t^2\, y_2 + \frac{2y_1}{\sqrt{y_1^2 + y_2^2}}\,, y_2(0) = 0, \quad y_2'(0) = 0\,,$$

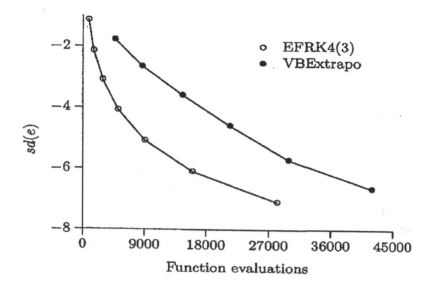

Figure 6.1.   Linear problem with variable coefficients: $\mu = i10$, $t_{end} = 10$.

is considered, whose solution is given by

$$y_1(t) = \cos(t^2), \qquad y_2(t) = \sin(t^2) \, .$$

This solution represents a periodic motion with variable frequency. We use $t_{end} = 10$ and the numerical results plotted in Figure 6.3 have been computed with error tolerances $Tol = 10^i$, $i \geq 2$ and $\mu = i\omega = it_n (n \geq 1)$ at each step.

In view of the numerical results obtained we may conclude that the code EFRK4(3) is clearly more efficient than the code VBExtrapo. This conclusion is not surprising because the technique based on embedded pairs for the estimation of the LTE requires a smaller computational cost than the technique based on Richardson's extrapolation.

2.6.4   Frequency determination and steplength control

In Section 2.6.1 a heuristic way of determining μ is introduced and discussed. In the present section a more reliable technique is developed for the determination of the μ-values, which is based on a combined use of embedded classical explicit and exponential-fitted (explicit) Runge-Kutta (EFRK) methods (see also [31]).

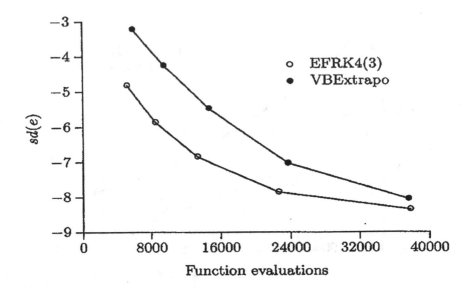

Figure 6.2. Undamped Duffing's equation: $\mu = i$, $t_{end} = 100$.

The methods. For the description of EFRK methods we use the notation introduced in Section 2.2. As an example we give below the form of an embedded EFRK method with 6 stages.

$$
\begin{aligned}
y_{n+1} &= y_n + h \sum_{i=1}^{5} b_i f(x_n + c_i h, Y_i)\,, \\
\hat{y}_{n+1} &= y_n + h \sum_{i=1}^{6} \hat{b}_i f(x_n + c_i h, Y_i)\,, \\
Y_1 &= y_n\,, \\
Y_i &= \gamma_i y_n + h \sum_{j=1}^{i-1} a_{ij} f(x_n + c_j h, Y_j)\,,
\end{aligned}
\tag{6.53}
$$

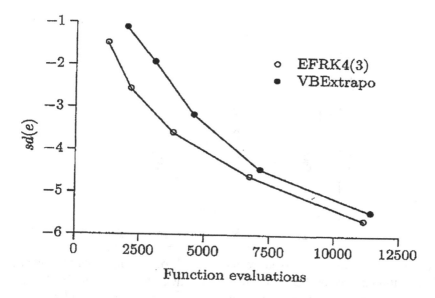

Figure 6.3. Nonlinear system: $\mu = it_n, t_{end} = 10$.

with $i = 2, \ldots, 6$, or in tableau form

c_1	1					
c_2	γ_2	a_{21}				
c_3	γ_3	a_{31}	a_{32}			
c_4	γ_4	a_{41}	a_{42}	a_{43}		
c_5	γ_5	a_{51}	a_{52}	a_{53}	a_{54}	
c_6	γ_6	a_{61}	a_{62}	a_{63}	a_{64}	a_{65}
		b_1	b_2	b_3	b_4	
		$\hat{b}_1$	$\hat{b}_2$	$\hat{b}_3$	$\hat{b}_4$	$\hat{b}_5$

A classical embedded Runge-Kutta method has all $\gamma_i = 1$ ($i = 1, \ldots, 6$) and all A, b and $\hat{b}$ values are just numbers. For EFRK methods these quantities can be Z-dependent functions. For the determination of the μ values we shall make use of an embedded classical RK-method in combination with an explicit EFRK-method; in particular we consider the well-known (4,5) England's method [9], i.e.

$$
\begin{array}{c|c|cccccc}
0 & 1 \\
\frac{1}{2} & 1 & \frac{1}{2} \\
\frac{1}{2} & 1 & \frac{1}{4} & \frac{1}{4} \\
1 & 1 & 0 & -1 & 2 \\
\frac{2}{3} & 1 & \frac{7}{27} & \frac{10}{27} & 0 & \frac{1}{27} \\
\frac{1}{5} & 1 & \frac{28}{625} & -\frac{1}{5} & \frac{546}{625} & \frac{54}{625} & -\frac{378}{625} \\
\hline
& & \frac{1}{6} & 0 & \frac{2}{3} & \frac{1}{6} & 0 \\
& & \frac{1}{24} & 0 & 0 & \frac{5}{48} & \frac{27}{56} & \frac{125}{336}
\end{array}
$$

The numerical solution derived from the fourth-order component in the embedded method is used as the inital value for the next step and it will be denoted as $y_{n+1}^{class.}$. A feature of this method is that the two last elements of b^T are zero, implying that if the error estimate is not required then only four stages need to be computed. The local truncation error of the fourth-order component of the embedded method has the form

$$
LTE^{class.} = h^5 \psi_1(x, y, f) + \mathcal{O}(h^6) , \tag{6.54}
$$

and it can be estimated numerically by

$$
h \sum_{i=1}^{6} (\hat{b}^T - b^T) k_i , \tag{6.55}
$$

with

$$
k_i = f\left(x_n + c_i h, y_n + \sum_{j=1}^{i-1} a_{ij} k_j\right) .
$$

As an EFRK method we have constructed the exponential-fitted analogue of the fourth-order component of England's method. Due to Theorem 6.2 an explicit Runge Kutta method can exactly integrate only a set of two linearly independent functions; for EFRK methods one can choose either the set $\exp(\mu x), \exp(-\mu x)$, or equivalently $\sin(\omega x), \cos(\omega x)$ $\mu = i\omega$, instead of the classical set $1, x$. This means that every stage equation and the final step b-dependent equation in (6.53) has to integrate exactly this set of functions. We also want to preserve the components of c at their classical values in the fourth-order Runge-Kutta scheme, i.e. $c_1 = 0, c_2 = c_3 = 1/2$ and $c_4 = 1$. The above conditions give rise to equations of the type (6.27-6.30) with $s = 4$.

It is clear that a lot of freedom in the determination of the A, b and γ components is still left, but we choose to accept just the classical values for some of them viz. (this choice does not give a special advantage; other choices are possible as well):

$$a_{43} = 2, \quad b_2 = 0, \quad \gamma_3 = 1, \quad \gamma_4 = 1 . \tag{6.56}$$

By solving the set of equations (6.27-6.30) and (6.31) the following Z-dependent coefficients result:

$$
\begin{aligned}
a_{21} &= \frac{\eta_0(Z/4)}{2}, & \gamma_2 &= \eta_{-1}(Z/4), & a_{41} &= 0, \\
a_{31} &= a_{32} = \frac{\eta_0(Z/4)}{2(\eta_{-1}(Z/4)+1)}, & a_{42} &= \eta_0(Z/4) - 2, & & (6.57) \\
b_1 &= b_4 = \frac{\eta_0(Z/4) - 1}{2(\eta_{-1}(Z/4) - 1)}, & b_3 &= \frac{\eta_{-1}(Z/4) - \eta_0(Z/4)}{(\eta_{-1}(Z/4) - 1)} .
\end{aligned}
$$

These expressions can be re-written by (3.35) (see for equivalent operations Tables 6.9 and 6.10).

The local truncation error of this EFRK-method has the following form:

$$LTE^{EFRK} = h^5(\psi_1(x,y,f) + \psi_2(x,y,f,\mu)) + \mathcal{O}(h^6) , \tag{6.58}$$

where the function $\psi_1(x,y,f)$ is defined in (6.54) and where

$$\psi_2(x,y,f,\mu) = -\mu^2 \psi_3(x,y,f) + \mathcal{O}(\mu^4) . \tag{6.59}$$

Typical expressions for the functions $\psi_1(x,y,f)$ and $\psi_3(x,y,f)$ for different EFRK-methods with 2, 3 and 4 stages can be found in (6.33–6.35). The numerical value obtained by the EFRK method (6.56–6.57) will be denoted as y_{n+1}^{EFRK}. A estimate for $h^5 \psi_2(x,y,f,\mu)$ is obtained from

$$y_{n+1}^{EFRK} - y_{n+1}^{class.} . \tag{6.60}$$

REMARK 6.7 *The splitting up of LTE^{EFRK} into the form (6.58–6.59) is valid only for scalar equations. It can be used for systems, when each component of the solution is characterized by one and the same μ. In other cases the μ present in the expressions of the $\psi_2(x,y,f,\mu)$ functions has to be considered as a diagonal matrix and has to be placed immediately in front of the occurring derivative form $y^{(i)}$. This problem will be considered in detail when implicit exponential-fitted Runge-Kutta methods are discussed.*

μ-determination and steplength control. On each integration step the following operations are performed:

1 The system is integrated by the classical (4,5) England's method.

2 The $LTE^{class.}$ is determined by (6.55).

3 The system is integrated by the constructed EFRK-method, where for each equation in the system an arbitrary μ-value is used. This is denoted by μ_0.

4 The value of $h^5\psi_2(x, y, f, \mu_0)$ is calculated by equation (6.60).

5 For each equation in the system equation (6.59) furnishes $h^5\psi_3(x, y, f)$ as

$$h^5\psi_3(x, y, f) = -\frac{h^5\psi_2(x, y, f, \mu_0)}{\mu_0^2} .$$

6 Again for each equation an optimal μ value is obtained by making the leading order term in LTE^{EFRK} as small as possible in terms of $\alpha = -\psi_1(x, y, f)/\psi_3(x, y, f)$, i.e.

- $\mu = \sqrt{\alpha}$ if $\alpha > 0$,

- $\mu = i\omega = i\sqrt{|\alpha|}$ if $\alpha < 0$. In this case the trigonometric functions instead of the exponential ones are integrated.

7 The system is re-integrated with the EFRK-method where for each equation the calculated μ-value is used for the determination of the A- and b-coefficients under (6.57). By the described μ determination the order of the EFRK-method is expected to increase by one unit. This can easily be tested in the following way by considering the equation $y' = -4y$ with $y(0) = 1$, having an exact solution $y(x) = \exp(-4x)$. It is solved twice over the interval $[0, 1]$ by the EFRK-method, the first time with a fixed steplength $h = 0.01$ and a second time with $h = 0.02$. In each knot point the global truncation error $|y_n - y(x_n)|$ is calculated and divided by h^4 and by h^5 in order to determine the x-dependent error form factor of the method, which should be independent of the chosen fixed steplength. The results are plotted in Figures 6.4 and 6.5.

One should usually expect that in so much as the order is assigned correctly the form factor should be one and the same, irrespective of the value of h. In Figure 6.4 it is assumed that the order is four and we see that the expectation is violated. More than that, the form factor corresponding to the halved stepsize is clearly half of the other. However, if the order is assumed as five the form factor is one and the same for the two stepsizes (see Figure 6.5). This indicates that five is the correct value of the order.

8 In order to find an estimate for the leading order term of the LTE corresponding to the calculation with the EFRK-method with adjusted μ-values we used a Richardson extrapolation technique. We apply the EFRK-method

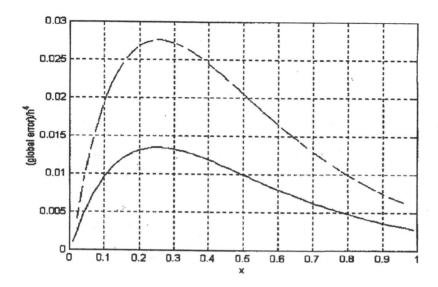

Figure 6.4. Plot of the error form factor when the order of the EFRK-method is assumed to be four. The solid line are the results obtained with $h = 0.02$, the dashed line the ones obtained with $h = 0.01$.

to obtain the solution y_{n+1} at x_{n+1}. Under the usual localizing assumption that $y_n = y(x_n)$ it is clear that the LTE^{EFRK} can be written in the form

$$LTE^{EFRK} = y(x_{n+1}) - y_{n+1} = C(y, f, \mu)h^6 + \mathcal{O}(h^7) ,$$

where $C(y, f, \mu)$ is some complicated function of y, its derivatives, $f(x, y)$ and its partial derivatives with respect to x and y, all evaluated at the point (x_n, y_n), and of μ.

A second value of the numerical solution at x_{n+1}, hopefully better, is calculated by applying the same EFRK-method twice with steplength $h/2$ and also starting in x_n; let this be denoted by z_{n+1}. The same set of μ-values can be used for both calculations. Under the usual localizing assumption the local truncation error is now given by

$$LTE^{EFRK} = y(x_{n+1}) - z_{n+1} = 2C(y, f, \mu)(h/2)^6 + \mathcal{O}(h^7) ,$$

which, at its turn, produces an estimate of the error in the second calculation, viz.:

$$error = 2C(y, f, \mu)(h/2)^6 \approx (z_{n+1} - y_{n+1})/31$$

If the user asks for a given tolerance *tol*, he can control the steplength and the error in the same way as described at the end of part 2.6.1.

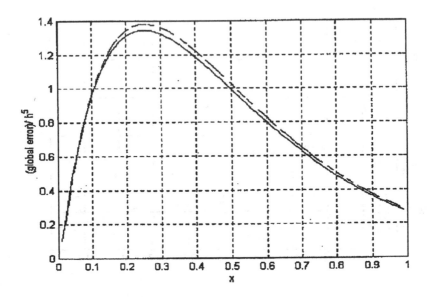

Figure 6.5. Plot of the error form factor when the order of the EFRK-method is assumed to be five. The solid line are the results obtained with $h = 0.02$, the dashed line the ones obtained with $h = 0.01$.

Numerical experiments. In this section we solve some initial value problems (single first-order equations as well as systems of first-order equations) having as solution a combination of sine, cosine or exponential functions. In the case of high-order systems the equations are presented as equivalent first-order systems. These systems have been solved with the described EFRK method and the considered classical embedded RK method; in the EFRK case we applied the above described Richardson extrapolation technique. For the classical method the error was calculated by the formula (6.55). Tolerance values, *tol* of 10^{-5}, 10^{-7} and 10^{-9} have been considered. For each equation an arbitrary μ_0-value has been used. For each example we have calculated the Euclidean norm of the error vector as defined by (6.50). These data together with the number of accepted and rejected steps and the number of evaluations of the r.h.s. (numf) of the systems are presented in Table 6.14. The following cases have been considered:

- **Example 1**

$$y' = x + y, \quad y(0) = 2 . \tag{6.61}$$

Its exact solution is $y(x) = 3\exp(3x) - x - 1$. Equation (6.61) has been solved in the interval $0 \le x \le 4$ with $\mu_0 = 0.5$.

- **Example 2**

$$y' = -4y, \quad y(0) = 1 , \tag{6.62}$$

with exact solution $y(x) = \exp(-4x)$. Equation (6.62) has been solved in the interval $0 \leq x \leq 2$ with $\mu_0 = 0.5$.

- **Example 3**

$$y' = 15\cos(15x), \quad y(0) = 0 , \tag{6.63}$$

with exact solution $y(x) = \sin(15x)$. Equation (6.63) has been solved in the interval $0 \leq x \leq 3\pi/2$ with $\lambda_0 = 0.2$.

- **Example 4**

$$y' = y\cos(x), \quad y(0) = 1, \tag{6.64}$$

with exact solution $y(x) = \exp(\sin(x))$ Equation (6.64) has been solved in the interval $0 \leq x \leq 10$ with $\lambda_0 = 0.5$.

- **Example 5**

$$\begin{cases} y_1' &= -y_1 + y_2, \quad y_1(0) = 3 \\ y_2' &= y_1 - y_2, \quad y_2(0) = 1 \end{cases} \tag{6.65}$$

with exact solution $y_1(x) = 2 + \exp(-2x)$ and $y_2(x) = 2 - \exp(-2x)$. The system (6.65) has been solved in the interval $0 \leq x \leq 2$ with $\mu_0 = 0.5$ for each component of the solution as a seed value.

It is clear from Table 6.14 that for all examples the number of steps is lower in the EFRK than in the classical method. The higher the asked tolerance the better the EFRK method scores over the classical method in the number of r.h.s. evaluations. In all cases, the Euclidean norm which in essence represents the absolute error, is compatible with the given tolerance. In Figure 6.6 we plot the calculated values for ω in each knot point for Example 3 for $tol = 10^{-5}$. In this case the exact value of ω is just 15 in all points. As far the calculated values are regarded they only deviate within 4 percent from that value. As a matter of fact in all cases where calculated and theoretical values could be compared, analogous results are obtained.

In Figure 6.7 we give the results obtained for Example 4. The solid line gives the exact solution, while the circles and the crosses represent the results at the mesh points corresponding to the EFRK and the classical method, respectively. The figure clearly demonstrates that the number of steps to obtain the solution is much smaller with the EFRK than with the classical method.

3. Exponential-fitted implicit methods

Up to now only explicit EFRK-methods have been discussed. In the present section implicit EFRK methods are constructed. The structure of the local truncation error (LTE) is studied. A technique is developed to determine an appropriate value for the occurring frequency in every step of the calculation.

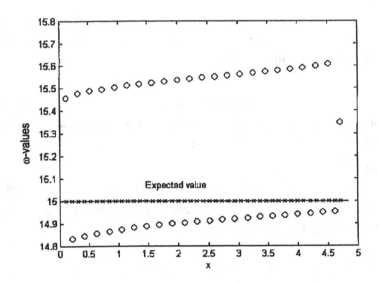

Figure 6.6. Plot of the calculated ω-values for the Example 3 with $tol = 10^{-5}$. The theoretically expected value 15 is also given.

3.1   Classical implicit Runge-Kutta methods

For the description of implicit EFRK methods we use the classical Butcher notation [4]:

$$y_{n+1} = y_n + h\sum_{i=1}^{s} b_i f(x_n + c_i h, Y_i) , \qquad (6.66)$$

$$Y_i = y_n + h\sum_{j=1}^{s} a_{ij} f(x_n + c_j h, Y_j) , \qquad (6.67)$$

with $i = 1, \ldots, s$, or in tableau form

$$
\begin{array}{c|cccc}
c_1 & a_{11} & a_{12} & \cdots & a_{1s} \\
c_2 & a_{21} & a_{22} & \cdots & a_{2s} \\
   &   & \cdots &   &   \\
c_s & a_{s1} & a_{s2} & \cdots & a_{ss} \\
\hline
   & b_1   & b_2   & \cdots & b_s
\end{array}
\qquad (6.68)
$$

There are several ways to introduce classical implicit Runge-Kutta methods.

Table 6.14. Comparison of the Euclidean norms of the end-point global errors obtained by using the fourth-order EFRK and the classical embedded RK-method with steplength control.

		$\log(tol)$	acc. steps	rej. steps	numf	error
Example 1	EFRK	−5	12	0	221	$9.33E - 4$
		−7	23	0	430	$2.40E - 5$
		−9	48	0	905	$5.70E - 7$
	Class.	−5	32	0	190	$2.05E - 3$
		−7	81	1	490	$5.32E - 5$
		−9	205	2	1240	$1.31E - 6$
Example 2	EFRK	−5	9	0	164	$4.82E - 6$
		−7	17	1	335	$5.89E - 8$
		−9	34	2	677	$4.04E - 9$
	Class.	−5	20	1	124	$3.50E - 6$
		−7	46	2	286	$1.23E - 7$
		−9	111	4	688	$3.66E - 9$
Example 3	EFRK	−5	45	21	1247	$5.96E - 5$
		−7	91	23	2159	$2.42E - 7$
		−9	189	33	4211	$7.10E - 9$
	Class.	−5	149	33	1090	$1.86E - 5$
		−7	362	32	2362	$4.29E - 7$
		−9	901	49	5698	$1.06E - 8$
Example 4	EFRK	−5	18	5	430	$6.88E - 6$
		−7	35	8	810	$4.51E - 8$
		−9	71	9	1513	$3.13E - 9$
	Class.	−5	36	6	250	$2.88E - 5$
		−7	82	8	538	$6.64E - 7$
		−9	196	11	1240	$8.15E - 9$
Example 5	EFRK	−5	7	0	126	$5.84E - 6$
		−7	12	0	221	$1.61E - 7$
		−9	24	1	468	$4.81E - 9$
	Class.	−5	14	0	82	$1.09E - 5$
		−7	33	2	208	$3.27E - 7$
		−9	81	3	502	$8.60E - 9$

- One can see them as the methods for which the components of the A-matrix and the c- and b-vectors are the solution of order conditions of a form similar to equations (6.12a-h), i.e. (see [4, 18]

  - Order 1:

$$\sum_{i=1}^{s} b_i = 1 \qquad\qquad (6.69a)$$

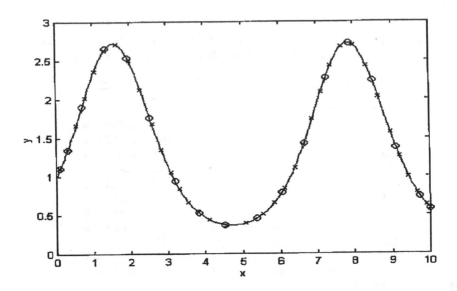

Figure 6.7.   Plot of the exact solution (solid line) and the numerical results obtained for Example 4 with $tol = 10^{-5}$ with the EFRK (circles) and with the classical Runge-Kutta method (crosses).

- Order 2:

$$\sum_{i=1}^{s} b_i c_i = \frac{1}{2} \qquad\qquad (6.69\text{b})$$

- Order 3:

$$\sum_{i=1}^{s} b_i c_i^2 = \frac{1}{3}, \qquad\qquad (6.69\text{c})$$

$$\sum_{i=1}^{s} b_i a_{ij} c_j = \frac{1}{6} \qquad\qquad (6.69\text{d})$$

- Order 4:

$$\sum_{i=1}^{s} b_i c_i^3 = \frac{1}{4}, \qquad\qquad (6.69\text{e})$$

$$\sum_{i=1,j=1}^{s} b_i c_i a_{ij} c_j = \frac{1}{8}, \qquad\qquad (6.69\text{f})$$

$$\sum_{i=1,j=1}^{s} b_i a_{ij} c_j^2 = \frac{1}{12}, \qquad\qquad (6.69\text{g})$$

$$\sum_{i=1,j=1,k=1}^{s} b_i a_{ij} a_{jk} c_k = \frac{1}{24} .\qquad (6.69\text{h})$$

- Fully implicit Runge-Kutta methods can be categorized by the class of quadrature formulae to which they revert when we put $f(x, y) = f(x)$. Three typical categories are known in the literature (see for example [18]):

  - **The Gauss or Gauss-Legendre methods.** They squeeze out the highest possible order, and the s-stage Gauss method has order $2s$.

  - **The Radau methods.** They are characterized by the requirement that the ordinates include one or other of the ends of the interval of integration. This means that the corresponding implicit Runge-Kutta methods have either $c_1 = 0$ (*Radau I*) or $c_s = 1$ (*Radau II*). The maximum attainable order of such an s-stage method is $2s - 1$.

  - **The Lobatto methods.** In that category the ordinates include both ends of the interval of integration. The corresponding implicit methods have $c_1 = 0$ and $c_s = 1$. The maximum attainable order is now $2s - 2$. According to a classification of Butcher [3] there are several types denoted as *LobattoIIIA, B* and *C*

- Implicit Runge–Kutta methods can also be categorized according to whether or not they are *collocation methods*. Collocation consists of choosing a function (usually a polynomial) and a set of mesh points, and then demanding that at the nodes the function does whatever is neccesary to make it mimic the behaviour of the unknown function we are trying to approximate numerically. In the context of solving initial value problems, $y' = f(x, y)$, $y(a) = \eta$, one can advance the numerical solution from x_n to x_{n+1} by choosing a polynomial P of degree s and a set of distinct collocation points $\{x_n + c_i\, h, i = 1, 2, \ldots, s\}$ and demanding that

$$\begin{aligned} P(x_n) &= y_n , \\ P'(x_n + c_i\, h) &= f(x_n + c_i\, h, P(x_n + c_i h)), \qquad i = 1, 2, \ldots, s . \end{aligned}$$

One then completes the step by taking $y_{n+1} = P(x_n + h)$. It was originally shown by Wright [36] that this process is identical with an s-stage implicit Runge-Kutta method. This way of working is closely related to the exponential-fitted approach where, instead of a polynomial basis, an exponential fitting set

$$\exp(\pm\mu x), x \exp(\pm\mu x), x^2 \exp(\pm\mu x), \ldots$$

is considered. Several important theorems concerning collocation methods have been proved. For further reference we just cite them here. Proofs are given by Hairer *et al.* [12].

THEOREM 6.3 *A collocation method is equivalent to the s-stage implicit Runge-Kutta method with*

$$a_{ij} = \int_0^{c_i} l_j(t)dt, \qquad b_j = \int_0^1 l_j(t)dt, \qquad i,j = 1,\ldots,s, \qquad (6.70)$$

where the $l_j(s)$ are the Lagrange polynomials

$$l_j(t) = \frac{\Pi_{k\neq j}(t - c_k)}{\Pi_{k\neq j}(c_j - c_k)}.$$

THEOREM 6.4 *An implicit Runge-Kutta method with all c_i different and of order at least s is a collocation method iff (6.15) is true for $\eta = s$.*

THEOREM 6.5 *The collocation polynomial $P(x)$ gives rise to a continuous implicit Runge-Kutta method of order s, i.e. for all $x_n \le x \le x_n + h$ one has*

$$\|y(x) - P(x)\| \le Ch^{s+1}.$$

Moreover for the derivatives of $P(x)$ one has

$$\|y^{(k)}(x) - P^{(k)}(x)\| \le Ch^{s+1-k}, \quad k = 0,\ldots,s.$$

The families of implicit Runge-Kutta methods quoted in this section split into two groups: the Gauss, RadauIIA and LobattoIIIA methods are collocation methods, while the RadauIA, LobattoIIIB and LobattoIIIC are not. Some one- and two-stage implicit collocation Runge-Kutta methods are listed in Tables 6.15–6.19.

Table 6.15. Gauss Method
(stage-order=1,order=2).

$$\begin{array}{c|c} \frac{1}{2} & \frac{1}{2} \\ \hline & 1 \end{array}$$

Table 6.16. Gauss Method
(stage-order=2, order=4).

$$\begin{array}{c|cc} \frac{3-\sqrt{3}}{6} & \frac{1}{4} & \frac{3-2\sqrt{3}}{12} \\ \frac{3+\sqrt{3}}{6} & \frac{3+2\sqrt{3}}{12} & \frac{1}{4} \\ \hline & \frac{1}{2} & \frac{1}{2} \end{array}$$

- Implicit Runge-Kutta methods can also be constructed following Albrecht's approach [1, 2, 18] and Section 2.2. In this approach each of the s stages and the final stage are considered as being a generalized linear multistep

Table 6.17. RadauIIA Method
(stage-order=1,order=1).

$$
\begin{array}{c|c}
1 & 1 \\
\hline
 & 1
\end{array}
$$

Table 6.18. RadauIIA Method
(stage-order=2, order=3).

$$
\begin{array}{c|cc}
\dfrac{1}{3} & \dfrac{5}{12} & -\dfrac{1}{12} \\[2mm]
1 & \dfrac{3}{4} & \dfrac{1}{4} \\[2mm]
\hline
 & \dfrac{3}{4} & \dfrac{1}{4}
\end{array}
$$

Table 6.19. LobattoIIIA Method
(stage-order=2,order=2).

$$
\begin{array}{c|cc}
0 & 0 & 0 \\[1mm]
1 & \dfrac{1}{2} & \dfrac{1}{2} \\[2mm]
\hline
 & \dfrac{1}{2} & \dfrac{1}{2}
\end{array}
$$

method on a nonequidistant grid and associate with it a linear operator in exactly the same way as described in the previous chapters, i.e.

$$
\mathcal{L}_i[h, \mathbf{a}]y(x) = y(x + c_i h) - y(x) - h \sum_{j=1}^{s} a_{ij}\, y'(x + c_j h)\ i = 1, 2, \ldots, s
$$

$$(6.71)$$

and

$$
\mathcal{L}[h, \mathbf{b}]y(x) = y(x + h) - y(x) - h \sum_{i=1}^{s} b_i y'(x + c_i h)\ . \tag{6.72}
$$

We describe below the derivation of the classical implicit Runge–Kutta methods. The set of power functions

$$
1, x, x^2, x^3 \ldots,
$$

is appropriate and then the subsequent treatment first implies the determination of the starred classical moments $L_{ij}^*(\mathbf{a})$ and $L_i^*(\mathbf{b})$, see the six-step flow chart of Chapter 3 and also [14].

Once these moments have been constructed we have to examine the linear systems

$$
\begin{aligned}
L_{ij}^*(\mathbf{a}) &= 0, \quad i = 1, \ldots .s,\ j = 0, 1, 2, \ldots, M - 1 \quad &(6.73)\\
L_i^*(\mathbf{b}) &= 0, \quad i = 0, 1, 2, \ldots, M' - 1 \quad &(6.74)
\end{aligned}
$$

to find out the maximal M and M' values for which they are compatible. In what follows it is obvious to choose $M = s + 1$; the RK-methods related to that choice are again the collocation methods with stage-order s. For this particular choice M' can vary between $s + 1$ and $2s + 1$ (Gauss methods). From the structure itself of the relations (6.71)–(6.72) it is clear that for every $i = 1, 2 \ldots, s$ one has

$$L_{i0}^*(\mathbf{a}) = 0 \text{ and } L_0^*(\mathbf{b}) = 0 .$$

In other words a constant is always exactly integrated by a RK-method. The conditions $L_{ij}^*(\mathbf{a}) = 0$ give rise to the equations

$$c_i^j - j \sum_{k=1}^{s} a_{ik} c_k^{j-1} = 0 , \quad j = 1, \ldots, s , i = 1, \ldots, M - 1 , \quad (6.75)$$

which are the well-known stage-order conditions or simplifying assumptions. The $j = 1$ case just represents the well-known row-sum condition (see (6.15) and Definition 6.4). The conditions $L_i^*(\mathbf{b}) = 0$ give rise to the equations

$$1 - i \sum_{j=1}^{s} b_j c_j^{i-1} = 0 , i = 1, \ldots, M' - 1 , \quad (6.76)$$

which represent the quadrature order conditions which are denoted in [12] as $B(M')$ (see also (6.17) and Definition 6.5). This is a sufficient condition for the method to have order $M' - 1$ if the restrictions $M = s + 1$ and $s + 1 \leq M' \leq 2s + 1$ are valid.

As an example of the use of the above theory we consider here the two-stage case (see also [32]). It is obvious to put in first instance $M = M' = s + 1 = 3$, i.e. if one considers the hybrid reference set (see also (3.54))

$$1, x, x^2, \ldots, x^K \text{ or } x^{K'}, \quad (6.77)$$
$$\exp(\pm \mu x), x \exp(\pm \mu x), \ldots, x^P \exp(\pm \mu x) \text{ or } x^{P'} \exp(\pm \mu x) ;$$

this choice is equivalent with $(K = K' = 2, P = P' = -1)$. The solution of the systems (6.73-6.74) with $(i = 1, 2; j = 1, 2)$ is then given by

$$a_{11} = \frac{c_1(-c_1 + 2c_2)}{2(-c_1 + c_2)} , \quad a_{12} = -\frac{c_1^2}{2(-c_1 + c_2)} , \quad a_{21} = \frac{c_2^2}{2(-c_1 + c_2)} ,$$
$$a_{22} = \frac{c_2(-2c_1 + c_2)}{2(-c_1 + c_2)} , \quad b_1 = \frac{(2c_2 - 1)}{2(-c_1 + c_2)} , \quad b_2 = -\frac{(2c_1 - 1)}{2(-c_1 + c_2)} .$$
$$(6.78)$$

As well-known, the particular choice of c_1 and c_2 determines the order of the method.

- For arbitrary c_i's the order is 2. In particular for $c_1 = 0$ and $c_2 = 1$ one obtains the LobattoIIIA method of order 2 (see also Table 6.19);

- Adding $L_3^*(\mathbf{b}) = 0$ (this means $M' = 4, K' = 3, P' = -1$) to the equations previously considered we find $c_1 = (3c_2 - 2)/3(2c_2 - 1)$; the special choice $c_2 = 1$ and then $c_1 = 1/3$ results in the two-stage RadauIIA method (see also Table 6.18);

- Adding $L_3^*(\mathbf{b}) = 0$ and $L_4^*(\mathbf{b}) = 0$ (this means $M' = 5, K' = 4, P' = -1$) to the equations considered delivers

$$c_1 = \frac{3 - \sqrt{3}}{6} \quad \text{and} \quad c_2 = \frac{3 + \sqrt{3}}{6} .$$

Note that the simplifying assumptions $C(s)$ and the quadrature order conditions $B(2s)$ are fulfilled. Due to the theorem 342A and 342D in [4] one can prove that the $D(s)$ simplifying conditions (see Definition 6.3)

$$\sum_{i=1}^{s} b_i c_i^{q-1} a_{ij} = \frac{b_j}{q}(1 - c_j^q), \quad j = 1, \ldots, s, \quad q = 1, \ldots, s ,$$

are also fulfilled, resulting in the fact that the maximal algebraic order $2s = 4$ is obtained (Gauss method) (see also Table 6.16).

In [18] the leading term of the local truncation error (*plte*) is found of the form

$$plte = \frac{h^{p+1}}{(p+1)!} \sum_{r(t)=p+1} \alpha(t)[1 - \gamma(t)\psi(t)]F(t) \tag{6.79}$$

where p is the order of the method (varying between M and M'); for the explanation of the used symbols we refer to [18]. When $p = M' - 1 = s$ and $q = M - 1 = s$ it is easy to show that for every occurring t one has

$$\gamma(t)\psi(t) = \gamma([\tau^{s+1}])\psi([\tau^{s+1}]) , \tag{6.80}$$

such that in that case

$$plte = \frac{h^{s+1}}{(s+1)!} (1 - \gamma([\tau^{s+1}])\psi([\tau^{s+1}])) \sum_{r(t)=s+1} \alpha(t)F(t) ,$$

which simplifies to

$$plte = \frac{h^{s+1}}{(s+1)!} (1 - \gamma([\tau^{s+1}])\psi([\tau^{s+1}]))y^{(s+1)} .$$

For the LobattoIIIA case with $s = 2$, this last expression reads

$$plte(LobattoIIIA, s = 2) = \frac{h^3}{3!}(1 - 3\sum_{i=1}^{2} b_i c_i^2)y^{(3)}$$

$$= -\frac{h^3}{12}y^{(3)} . \tag{6.81}$$

In the case $p = M' - 1 = s + 1$ and $q = M - 1 = s$ (6.79) does not simplify to a very simple expression, depending only on one total derivative of the solution y. One can use the simplifying assumptions (in fact the moments L_{ij}^*) to derive the *lte* expression. For the case $s = 2$ one finds

$$plte = \frac{h^4}{4!}\left[(1 - 4\sum_{i=1}^{2} b_i c_i^3)y^{(4)} + (4\sum b_i c_i^3 - 12\sum b_i a_{ij} c_j^2)f_y y^{(3)}\right] .$$

This expression clearly shows that the methods considered have an order 3 and a stage-order 2. In case of the RadauIIA method this form reduces to (we give the expression for the case of a scalar ODE):

$$plte(RadauIIA, s = 2) = -\frac{h^4}{216}(y^{(4)} - 4f_y y^{(3)}) . \tag{6.82}$$

In the same spirit one can construct the *lte* for the $s = 2$ Gauss method taking into account the simplifying conditions. One finds

$$plte(Gauss, s = 2) = \frac{h^5}{5!}\left[A'y^{(5)} + (C' - A')f_y y^{(4)} \right.$$
$$\left. + [(D' - C')f_y^2 + 4(B' - A')(f_{xy} + f_{yy}f)]y^{(3)}\right]$$

showing again the order 4 and stage-order 2 properties. The expressions of A', B', C' and D' are as it follows:

$$A' = (1 - 5\sum b_i c_i^4) ,$$
$$B' = (1 - 15\sum b_i c_i a_{ij} c_j^2) ,$$
$$C' = (1 - 20\sum b_i a_{ij} c_j^3) ,$$
$$D' = (1 - 60\sum b_i a_{ij} a_{jk} c_k^2) .$$

On substituting here the numerical values of the coefficients the expression of the *plte* simplifies to

$$plte(Gauss, s = 2) = \frac{h^5}{4320}(y^{(5)} - 5f_y y^{(4)} + 10(f_y^2 - f_{xy} - f_{yy}f)y^{(3)}) . \tag{6.83}$$

3.2 A detailed study of the exponential-fitted two-stage implicit Runge-Kutta methods

In the spirit of the six-step flow chart and in the same way as we did in Section 2.2 for the explicit Runge-Kutta schemes we have to construct the formal expressions $G_i^{\pm}(Z, \mathbf{a})$ and $G^{\pm}(Z, \mathbf{b})$, but they are already known. In fact they are given by (6.27-6.30) with all $\gamma_i = 1$. We also consider, if appropriate, their derivatives with respect to Z, which are denoted, as usual, as $G_{ij}^{\pm(p)}(Z, \mathbf{a})$ and $G^{\pm(p)}(Z, \mathbf{b})$ for $p = 1, 2, \ldots$ (see also [32]).

We choose as reference set of M and M' functions the hybrid set (6.77) with either $K + 2P = M - 3$ or $K' + 2P' = M' - 3$ .This means that the parameters P or P' and K' or K' can have different values in the internal stages and in the final stage. The two parameters $K(K')$ and $P(P')$ characterise the reference set. The set in which there is no classical component (except for 1) is identified by $K = 0$ and $K' = 0$ while the set in which there is no exponential fitting component is identified by $P = -1$ and $P' = -1$ (see also Section 3.1). Parameters P' and P are called the levels of tuning for the final and internal stages. Note that in contrast to the multistep case ([15] and Chapter 5) the smallest value here attainable for $K(K')$ is 0 and not -1.

Having chosen the reference set one solves formally the linear systems

$$L_{ij}^*(\mathbf{a}) = 0, \ G_i^{\pm}(Z, \mathbf{a}) = 0, \ G_i^{\pm(p)}(Z, \mathbf{a}) = 0, \ p = 1, \ldots, P ,$$
$$\text{with } i = 1, 2, \ldots, s, \ j = 1, \ldots, K ,$$
$$L_i^*(\mathbf{b}) = 0, \ G^{\pm}(Z, \mathbf{b}) = 0, \ G^{\pm(p)}(Z, \mathbf{b}) = 0, \ p = 1, \ldots, P' ,$$
$$\text{with } i = 1, 2, \ldots, K' ,$$

with Z dependent coefficients. The numerical values of a_{ij} and b_i are computed either for real μ-values (exponential case) or for pure imaginary values $\mu = i\omega$ (oscillatory case).

Once the Butcher tableau is determined, the leading order term of the local truncation error can be written down both for the pure exponential fitted and for the mixed case.

3.2.1 Order 2 methods

As for the classical case we start again with choosing $M = M' = s+1 = 3$. In order to have $\exp(\pm\mu x)$ as basic functions we select $K = 0$ and $K' = 0$ which guarantees that only one classical component, that is 1, is present. We consider the solution $\mathbf{b}$ of the equations $G^{\pm}(Z, \mathbf{b}) = 0$ and the solutions $\mathbf{a}$ of $G_i^{\pm}(Z, \mathbf{a}) = 0, i = 1, 2$, i.e.

$$G^+(Z, \mathbf{b}) = \eta_{-1}(Z) - 1 - Z(b_1 c_1 \eta_0(c_1^2 Z) + b_2 c_2 \eta_0(c_2^2 Z)) = 0 ,$$

(6.84)

$$G^-(Z, \mathbf{b}) = \eta_0(Z) - (b_1 \eta_{-1}(c_1^2 Z) + b_2 \eta_{-1}(c_2^2 Z)) = 0 ,$$ (6.85)

and

$$G_i^+(Z, \mathbf{a}) = \eta_{-1}(c_i^2 Z) - 1 - Z(a_{i1}c_1\eta_0(c_1^2 Z) + a_{i2}c_2\eta_0(c_2^2 Z)) = 0 ,$$
(6.86)

$$G_i^-(Z, \mathbf{a}) = c_i\eta_0(c_i^2 Z) - (a_{i1}\eta_{-1}(c_1^2 Z) + a_{i2}\eta_{-1}(c_2^2 Z)) = 0 , \quad (6.87)$$

for $i = 1, 2$. The following results are obtained:

$$\begin{aligned}
D &= Z(-\eta_{-1}(c_1^2 Z)c_2\eta_0(c_2^2 Z) + c_1\eta_0(c_1^2 Z)\eta_{-1}(c_2^2 Z)) , \\
a_{11} &= (\eta_{-1}(c_1^2 Z)\eta_{-1}(c_2^2 Z) - \eta_{-1}(c_2^2 Z) - Zc_1\eta_0(c_1^2 Z)c_2\eta_0(c_2^2 Z))/D , \\
a_{12} &= (-\eta_{-1}^2(c_1^2 Z) + \eta_{-1}(c_1^2 Z) + Zc_1^2\eta_0^2(c_1^2 Z))/D , \\
a_{21} &= (\eta_{-1}^2(c_2^2 Z) - \eta_{-1}(c_2^2 Z) - Zc_2^2\eta_0^2(c_2^2 Z))/D , \quad (6.88) \\
a_{22} &= (Zc_1\eta_0(c_1^2 Z)c_2\eta_0(c_2^2 Z) - \eta_{-1}(c_1^2 Z)\eta_{-1}(c_2^2 Z) + \eta_{-1}(c_1^2 Z))/D , \\
b_1 &= (\eta_{-1}(Z)\eta_{-1}(c_2^2 Z) - \eta_{-1}(c_2^2 Z) - \eta_0(Z)Zc_2\eta_0(c_2^2 Z))/D , \\
b_2 &= (\eta_0(Z)Zc_1\eta_0(c_1^2 Z) - \eta_{-1}(Z)\eta_{-1}(c_1^2 Z) + \eta_{-1}(c_1^2 Z))/D .
\end{aligned}$$

These general expressions reduce to very simple ones for particular values of c_1 and c_2. For example for the LobattoIIIA knot points $c_1 = 0$ and $c_2 = 1$ the following results emerge:

$$a_{11} = 0; \ a_{12} = 0; \ a_{21} = a_{22} = b_1 = b_2 = \frac{\eta_{-1}(Z) - 1}{Z\eta_0(Z)} . \quad (6.89)$$

As for the error, the differential equation $y^{(3)} - \mu^2 y' = 0$ is the one which has the three functions $1, \exp(\pm\mu x)$ as its linearly independent solutions; therefore the leading term should be of the form

$$plte = X(-\mu^2 y' + y^{(3)}) . \quad (6.90)$$

The factor X is fixed in terms of (6.6). Indeed the coefficient of y' should be the same in (6.90) and (6.6), i.e.

$$X = -\frac{1}{\mu^2} L_1(h, \mathbf{b}) .$$

For the LobattoIIIA case ($c_1 = 0, c_2 = 1$) this leads to the result:

$$plte(LobattoIIIA, s = 2, exp) = -h^3 \frac{Z\eta_0(Z) + 2 - 2\eta_{-1}(Z)}{Z^2\eta_0(Z)}(-\mu^2 y' + y^{(3)}) . \quad (6.91)$$

Note that the Z-dependent factor in this $plte$ can be re-expressed by (3.35) in the following form, which also can be used when Z tends to zero, as

$$\frac{\eta_0(Z/4) + 1}{8\eta_0(Z)}(\eta_0^2(Z/16) - 2\eta_1(Z/4)) - \frac{\eta_0^2(Z/4) - 2\eta_1(Z)}{2\eta_0(Z)} .$$

In the limit $\mu \to 0$ (this means implicitly that z or $Z \to 0$) all new formulae tend to the classical formulae. This is not immediately visible from (6.89) and (6.91). However, developing the expressions for the elements of the Butcher array and the *plte* one obtains

$$b_1 = b_2 = a_{21} = a_{22} = \frac{1}{2} - \frac{1}{24}Z + \frac{1}{240}Z^2 + \mathcal{O}(Z^3)$$

and

$$-\frac{Z\eta_0(Z) + 2 - 2\eta_{-1}(Z)}{Z^2\eta_0(Z)} = -\frac{1}{12} + \frac{1}{120}Z - \frac{17}{20160}Z^2 + \mathcal{O}(Z^3).$$

We see that for small $|Z|$ the first term at the right side is a good approximation of the whole expression in the l.h.s.. This will be the case also for the other exponential fitting versions and then in the following we shall use in the formulae of *plte* only this limit. On this basis it results that

$$plte(LobattoIIIA, s = 2, exp) = -\frac{h^3}{12}(-\mu^2 y' + y^{(3)}),$$

is a realistic expression for the *plte*, confirming the algebraic order 2.

3.2.2   Order 3 methods

Two strategies can be followed in order to increase the algebraic order . One can either introduce into (6.88) the classical constant c_i-values of the corresponding RadauIIA or Gauss methods or one can add one or two additional equations to the equations (6.84)-(6.85), by increasing M' up to 4 or 5. In this section we discuss the case $M' = 4$, while in Section 3.2.3 we consider the case $M' = 5$.

CASE 1 : FIXED c-VALUES

To obtain methods of algebraic order 3 we introduce $c_1 = 1/3$ and $c_2 = 1$ into (6.88). Considering the Taylor expansion with respect to Z one obtains (RadauIIA, case 1):

$$a_{11} = \frac{5}{12} + \frac{25}{1296}Z - \frac{5}{23328}Z^2 + \mathcal{O}(Z^3),$$

$$a_{12} = -\frac{1}{12} + \frac{7}{1296}Z - \frac{31}{116640}Z^2 + \mathcal{O}(Z^3),$$

$$a_{21} = b_1 = \frac{3}{4} + \frac{1}{144}Z + \frac{13}{38880}Z^2 + \mathcal{O}(Z^3),$$

$$a_{22} = b_2 = \frac{1}{4} - \frac{1}{144}Z + \frac{11}{38880}Z^2 + \mathcal{O}(Z^3).$$

To check the algebraic order we consider the classical order conditions in which we introduce the μ-dependent coefficients; the following results are obtained (for the case where the row-sum condition is not satisfied):

$$\sum_i b_i = 1 \qquad\qquad + \frac{1}{1620}Z^2 + \mathcal{O}(Z^3)$$

$$\sum_i b_i c_i = \frac{1}{2} - \frac{1}{216}Z + \frac{23}{58320}Z^2 + \mathcal{O}(Z^3)$$

$$\sum_{i,j} b_i a_{ij} = \frac{1}{2} + \frac{1}{72}Z + \frac{7}{19440}Z^2 + \mathcal{O}(Z^3)$$

$$\sum_i b_i c_i^2 = \frac{1}{3} - \frac{1}{162}Z + \frac{7}{21870}Z^2 + \mathcal{O}(Z^3)$$

$$\sum_{i,j,k} b_i a_{ij} a_{ik} = \frac{1}{3} + \frac{1}{162}Z + \frac{7}{7290}Z^2 + \mathcal{O}(Z^3)$$

$$\sum_{i,j} b_i a_{ij} c_i = \frac{1}{3} \qquad\qquad - \frac{1}{2430}Z^2 + \mathcal{O}(Z^3)$$

$$\sum_{i,j,k} b_i a_{ij} a_{jk} = \frac{1}{6} + \frac{11}{648}Z + \frac{13}{58320}Z^2 + \mathcal{O}(Z^3)$$

$$\sum_{i,j} b_i a_{ij} c_j = \frac{1}{6} + \frac{1}{216}Z + \frac{7}{58320}Z^2 + \mathcal{O}(Z^3),$$

confirming the algebraic order 3. For the *plte* we then find

$$plte(RadauIIA, s = 2, exp, \text{case } 1) =$$

$$-\frac{h^4}{216}\left(-\mu^2 y^{(2)} + y^{(4)} - 4f_y(-\mu^2 y' + y^{(3)})\right). \qquad (6.92)$$

CASE 2 : μ-DEPENDENT c-VALUES

By considering $M' = 4$, (i.e. $K' = 1, P' = 0$) we look for the solution **b** of the equations $L_1^*(\mathbf{b}) = 0$, i.e.

$$L_1^*(\mathbf{b}) = b_1 + b_2 - 1 = 0 \qquad (6.93)$$

and $G^{\pm}(Z, \mathbf{b}) = 0$, as defined in (6.84), (6.85). Solving this system for b_1 and b_2 we obtain

$$b_1 = \frac{\eta_0(Z) - \eta_{-1}(c_2^2 Z)}{\eta_{-1}(c_1^2 Z) - \eta_{-1}(c_2^2 Z)}, \qquad b_2 = \frac{\eta_{-1}(c_1^2 Z) - \eta_0(Z)}{\eta_{-1}(c_1^2 Z) - \eta_{-1}(c_2^2 Z)}, \qquad (6.94)$$

and a transcendental relation in the unknown c's. Defining $d_1 := (c_1 - c_2)/2$ and $d_2 := (c_1 + c_2)/2$ this relation can be written as

$$\eta(d_1^2 Z)\left(\eta_{-1}(d_1^2 Z) - (1 - d_2)\eta((1 - d_2)^2 Z) - d_2 \eta(d_2^2 Z)\right) = 0. \quad (6.95)$$

Guided by the classical RadauIIA case, one can choose either $c_1 = 1/3$ (RadauIIA, case 2a) or $c_2 = 1$ (RadauIIA, case 2b).

case 2a : $c_1 = \frac{1}{3}$

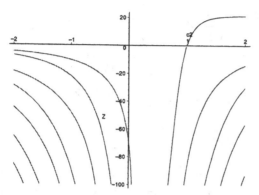

Figure 6.8. The different solutions of (6.95) in case $c_1 = 1/3$.

The equation (6.95) is now a transcendental equation in c_2. A contour plot as in Figure 6.8 (showing where the l.h.s. of (6.95) is equal to zero) reveals that there are several solutions, however, the curve intersecting the real axis (there is only one such curve), is the one in which we are interested. In practice we are interested in that part of the curve for which $|Z|$ is of moderate size (say smaller than 5). For such values of Z, one can solve (6.95) explicitly for c_2 to obtain

$$
c_2 = \begin{cases}
\dfrac{2}{3} + \dfrac{1}{\sqrt{Z}} \log\left(\dfrac{-G_1 + 1 + \sqrt{Z} G_1^{1/3}}{-1 + G_1 - \sqrt{Z} G_1^{2/3}} \right) & Z > 0 \\[2ex]
1 & Z = 0 \\[2ex]
\dfrac{2}{3} - \dfrac{i}{\sqrt{-Z}} \log\left(\dfrac{-G_2 + 1 + i\sqrt{-Z} G_2^{1/3}}{-1 + G_2 - i\sqrt{-Z} G_2^{2/3}} \right) & Z < 0,
\end{cases} \tag{6.96}
$$

where

$$
G_1 = \exp(\sqrt{Z}) = \eta_{-1}(Z) + \sqrt{Z}\eta_0(Z) \tag{6.97}
$$

and

$$
G_2 = \exp(i\sqrt{-Z}) = \eta_{-1}(Z) + i\sqrt{-Z}\eta_0(Z). \tag{6.98}
$$

For very small values of $|Z|$ we also have

$$
c_2 = 1 + \frac{1}{135}Z + \frac{19}{102060}Z^2 + \frac{1}{157464}Z^3 + \mathcal{O}(Z^4),
$$

showing that, in the limit as $\mu \rightarrow 0$, the value of c_2 tends to the classical value of the RadauIIA method. Introducing this expression for c_2 into (6.88) and (6.94) and considering the Taylor series with respect to Z one obtains:

$$b_1 = \frac{3}{4} + \frac{7}{720}Z + \frac{41}{453600}Z^2 + \mathcal{O}(Z^3),$$

$$b_2 = \frac{1}{4} - \frac{7}{720}Z - \frac{41}{453600}Z^2 + \mathcal{O}(Z^3),$$

$$a_{11} = \frac{5}{12} + \frac{119}{6480}Z - \frac{403}{4082400}Z^2 + \mathcal{O}(Z^3),$$

$$a_{12} = -\frac{1}{12} + \frac{41}{6480}Z - \frac{239}{1360800}Z^2 + \mathcal{O}(Z^3),$$

$$a_{21} = \frac{3}{4} + \frac{7}{240}Z + \frac{41}{453600}Z^2 + \mathcal{O}(Z^3),$$

$$a_{22} = \frac{1}{4} - \frac{7}{240}Z - \frac{41}{453600}Z^2 + \mathcal{O}(Z^3).$$

case 2b : $c_2 = 1$

We now have in (6.95) a transcendental equation in c_1 and a contour plot as in Figure 6.9 shows that there are again several solutions, however, we are primarily interested in the curve intersecting the real axis. There is only one such situation (see Figure 6.9). For small values of Z, one can solve (6.95) explicitly for c_1 to obtain

$$c_1 = \begin{cases} \dfrac{1}{\sqrt{Z}} \log\left(\dfrac{-G_1 + 1 + \sqrt{Z}G_1}{-1 + G_1 - \sqrt{Z}} \right) & Z > 0 \\[2mm] \dfrac{1}{3} & Z = 0 \\[2mm] \dfrac{-i}{\sqrt{-Z}} \log\left(\dfrac{-G_2 + 1 + i\sqrt{-Z}G_2}{-1 + G_2 - i\sqrt{-Z}} \right) & Z < 0. \end{cases} \qquad (6.99)$$

A series expansion gives

$$c_1 = \frac{1}{3} - \frac{1}{405}Z + \frac{1}{34020}Z^2 - \frac{1}{3061800}Z^3 + \mathcal{O}(Z^4). \qquad (6.100)$$

and introducing this c_1 in (6.94) and (6.88) a Taylor series development with respect to Z gives

$$a_{11} = \frac{5}{12} + \frac{11}{720}Z - \frac{23}{50400}Z^2 + \mathcal{O}(Z^3),$$

$$a_{12} = -\frac{1}{12} + \frac{1}{144}Z - \frac{17}{50400}Z^2 + \mathcal{O}(Z^3),$$

$$a_{21} = b_1 = \frac{3}{4} + \frac{1}{240}Z - \frac{1}{16800}Z^2 + \mathcal{O}(Z^3),$$

$$a_{22} = b_2 = \frac{1}{4} - \frac{1}{240}Z + \frac{1}{16800}Z^2 + \mathcal{O}(Z^3),$$

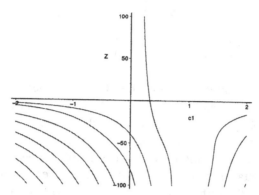

Figure 6.9. The different solutions of (6.95) in case $c_2 = 1$.

showing again that this result is a real extension of the (polynomial based) RadauIIA method.

In the cases 2a and 2b the final equation integrates exactly the functions $1, x$ and $\exp(\pm\mu x)$, which are the linearly independent solutions of the differential equation $y^{(4)} - \mu^2 y'' = 0$; on the other hand the internal stages still only integrate the functions $1, \exp(\pm\mu x)$ which are the linearly independent solutions of the linear differential equation $y^{(3)} - \mu^2 y'$. Taking this into account and the expression (6.82) the *plte* should be of the form:

$$plte(RadauIIA, s = 2, exp, \text{case } 2) =$$
$$-\frac{h^4}{216}\left(-\mu^2 y^{(2)} + y^{(4)} - 4f_y(-\mu^2 y' + y^{(3)})\right), \qquad (6.101)$$

which is exactly the same as in case 1.

3.2.3 Order 4 methods

CASE 1 : FIXED c-VALUES

In order to obtain an algebraic order 4 method we introduce $c_1 = (3-\sqrt{3})/6$ and $c_2 = (3+\sqrt{3})/6$ into (6.88). Considering the Taylor expansion with respect to Z one obtains (Gauss, case 1):

$$b_1 = b_2 = \frac{1}{2} + \frac{Z^2}{8640} + \mathcal{O}(Z^3),$$
$$a_{11} = \frac{1}{4} + \frac{\sqrt{3}}{288}Z + \frac{3 - 4\sqrt{3}}{51840}Z^2 + \mathcal{O}(Z^3),$$

$$a_{12} = \frac{1}{4} - \frac{1}{6}\sqrt{3} + \frac{\sqrt{3}}{864}Z + \frac{3 - 4\sqrt{3}}{51840}Z^2 + \mathcal{O}(Z^3),$$

$$a_{21} = \frac{1}{4} + \frac{1}{6}\sqrt{3} - \frac{\sqrt{3}}{864}Z + \frac{3 + 4\sqrt{3}}{51840}Z^2 + \mathcal{O}(Z^3),$$

$$a_{22} = \frac{1}{4} - \frac{\sqrt{3}}{288}Z + \frac{3 + 4\sqrt{3}}{51840}Z^2 + \mathcal{O}(Z^3).$$

Here again one can check (by considering the algebraic order conditions in which the μ-dependent coefficients are plugged in) that the order is four. The special constant choice of c_1 and c_2 makes it impossible to derive in a straightforward way the form of the *plte* in the exponential-fitted case from the classical form (6.83) of the *plte* . This situation is comparable with the results obtained for the explicit Runge-Kutta methods (6.33-6.35). As for these methods we also calculate here the *plte* by the difference $y(x_{n+1}) - y_{n+1}$ of the exact and the numerical solution in the point x_{n+1}. The following form for the *plte* emerges:

$$plte(Gauss, s = 2, exp, \text{case } 1) =$$
$$\frac{h^5}{4320}(-\mu^2 y^{(3)} + y^{(5)} + \mu^2(-\mu^2 y' + y^{(3)}))$$
$$-5f_y(-\mu^2 y^{(2)} + y^{(4)}) + 10(f_y^2 - f_{xy} - f_{yy}f)(-\mu^2 y' + y^{(3)})).$$

CASE 2 : μ-DEPENDENT c-VALUES

We will discuss two cases : one for which $P' = 0$ (case 2a) and one for which $P' = 1$ (case 2b).

case 2a : $P' = 0$

We take $P' = 0$ such that $K' = 2$ and solve the equations (6.84)–(6.85), (6.93) and

$$L_2^*(\mathbf{b}) = b_1 c_1 + b_2 c_2 - \frac{1}{2} = 0. \tag{6.102}$$

Equations (6.93) and (6.102) give simple expressions for b_1 and b_2 in terms of the knot points, i.e.

$$b_1 = \frac{1 - 2c_2}{2(c_1 - c_2)}, \quad b_2 = \frac{2c_1 - 1}{2(c_1 - c_2)}.$$

Due to symmetry reasons we can hope for a solution if $c_2 = 1 - c_1$, such that $b_1 = b_2 = 1/2$. For this choice the system of nonlinear equations considered has a solution, a fact which has been confirmed by [16] for Gauss quadrature rules for oscillatory functions. Introducing the new variable $d > 0$ as $d = \frac{1}{2} - c_1 = c_2 - \frac{1}{2}$, the resulting condition can be written as

$$\eta_{-1}(d^2 Z) - \eta_0(Z/4) = 0. \tag{6.103}$$

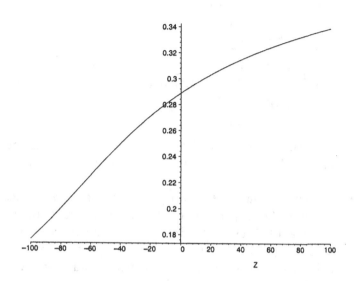

Figure 6.10. A plot of the solution d of (6.103) as a function of Z.

This equation can be solved for d as a function of Z. The graph of the solution is given in Figure 6.10.

The solution for c_2 can then be written as

$$
c_2 = \begin{cases}
\dfrac{1}{\sqrt{Z}} \log\left(\dfrac{G_1 - 1 + \sqrt{(G_1 - 1)^2 - ZG_1}}{\sqrt{Z}} \right) & Z > 0 \\[3ex]
\dfrac{3 + \sqrt{3}}{6} & Z = 0 \\[3ex]
\dfrac{1}{i\sqrt{-Z}} \log\left(\dfrac{G_2 - 1 + \sqrt{(G_2 - 1)^2 - ZG_2}}{i\sqrt{-Z}} \right) & Z < 0
\end{cases} \qquad (6.104)
$$

with G_1 and G_2 respectively defined in (6.97)–(6.98). A series expansion gives

$$
\begin{aligned}
c_1 &= \frac{1}{2} - \frac{\sqrt{3}}{6} - \frac{\sqrt{3}Z}{2160} + \frac{\sqrt{3}Z^2}{403200} + \mathcal{O}(Z^3)\,, \\
c_2 &= \frac{1}{2} + \frac{\sqrt{3}}{6} + \frac{\sqrt{3}Z}{2160} - \frac{\sqrt{3}Z^2}{403200} + \mathcal{O}(Z^3)\,,
\end{aligned}
$$

and

$$a_{11} = \frac{1}{4} + \frac{\sqrt{3}Z}{360} - \frac{19\sqrt{3}Z^2}{302400} + \mathcal{O}(Z^3),$$

$$a_{12} = \frac{1}{4} - \frac{1}{6}\sqrt{3} + \frac{\sqrt{3}Z}{720} - \frac{61\sqrt{3}Z^2}{1209600} + \mathcal{O}(Z^3),$$

$$a_{21} = \frac{1}{4} + \frac{1}{6}\sqrt{3} - \frac{\sqrt{3}Z}{720} + \frac{61\sqrt{3}Z^2}{1209600} + \mathcal{O}(Z^3),$$

$$a_{22} = \frac{1}{4} - \frac{\sqrt{3}Z}{360} + \frac{19\sqrt{3}Z^2}{302400} + \mathcal{O}(Z^3).$$

Again these results demonstrate the μ-dependence of (some of) the coefficients of the Butcher tableau and the fact that, as $\mu \to 0$, the classical c_i- and a_{ij}-values are recovered.

The final stage in the case 2a integrates exactly the functions 1, x, x^2 and $\exp(\pm\mu x)$, which are the linearly independent solutions of the differential equation $y^{(5)} - \mu^2 y^{(3)} = 0$; again we have to realize that the internal stages only integrates the functions 1, $\exp(\pm\mu x)$ which are linearly independent solutions of $y^{(3)} - \mu^2 y' = 0$. This means that a term $y^{(3)} - \mu^2 y'$ will naturally appear in the *plte*. Due to the Taylor series technique used in the derivation of the *plte* a component of the form $y^{(4)} - \mu^2 y^{(2)}$ will also appear. Taking all this into account and the classical form (6.83) the *plte* should be

$$plte(Gauss, s = 2, exp, \text{case 2a}) =$$
$$\frac{h^5}{4320}(-\mu^2 y^{(3)} + y^{(5)} - 5f_y(-\mu^2 y^{(2)} + y^{(4)})$$
$$+10(f_y^2 - f_{xy} - f_{yy}f)(-\mu^2 y' + y^{(3)})),$$

which is the same as in case 1.

case 2b : $P' = 1$

We consider the choice where $K' = 0$ and $P' = 1$, i.e. the full exponential fitting case. This means that in order to determine the b_i and c_i-values we combine the equations (6.84-6.85) with $G^{\pm(1)}(Z, \mathbf{b}) = 0$, which taking into account the properties of the η_{-1} and η_0 functions read

$$G^{+(1)}(Z, \mathbf{b}) = \eta_0(Z) - \sum_{i=1}^{2} b_i c_i \eta_0(c_i^2 Z) - \sum_{i=1}^{2} b_i c_i \eta_{-1}(c_i^2 Z) = 0,$$

$$G^{-(1)}(Z, \mathbf{b}) = -\eta_0(Z) + \eta_{-1}(Z) - Z\sum_{i=1}^{2} b_i c_i^2 \eta_0(c_i^2 Z) = 0.$$

The solution of these four equations can not anymore be given in a closed form as was the case for the other cases studied so far. It is however evident that again a solution can only be found provided the c_i-values are μ-dependent. The following results have been obtained in a series expansion form:

$$c_1 = \frac{1}{2} - \frac{\sqrt{3}}{6} - \frac{\sqrt{3}Z}{1080} - \frac{13\sqrt{3}Z^2}{2721600} + \mathcal{O}(Z^3) \,,$$

$$c_2 = \frac{1}{2} + \frac{\sqrt{3}}{6} + \frac{\sqrt{3}Z}{1080} + \frac{13\sqrt{3}Z^2}{2721600} + \mathcal{O}(Z^3) \,,$$

$$b_1 = \frac{1}{2} - \frac{Z^2}{8640} + \mathcal{O}(Z^3) \,,$$

$$b_2 = \frac{1}{2} - \frac{Z^2}{8640} + \mathcal{O}(Z^3) \,,$$

$$a_{11} = \frac{1}{4} + \frac{\sqrt{3}Z}{480} - \frac{(37\sqrt{3}+35)Z^2}{604800} + \mathcal{O}(Z^3) \,,$$

$$a_{12} = \frac{1}{4} - \frac{1}{6}\sqrt{3} + \frac{7\sqrt{3}}{4320}Z - \frac{(315+113\sqrt{3})Z^2}{5443200} + \mathcal{O}(Z^3) \,,$$

$$a_{21} = \frac{1}{4} + \frac{1}{6}\sqrt{3} - \frac{7\sqrt{3}}{4320}Z + \frac{(-315+113\sqrt{3})Z^2}{5443200} + \mathcal{O}(Z^3) \,,$$

$$a_{22} = \frac{1}{4} - \frac{\sqrt{3}Z}{480} + \frac{(37\sqrt{3}-35)Z^2}{604800} + \mathcal{O}(Z^3) \,.$$

The expression for the *plte* can now be derived as follows. The final stage integrates exactly the functions $1, \exp(\pm\mu x)$, $x\exp(\pm\mu x)$ which are linear independent solutions of the differential equation $y^{(5)} - 2\mu^2 y^{(3)} + \mu^4 y' = 0$; on the other hand the internal stages still only integrate the functions $1, \exp(\pm\mu x)$, and for this reason a term $y^{(3)} - \mu^2 y'$ will naturally appear in the *plte*. One can also expect, due to the Taylor technique used in the derivation of the *plte*, a component of the form $y^{(4)} - \mu^2 y''$.

This means that the *plte* should be of the form:

$$plte(Gauss, s = 2, exp, \text{case 2b}) =$$
$$\frac{h^5}{4320}\left[\mu^4 y' - 2\mu^2 y^{(3)} + y^{(5)} - 5f_y(-\mu^2 y'' + y^{(4)})\right.$$
$$\left. +10(f_y^2 - f_{xy} - f_{yy}f)(-\mu^2 y' + y^{(3)})\right] \,.$$

3.2.4 Numerical examples

As a first example we consider the problem

$$y' = y \,, \qquad y(0) = 1 \,, \tag{6.105}$$

and we will apply several versions of the RadauIIA method to this problem with fixed stepsize. The stepsizes used are given in the first column of Table 6.20. In the other columns the absolute values of the global errors in $x = 1$ are given.

First of all, the classical method is applied. In column two of Table 6.20, we can infer that this classical method indeed behaves like a third-order method, since halving the stepsize leads to a reduction of the error by a factor which is approximately 8.

Secondly, we apply the exponential fitted method (case 1) where $c_1 = 1/3$ and $c_2 = 1$. The value of μ in each step is obtained by annihilating the leading term in the local trunction error as computed in (6.92). For this problem we obtain the constant value $\mu = 1$. Since the analytical solution $y(x) = \exp(x)$ is in the space of functions which are integrated exactly with that choice for μ, we obtain in column 3 of Table 6.20 machine accuracy for all values of h.

This is also the case for the exponential fitted method where the $c_1 = 1/3$ and c_2 is determined by (6.96) (case 2a) and for the case 2b where $c_2 = 1$ and c_1 is given by (6.99). Indeed, for both methods the constant value $\mu = 1$ is obtained in each step if the leading term in the local trunction error is set equal to zero.

So from Table 6.20 we conclude that our experimental results agree with the theory.

h	classical	case 1	case 2a	case 2b
1	$5.16\ 10^{-2}$	$1.33\ 10^{-15}$	0.00	$1.78\ 10^{-15}$
$\frac{1}{2}$	$5.55\ 10^{-2}$	$4.44\ 10^{-16}$	$8.88\ 10^{-16}$	$3.11\ 10^{-15}$
$\frac{1}{4}$	$6.33\ 10^{-4}$	$8.88\ 10^{-16}$	$1.78\ 10^{-15}$	$1.69\ 10^{-14}$
$\frac{1}{8}$	$7.63\ 10^{-5}$	0.00	$4.44\ 10^{-16}$	0.00
$\frac{1}{16}$	$9.37\ 10^{-6}$	$1.33\ 10^{-15}$	$1.33\ 10^{-15}$	$4.44\ 10^{-16}$

Table 6.20.   Absolute values of global errors in $x = 1$ of several versions of the RadauIIA method applied to (6.105).

As a second example, we consider the problem

$$\begin{cases} y_1' = -y_2 + \cos x + \sin 2x \\ y_2' = y_1 + 2\cos 2x - \sin x \end{cases} \qquad \begin{cases} y_1(0) = 0 \\ y_2(0) = 0 \end{cases} \qquad (6.106)$$

Its solution is given by

$$\begin{cases} y_1(x) = \sin x \\ y_2(x) = \sin 2x \end{cases}$$

We apply the same methods and again we compare the absolute values of the global errors (for each of the components) at the endpoint $x = 1$. However, since we now have a system of equations, the expressions for the local truncation errors should be interpreted appropriately : f_y is now the jacobian and for the exponential fitted versions, μ^2 is a diagonal matrix with diagonal elements μ_1^2 and μ_2^2 where μ_1 is the frequency to which the first component of the solution is fitted and μ_2 the one to which the second component is fitted. To annihilate the leading term of the local trunction error we now solve a system and we obtain $\mu_1^2 = -1$, $\mu_2^2 = -4$ for all of the three exponential fitted methods, such that we fit to linear combinations of $\sin x$ and $\cos x$ for the first component and $\sin 2\,x$ and $\cos 2\,x$ for the second component.

The results in Table 6.21 for the classical method and the exponential fitted method with fixed collocation points (case 1) again confirm the theoretical results. However, the exponential fitted methods where one of the c's is μ dependent (cases 2a and 2b) do not integrate the system exactly. Although in each case the *plte* was annihilated, these methods still behave like third-order methods.

This is at first sight surprising in so much as it shows that there is a clear distinction between methods with fixed collocation points and methods with μ dependent collocation points. The analysis of this discrepancy will be discussed in the next section.

h	classical	case 1	case 2a	case 2b
1	$8.25 \ 10^{-2}$	0.00	$3.98 \ 10^{-3}$	$4.59 \ 10^{-3}$
	$2.60 \ 10^{-2}$	$1.11 \ 10^{-16}$	$9.97 \ 10^{-3}$	$3.07 \ 10^{-3}$
$\frac{1}{2}$	$8.91 \ 10^{-3}$	$1.11 \ 10^{-16}$	$2.57 \ 10^{-4}$	$1.12 \ 10^{-4}$
	$1.83 \ 10^{-3}$	$1.11 \ 10^{-16}$	$1.00 \ 10^{-3}$	$6.61 \ 10^{-4}$
$\frac{1}{4}$	$1.11 \ 10^{-3}$	$2.22 \ 10^{-16}$	$2.48 \ 10^{-5}$	$5.76 \ 10^{-6}$
	$2.08 \ 10^{-1}$	0.00	$1.21 \ 10^{-4}$	$9.82 \ 10^{-5}$
$\frac{1}{8}$	$1.40 \ 10^{-4}$	0.00	$2.77 \ 10^{-6}$	$1.69 \ 10^{-6}$
	$2.57 \ 10^{-5}$	$1.11 \ 10^{-16}$	$1.50 \ 10^{-5}$	$1.33 \ 10^{-5}$
$\frac{1}{16}$	$1.77 \ 10^{-5}$	$1.11 \ 10^{-16}$	$3.29 \ 10^{-7}$	$2.65 \ 10^{-7}$
	$3.24 \ 10^{-6}$	$1.11 \ 10^{-16}$	$1.83 \ 10^{-6}$	$1.73 \ 10^{-6}$

Table 6.21. Absolute values of the global errors in $x = 1$ of several versions of the RadauIIA method applied to (6.106).

3.2.5   Fixed versus μ-dependent knot points

There are several ways in which an exponential fitted Runge–Kutta method can be applied to a system of equations. A first approach is to use the same μ to fit to all the components of the solution. This approach may be well suited if all components exhibit the same behaviour, but if serious differences arise between components it is not possible to determine a unique value for μ for which the error in all components will be reduced significantly compared to the classical case. In that case a better approach is to use a separate μ for each component such that the EFRK method becomes a partitioned method [12].

A partitioned method can be applied to a system of the form

$$\begin{cases} y' = f(x,y,z) \\ z' = g(x,y,z) \end{cases} \tag{6.107}$$

where y and z can be vectors of different dimensions. The idea of a partitioned RK method is to treat the y-variables with one RK method and the z-variables with a second method. This idea can be extended to more than two methods.

It is well known from the theory of partitioned RK methods that, if a partitioned method is made up of 2 methods of order p, its order can not exceed p. The order conditions for partitioned RK methods can be divided in two categories: the usual order conditions for RK methods and additional so-called coupling order conditions. It is only when these additional conditions are fulfilled that the partitioned RK method also has order p.

Of course, the different methods which are used in our exponential fitted partitioned method are very similar : they all reduce to the same classical method if all μ's tend to 0. With this in mind, it is easy to verify by considering the different series expansions that for all the 2-stage methods (Lobatto, Radau, Gauss) considered in this chapter the order of the partitioned method coincide with the order of the originating RK method.

However, these same series expansions reveal that extra terms may occur in the expressions for the leading term of the local trunction errors of the partitioned RK methods. We obtain the following expressions for the exponential fitted Radau IIA methods when applied to the system (6.107) where y and z are scalars :

- case 1 :

$$plte(RadauIIA, s = 2, exp, \text{case } 1) =$$
$$-\frac{h^4}{216} \left\{ -\begin{pmatrix} \mu_1^2 & 0 \\ 0 & \mu_2^2 \end{pmatrix} \begin{pmatrix} y^{(2)} \\ z^{(2)} \end{pmatrix} + \begin{pmatrix} y^{(4)} \\ z^{(4)} \end{pmatrix} \right.$$
$$\left. -4 \begin{pmatrix} f_y & f_z \\ g_y & g_z \end{pmatrix} \left[-\begin{pmatrix} \mu_1^2 & 0 \\ 0 & \mu_2^2 \end{pmatrix} \begin{pmatrix} y' \\ z' \end{pmatrix} + \begin{pmatrix} y^{(3)} \\ z^{(3)} \end{pmatrix} \right] \right\}$$

- case 2a :

$$plte(RadauIIA, s = 2, exp, \text{case } 2a) =$$

$$-\frac{h^4}{216}\left\{-\begin{pmatrix} \mu_1^2 & 0 \\ 0 & \mu_2^2 \end{pmatrix}\begin{pmatrix} y^{(2)} \\ z^{(2)} \end{pmatrix} + \begin{pmatrix} y^{(4)} \\ z^{(4)} \end{pmatrix}\right.$$

$$-4\begin{pmatrix} f_y & f_z \\ g_y & g_z \end{pmatrix}\left[-\begin{pmatrix} \mu_1^2 & 0 \\ 0 & \mu_2^2 \end{pmatrix}\begin{pmatrix} y' \\ z' \end{pmatrix} + \begin{pmatrix} y^{(3)} \\ z^{(3)} \end{pmatrix}\right]$$

$$\left.-\frac{2}{5}\begin{pmatrix} (\mu_1^2 - \mu_2^2)\, f_z\, g \\ (\mu_2^2 - \mu_1^2)\, g_y\, f \end{pmatrix}\right\}$$

- case 2b :

$$plte(RadauIIA, s = 2, exp, \text{case } 2b) =$$

$$-\frac{h^4}{216}\left\{-\begin{pmatrix} \mu_1^2 & 0 \\ 0 & \mu_2^2 \end{pmatrix}\begin{pmatrix} y^{(2)} \\ z^{(2)} \end{pmatrix} + \begin{pmatrix} y^{(4)} \\ z^{(4)} \end{pmatrix}\right.$$

$$-4\begin{pmatrix} f_y & f_z \\ g_y & g_z \end{pmatrix}\left[-\begin{pmatrix} \mu_1^2 & 0 \\ 0 & \mu_2^2 \end{pmatrix}\begin{pmatrix} y' \\ z' \end{pmatrix} + \begin{pmatrix} y^{(3)} \\ z^{(3)} \end{pmatrix}\right]$$

$$\left.+\frac{2}{5}\begin{pmatrix} (\mu_1^2 - \mu_2^2)\, f_z\, g \\ (\mu_2^2 - \mu_1^2)\, g_y\, f \end{pmatrix}\right\}$$

In Case 1, it turns out that there are no extra terms. The values for $\mu_1^2 = -1$ and $\mu_2^2 = -4$ which annihilate the leading term were already correctly computed and since they are also the squares of the frequencies of the components of the exact solution of the problem, we have machine accuracy.

For the cases 2a and 2b there are extra terms, such that the true values of μ_1^2 and μ_2^2 which make the *plte* vanish do not coincide with -1 and -4. In fact, for the problem (6.106), these true values are x-dependent functions given by:

- case 2a :

$$\begin{cases} \mu_1^2(x) = -\dfrac{128\sin^3 x + 220\sin^2 x - 63\sin x - 120}{32\sin^3 x + 172\sin^2 x - 15\sin x - 96} \\[2mm] \mu_2^2(x) = -\dfrac{128\sin^3 x + 640\sin^2 x - 63\sin x - 360}{32\sin^3 x + 172\sin^2 x - 15\sin x - 96} \end{cases} \qquad (6.108)$$

- case 2b :

$$\begin{cases} \mu_1^2(x) = -\dfrac{128\sin^3 x - 60\sin^2 x - 63\sin x + 40}{32\sin^3 x - 108\sin^2 x - 15\sin x + 64} \\[2mm] \mu_2^2(x) = -\dfrac{128\sin^3 x - 480\sin^2 x - 63\sin x + 280}{32\sin^3 x - 108\sin^2 x - 15\sin x + 64} \end{cases} \qquad (6.109)$$

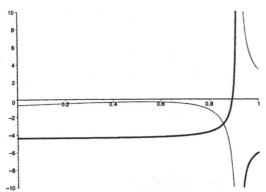

Figure 6.11. The values for μ_1^2 (thin line) and μ_2^2 (thick line) for case 2b applied to problem (6.106).

In Figure 6.11 the values for μ_1^2 and μ_2^2 are plotted for the case 2b (the case 2a is very similar). The optimal values for μ_1 and μ_2 are no longer constants, but rather defined by an x-dependent function. One notices that, already at the start of the integration interval, these computed values are rather big corrections to the true frequencies of the components of the solution. As the integration advances the computed μ values move further away (μ_2^2 even becomes positive such that we fit the second component to hyperbolic functions rather than trigonometric functions) and for $x \approx 0.917$ there is a discontinuity.

Since the computed values for the μ's do not coincide with the squares of the frequencies in the solutions, we can no longer hope for machine accuracy: we can only obtain an increase of the order of the partitioned method from 3 to 4. This is indeed what we find in Table 6.22 and in Figure 6.12. In Table 6.22 we show the absolute values of the errors at the endpoint $x = 1$ for different values of h. The upper/lower entry corresponds to the first/second component respectively. In Figure 6.12 we display $- \log_{10} h$ vs. $- \log_{10} GE$ where GE is the global error (in the Euclidean norm) at the endpoint. We notice that, for the exponential fitted methods, the shape of the dependence slightly deviates from a straight line. This is mainly due to the discontinuity in the expressions of the μ's. However, if we compute the slopes of the best fitting straight lines we obtain 3.009 for the classical RadauIIA method, while we have 4.040 for the EFRK 2a case and 3.823 for the EFRK 2b case. Clearly, these numbers confirm the raising of the order.

h	case 2a	case 2b
1	$7.11\ 10^{-3}$	$4.06\ 10^{-4}$
	$1.71\ 10^{-4}$	$8.30\ 10^{-3}$
$\frac{1}{2}$	$3.53\ 10^{-4}$	$1.02\ 10^{-4}$
	$5.07\ 10^{-5}$	$3.93\ 10^{-4}$
$\frac{1}{4}$	$2.91\ 10^{-5}$	$1.72\ 10^{-5}$
	$7.92\ 10^{-6}$	$1.78\ 10^{-5}$
$\frac{1}{8}$	$4.45\ 10^{-7}$	$3.00\ 10^{-6}$
	$2.88\ 10^{-7}$	$3.35\ 10^{-7}$
$\frac{1}{16}$	$8.73\ 10^{-8}$	$4.64\ 10^{-8}$
	$4.01\ 10^{-8}$	$9.64\ 10^{-8}$

Table 6.22. Absolute values of the global errors in $x = 1$ of the μ-dependent RadauIIA methods applied to (6.106). The values for μ are obtained from (6.108)-(6.109).

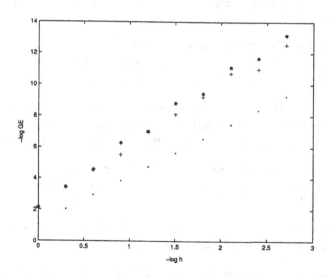

Figure 6.12. A log-log plot (base 10) of the step size versus the global error in the endpoint for the classical RadauIIA methode (dots) and its exponential fitted cases IIA (stars) and IIB (crosses).

3.2.6   Conclusion

In this section we have constructed several exponential fitted versions of 2-stage RK methods. These EFRK methods can be divided into two classes : methods for which all of the c-points are fixed constants (case 1) and methods for which some of the c-points are frequency dependent (case 2). From a theoretical point of view, both classes of methods are quite easy to construct. However, in practice, there is a big difference. We have shown that there are problems for which machine accuracy is obtained for case 1 but not for case 2. A closer examination of the local truncation error revealed that, unlike the methods of case 1, the methods of case 2 behave like partitioned EFRK methods. In that case, an accurate computation of the frequencies of the different components of the solution by annihilating the *plte* is no longer possible due to the presence of extra coupling terms. Therefore, the methods with fixed knot points should be preferred.

3.3   A specific implementation of implicit methods

As discussed in previous sections an s-stage Runge-Kutta method can integrate exactly at most a set of $s + 1$ linearly independent functions. Methods for which this maximum is reached are the collocation methods. In general these functions may be chosen as members of the hybrid set (6.77); however in the present section we restrict the discusion to $\exp(\mu_i x), i = 1, 2, \ldots, s + 1$, and consider only three relevant cases (see also 5.1.1 for an equivalent approach for linear multistep solvers and [30]):

- Algorithm A0. This corresponds to $\mu_i = 0$, $(i = 1, 2, \ldots, s + 1)$ and one obtains just the classical s-stage Runge-Kutta method of collocation type with fixed coefficients and the reference set $1, x, x^2, \ldots, x^s$.

- Algorithm A1. This corresponds to some μ which may be either given in advance or determined internally by the algorithn and it uses $\mu_1 = \mu$ and $\mu_i = 0$, $(i = 2, 3, \ldots, s + 1)$. Its coefficients depend on $z = \mu h$.

- Algorithm A2. For some given or internally determined μ this takes $\mu_1 = -\mu_2 = \mu$ and $\mu_i = 0$, $(i = 3, \ldots, s + 1)$. Its coefficients depend on $Z = z^2 = \mu^2 h^2$.

The above conditions and choices for the μ's give rise to the following type of equations, to be fulfilled by the A- and b- coefficients.

- Algorithm A0: The equations (6.75) and (6.76) have to be used here.

- Algorithm A1:

$$\exp(z) - 1 - z \sum_{i=1}^{s} b_i \exp(c_i z) = 0 \, ,$$

$$\sum_{i=1}^{s} b_i c_i^{k-1} = 1/k, \quad k = 1, \ldots, s-1,$$

$$\exp(c_i z) - 1 - v \sum_{j=1}^{s} a_{ij} \exp(c_j z) = 0,$$

$$\sum_{j=1}^{s} a_{ij} c_j^{k-1} = c_i^k/k, \quad k = 1, \ldots, s-1, \text{ and } i = 1, \ldots, s.$$

- Algorithm A2:

    1 As long as $s > 1$ the four equations (6.27–6.30) with $\gamma_i = 1$ have to be solved.

    2 For $s = 1$ a special situation arises; with the classical definition (6.68) it is not possible to construct an implicit EFRK method of type A2 with the defined properties. However using the notation of Section 2.2, where extra parameters γ's are introduced one can find a method integrating two exponential functions. For the case considered the occurring γ_1, a_{11} and b_1 become functions of $z = \mu h$ and are the solutions of the following equations:

$$\exp(\pm z) - 1 \mp z b_1 \exp(\pm c_1 v) = 0,$$
$$\exp(\pm c_1 z) - \gamma_1 \mp z a_{11} \exp(\pm c_1 z) = 0.$$

One can verify that this system has a nontrivial solution only when $c_1 = 1/2$.

3.3.1 The special case of a one-stage EFRK method

For the three types of algorithms considered two implicit EFRK methods can be constructed, one of algebraic order 1 and another of algebraic order 2. For the A0 case the equations (6.75,6.76) give rise to $c_1 = a_{11}$ and $b_1 = 1$, showing that for $c_1 = 1$ the method reduces to a RadauIIA method of order 1 and stage-order 1 (Table 6.17). For $c_1 = a_{11} = 1/2$ and $b_1 = 1$ a Gauss type method is obtained of algebraic order 2 and stage-order 1 (Table 6.15). In the first case the principal local trunction term (*plte*) reads

$$-\frac{h^2}{2} y^{(2)},$$

while for the latter case this term has the more complicated form

$$\frac{h^3}{24} \left(y^{(3)} - 3 f_y y^{(2)} \right),$$

where all occurring functions are evaluated at $x = x_n$ and $y = y_n$.

For the A1 case a_{11} and b_1 are expressed in terms of c_1 in the following way:

$$a_{11} = \frac{\exp(c_1 z) - 1}{z \exp(c_1 z)}, \quad b_1 = \frac{\exp(z) - 1}{z \exp(c_1 z)},$$

with series expansion

$$a_{11} = c_1 + \frac{c_1^2 z}{2} + \frac{c_1^3 z^2}{6} + \mathcal{O}(z^3),$$

$$b_1 = 1 + (\frac{1}{2} - c_1) z + (-\frac{1}{2} c_1^2 + \frac{1}{6} + (c_1 - \frac{1}{2}) c_1) z^2 + \mathcal{O}(z^3),$$

showing that the algebraic order is 1 as long as $c_1 \neq 1/2$, and 2 otherwise, while the stage-order is 1 for every choice of c_1. Analogously the *plte*, which has the form

$$plte = \frac{h^2}{2}(1 - 2 c_1)(y^{(2)}(x_n) - \mu y'(x_n)),$$

for $c_1 \neq 1/2$ and

$$plte = \frac{h^3}{24}(y^{(3)} - \mu y^{(2)} - 3f_y(y^{(2)} - \mu y')),$$

for $c_1 = 1/2$ reveals the same feature. For systems f_y is the Jacobian and μ represents a diagonal matrix with the frequencies related to each equation present on the diagonal position.

For the A2 case the γ_1 parameter has to be considered together with $c_1 = 1/2$ in order to obtain a nontrivial solution:

$$a_{11} = \frac{\exp(z) - 1}{z (\exp(z) + 1)} = \frac{1}{2} - \frac{1}{24} z^2 + \frac{1}{240} z^4 + \mathcal{O}(z^6),$$

$$\gamma_1 = \frac{2 \exp(z/2)}{\exp(z) + 1} = 1 - \frac{1}{8} z^2 + \frac{5}{384} z^4 + \mathcal{O}(z^6),$$

$$b_1 = \frac{\exp(z/2) - \exp(-z/2)}{z} = 1 + \frac{1}{24} z^2 + \frac{1}{1920} z^4 + \mathcal{O}(z^6),$$

resulting in an algebraic order 2 method. The *plte* has the following form:

$$plte = \frac{h^3}{24}(y^{(3)} - \mu^2 y' - 3f_y(y^{(2)} - \mu^2 y)).$$

3.3.2   A special two-stage implicit EFRK method

For the A0 algorithm the results (6.78) are obtained: as well-known, the particular choice of c_1 and c_2 determines the order of the method.

In the further construction of A1 and A2 algorithms we analyse the particular situation where the knot points are symmetrically distributed over the interval

$]0, 1[$, i.e. $c_1 = \dfrac{1}{3}$ and $c_2 = \dfrac{2}{3}$. This choice guarantees simple expressions for the principal local truncation error terms; they can be expressed as functions of total derivatives of the solution vector y, without the occurrence of partial derivatives with respect to y of the r.h.s. of the system. The related classical method is for these values A-stable and has a *plte*

$$plte(A0) = \frac{h^3}{36} y^{(3)}(x_n) \, .$$

For the A1 algorithm the following results are obtained:

$$a_{11} = \frac{(-\exp(c_1 z) + 1 + c_1 z \exp(c_2 z))}{z(-\exp(c_1 z) + \exp(c_2 z))} \, ,$$

$$a_{12} = -\frac{(-\exp(c_1 z) + 1 + z \exp(c_1 z) c_1)}{z(-\exp(c_1 z) + \exp(c_2 z))} \, ,$$

$$a_{21} = \frac{(-\exp(c_2 z) + 1 + c_2 z \exp(c_2 z))}{z(-\exp(c_1 z) + \exp(c_2 z))} \, ,$$

$$a_{22} = -\frac{(-\exp(c_2 z) + 1 + z \exp(c_1 z) c_2)}{z(-\exp(c_1 z) + \exp(c_2 z))} \, ,$$

$$b_1 = -\frac{(-\exp(z) + 1 + z \exp(c_2 z))}{z(\exp(c_1 z) - exp(c_2 z))} \, ,$$

$$b_2 = \frac{(-\exp(z) + 1 + z \exp(c_1 z))}{z(\exp(c_1 z) - \exp(c_2 z))} \, .$$

This method is of algebraic order 2 and has for the symmetrical placed knot points the following *plte*:

$$plte(A1) = \frac{h^3}{36} (y^{(3)}(x_n) - \mu y^{(2)}(x_n)) \, . \tag{6.110}$$

For the A2 case the A- and b coefficients are given by (6.88); this method is also of algebraic order 2 and has for the symmetrical placed knot points a *plte*

$$plte(A2) = \frac{h^3}{36} (y^{(3)}(x_n) - \mu^2 y'(x_n)) \, . \tag{6.111}$$

We see that in the three cases the *plte* consists of a product of three factors, i.e. a general h^3 factor, a numerical factor $1/36$ and a factor which involves one or two derivatives of the solution vector. The different behaviour of this third, differential factor can make a difference in the accuracy. Let us consider as an example the A1-algorithm. For that purpose we introduce the two operators

$$\mathcal{D}_{A0}[y(x); \mu] := y^{(3)}(x), \quad \mathcal{D}_{A1}[y(x); \mu] := y^{(3)}(x) - \mu y^{(2)}(x) \, .$$

The problem consists in comparing the behaviours of these two operators in the interval $x_n < x < x_n + h = x_{n+1}$ when $y(x)$ is the exact solution of the considered problem. When $\mathcal{D}_{A0}$ identically vanishes the classical A0 algorithm is exact and there is no need of improving it. This is however an unlikely situation in most applications. In general the replacement of A0 by A1 or A2 is expected to improve the accuracy in so far as μ is properly selected. An analogous situation has been discussed for multistep algorithms in [15]. If a μ exists such that $\mathcal{D}_{A1}$ identically vanishes on the quoted interval then the version A1 corresponding to that μ will be exact. The reason is that $\mathcal{D}_{A1}$ identically vanishing is equivalent to looking at the differential equation

$$y^{(3)}(x) - \mu \, y^{(2)} = 0$$

and, indeed, $\exp(\mu \, x)$ is a solution. In general, no constant μ can be found such that $\mathcal{D}_{A1}$ identically vanishes but it makes sense to address the problem of finding that value of μ which ensures that the values of $\mathcal{D}_{A1}$ are kept as close to zero as possible for x in the considered interval. Upon taking

$$\mu(A1) = \frac{y^{(3)}(x_n)}{y^{(2)}(x_n)} \qquad (6.112)$$

and using the Taylor series expansion for $y^{(3)}(x)$ and $y^{(2)}(x)$ around x_n we get

$$\mathcal{D}_{A1}[y(x); \mu] = (x - x_n)(y^{(4)}(x_n) - \mu y^{(3)}(x_n)) + \mathcal{O}((x - x_n)^2) \, .$$

This indicates that the bound of $\mathcal{D}_{A1}$ behaves as h and consequently the bound of $plte(A1)$ given by (6.110) behaves as h^4. This means that the order of version A1 corresponding to the choice (6.112) for μ will be three, that is by one unit higher than the order of the classical A0 or of A1, when no careful selection of μ is operated for the latter. The same considerations can be repeated for algorithm A2 with the result that, if

$$\mu^2(A2) = \frac{y^{(3)}(x_n)}{y'(x_n)} \qquad (6.113)$$

then the order of A2 becomes three.

It is known that for a Runge-Kutta method of order p the global error can be written as

$$y(x_n) - y_n = h^p E(x_n) \, ,$$

where some comments are given in [4] concerning the typical structure of $E(x)$. This fact will be used later on to obtain graphical confirmation of the order of various versions. It will be seen that the absolute value of the scaled errors $|y(x_n) - y_n|/h^2$, $n = 1, 2, \ldots$ corresponding to runs with A0 on different h lie along one and the same curve and the same is true for $(y(x_n) - y_n)/h^3$, $n = 1, 2, \ldots$ when the results from the optimal A1 and A2 are used. This confirms that A0 is a second order method while the new methods are of order three. Analogous techniques have been used in Chapter 5 for the experimental determination of the order for linear multistep methods and in Section 2.6.4 for the same of explicit EFRK methods.

3.3.3   Technical details for implementation

The choice between A1 and A2 and the determination of the optimal μ require some ad-hoc techniques.

Calculation of the derivatives.   The calculation of μ by (6.112) or (6.113) requires the knowledge of first-, second- and third-order derivatives of the exact solution. For the first-order derivative we have used the r.h.s. $f(x, y(x))$ of the equations. One could expect that a straightforward procedure for the computation of the second- and third-order derivatives consists in using the analytical expressions of the total derivatives with respect to x of $f(x, y(x))$. In [15] it has been explained that the A1 and A2 algorithms, combined with an μ-determination based on these analytically based derivatives performed badly on stiff ODEs. It has been also shown that an optimal use of the above algorithms is guaranteed when these derivatives are evaluated by the usual difference formulae. In our implementation we have used finite-difference formulae for the second- and third-order derivatives which have an accuracy of $\mathcal{O}(h)$. We use data at the knot points x_{n-2}, x_{n-1}, x_n and x_{n+1} for input. The first three data were the values of the solution just as resulted from the previous two runs. For the latter we used a value y_{n+1} determined by

$$y_{n+1} = y_{n-1} + \frac{h}{3} \left(f(x_{n-1}, y_{n-1}) + 4f(x_n, y_n) + f(x_{n+1}, y_{n+1}) \right) , \quad (6.114)$$

which is known as the generalized Milne-Simpson two-step formula with classical weights. This method is more accurate than A1 or A2, but has bad stability properties. We only use it for the calculation of the derivatives and not for the propagation of the solution.

Starting procedure.   It is clear that the above strategy for the calculation of derivatives can only be started from the moment that besides the initial conditions at the point x_0 the solution is also known in one additional point x_1. As starting algorithm we use a collocation method of order 5 with five stages and equidistant knot points $c_j = x_0 + jh/4$, $j = 0, 1, \ldots, 5$, with evidently stage-order 5, i.e. a method with the following Butcher tableau:

$$
\begin{array}{c|ccccc}
0 & 0 & 0 & 0 & 0 & 0 \\
1/4 & 251/2880 & 323/1440 & -11/120 & 53/1440 & -19/2880 \\
1/2 & 29/360 & 31/90 & 1/15 & 1/90 & -1/360 \\
3/4 & 27/320 & 51/160 & 9/40 & 21/160 & -3/320 \\
1 & 7/90 & 16/45 & 2/15 & 16/45 & 7/90 \\
\hline
 & 7/90 & 16/45 & 2/15 & 16/45 & 7/90
\end{array}
\tag{6.115}
$$

This method is A-stable and the fact that the stage-order is 5 allows us, even for stiff problems, to construct very accurate finite difference approximations based on the internal stage vectors Y_i, $i = 1, \ldots, 5$ for $y'(x_1), y^{(2)}(x_1)$ and $y^{(3)}(x_1)$. With these values and equations (6.112) and (6.113) an optimal value for μ is calculated in x_1. For the next step either the A1 or A2 algorithm is activated in order to determine the solution at the point x_2. For the numerical integration at the other points $x_0 + jh$, $j = 3, 4, \ldots$ the general strategy described below is followed.

Choosing between A1 and A2. One must realize that algorithm A1 is not defined if x_n is a root of $y^{(2)}(x)$, (see (6.112)), while A2 is not defined when x_n is a root of $y'(x)$, (see (6.113)). It follows that at the very special case when x_n is a root of both $y'(x)$ and $y^{(2)}(x)$ the classical A0 cannot be replaced by an EFRK method of the type A1 or A2. However this situation is quite exceptional. In general, a reasonable way of selecting between A1 and A2 consists in comparing $|y'(x_n)|$ and $|y^{(2)}(x_n)|$. If

$$
|y'(x_n)| < |y^{(2)}(x_n)| ,
\tag{6.116}
$$

version A1 is selected. If

$$
|y'(x_n)| \geq |y^{(2)}(x_n)| ,
\tag{6.117}
$$

A2 is selected. By this procedure one avoids the occurrence of extreme large values for μ. For systems of differential equations this selection procedure should be repeated for each component separately.

A flow chart of the procedure.

- Starting the calculation: By means of (6.115) the solution as well as an appropriate value of μ is calculated in the point x_1 and depending on the conditions (6.116) and (6.117) the solution in x_2 is determined either with A1 or A2.

- The procedure for all other knot points:

– Choosing the version:

    * The values at x_{j-2}, x_{j-1} and at x_j, $j = 2, \ldots$, resulting from the previous runs are accepted as reasonable approximations of the exact solution.

    * A guess for the solution at the point x_{j+1} is calculated by (6.114).

    * These values are used for the computation of $y'(x_n), y^{(2)}(x_n)$ and $y^{(3)}(x_n)$ by means of finite difference methods with at least an accuracy of $\mathcal{O}(h)$.

    * Algorithm A1 or A2 is chosen on the basis of the above criterion.

– Evaluation of μ. If A1 is chosen μ is calculated by (6.112); its value is real. If A2 is chosen the two frequencies are

$$\mu_{1,2} = \pm\sqrt{y^{(3)}(x_n)/y'(x_n)} \ .$$

In the latter case the two μ's are either both real or both imaginary.

– Construction of the algorithm weights. If $|\mu|\ h > 0.2$ the analytical expressions given above for the A and b elements are used for the calculation; otherwise a Taylor series expansion of these expressions is introduced.

    For systems the above stages are performed on each component.

– Calculation of the solution at x_{n+1}. To solve the implicit system, related to either A1 or A2 a Newton-Raphson iteration procedure is used.

Numerical illustration. As a test case we consider again the system of two equations (see also 5.81):

$$\begin{cases} y_1' &= -2y_1 + y_2 + 2\sin(x) \\ y_2' &= -(\beta+2)y_1 + (\beta+1)(y_2 + \sin(x) - \cos(x)) \ . \end{cases} \tag{6.118}$$

This system has been used in [18] with $\beta = -3$ and $\beta = -1000$ in order to illustrate the phenomenon of stiffness. If the initial conditions are $y_1(0) = 2$ and $y_2(0) = 3$ the exact solution is β-independent and reads:

$$y_1(x) = 2\exp(-x) + \sin(x), \quad y_2(x) = 2\exp(-x) + \cos(x) \ .$$

We adopted the same initial conditions and carried out separate runs with the classical A0 and with the optimal EFRK-method, whose algorithm is described in the above flow chart. The interval $[0, 10]$ was considered and the stepsizes used were $h = 1/10$, $1/20$ and $1/40$. In Table 6.23 we give the absolute errors from the two methods at $x = 10$, for $\beta = -3$ (a nonstiff case) and $\beta = -1000$

Table 6.23. Errors $\Delta y_i(x) = |y_i(x) - y_i^{comput}(x)|$ at $x = 10$ for the two components $i = 1, 2$ from the classical two-stage algorithm A0 and from its optimal exponential fitting extension for system (6.118).

	$\beta = -3$		$\beta = -1000$	
h	Δy_1	Δy_2	Δy_1	Δy_2
classical A0				
0.1000	0.171(−03)	0.062(−03)	0.192(−03)	0.002(−03)
0.0500	0.425(−04)	0.157(−04)	0.479(−04)	0.168(−03)
0.0250	0.106(−04)	0.039(−04)	0.119(−04)	0.098(−04)
optimal EFRK-method				
0.1000	0.021(−04)	0.105(−04)	0.603(−05)	0.457(−05)
0.0500	0.034(−05)	0.135(−05)	0.666(−06)	0.191(−06)
0.0250	0.046(−06)	0.169(−06)	0.800(−07)	0.596(−07)

(a moderately stiff case). We see that the EFRK-method works much better than A0, irrespective of whether the system is stiff or nonstiff.

The fact that the EFRK-method is of the third order is also visible but, for additional confirmation, we present some graphs for the case $\beta = -1000$. (The graphs corresponding to $\beta = -3$ are similar.) In Figure 6.13 we give the absolute value of the scaled errors $|y(jh) - y_j|/h^2$, $j = 1, 2, 3, \ldots$ obtained by the classical A0 for the component y_1 and this confirms the fact that A0 is a second order method. The graph corresponding to the component y_2 is pretty much the same.

In Figure 6.14 the absolute value of the scaled errors $|y(jh) - y_j|/h^3$, $j = 1, 2, 3, \ldots$ for the first component obtained by the optimal EFRK-method is given, confirming the third-order behaviour. In Figure 6.15 we depict the variation of μ for the first component. On this graph one or two curves are alternatively shown. The regions of x where only one curve is shown represent the regions where A1 was chosen while when two curves appear (these are always symmetric) it means that A2 was activated and the two curves give the values of the two μ's. In the latter case, the circles indicate that the two μ's are real while the crosses indicate that they are imaginary.

In Figure 5.8 and in [15] quite analogous results were obtained for the same problem with an exponential-fitted bdf method of order three. What has been remarked in [15] about the determined μ-values can be repeated here, i.e. that the exact solution of the system has three characteristic frequencies $\Lambda_1 = -1$, $\Lambda_2 = i$ and $\Lambda_3 = -i$ because each of the two components of the

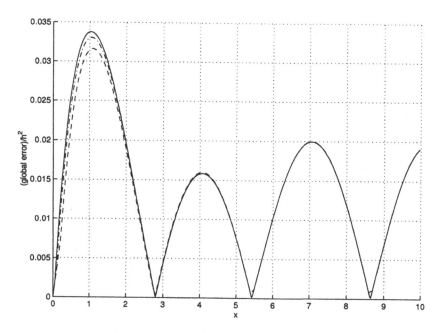

Figure 6.13. Absolute values of the scaled errors $|y(jh) - y_j|/h^2$, $j = 1, 2, \ldots$ for the first component obtained by the classical RK-method of order two. The dashed line displays the results obtained with $h = 0.1$, the dash-dot line the ones obtained with $h = 0.05$ and the solid one those calculated with $h = 0.025$. Observe that the three lines cover each other completely in the r.h.s. of the figure.

exact solution is some linear combination of $\exp(\Lambda_i x)$, $i = 1, 2, 3$. It follows that the detected values of μ for A1 and of $\mu_1 = -\mu_2$ for A2 may be seen as some compact representatives of the three true Λ's of the problem, but this does not imply necessarily that their values should be some average values of the latter. Close to the origin the detected μ is clearly outside the range of the Λ's (at least for the first component) but at larger x and when A2 is chosen, the two μ's are close to Λ_2 and Λ_3. The latter can be interpreted as reflecting the fact that in these intervals the trigonometric terms dominate in the involved derivatives of the solution.

Conclusions. In this section we considered one particular second order implicit Runge-Kutta method for first order ODEs and examined the problem of how the frequencies should be tuned in order to obtain the maximal benefit from the exponential fitting versions of such algorithms. We have shown that the key to the answer consists in analysing the behaviour of the error. On this

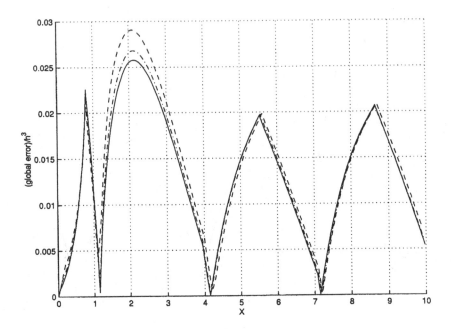

Figure 6.14. Absolute values of the scaled errors $|y(jh) - y_j|/h^3$, $j = 1, 2, \ldots$ for the first component obtained with the optimal EFRK-method of order three. The dashed line are the results obtained with $h = 0.1$, the dash-dot line the ones obtained with $h = 0.05$ and the solid ones calculated with $h = 0.025$.

basis we were able to propose formulae for the best μ-values and to show that, under these conditions, the order of the method is increased by one unit. Such methods are called optimal implicit EFRK-methods. The whole analysis is restricted to a simple two-stage method with knot points symmetrically placed in the interval $]0, 1[$.

4.   Some related research: Runge-Kutta-Nyström methods

The construction of Runge-Kutta-Nyström methods with frequency dependent coefficients is a natural extension of the theory described above for Runge-Kutta methods. Three different research groups have contributed to this problem: Coleman and Duxbury [7], Ozawa [21] and Paternoster [22]. The last author tangentially investigates Runge-Kutta methods; her results confirm the ones we have presented in the above chapter.

The techniques proposed by the three groups for the derivation of such Runge-Kutta-Nyström (RKN) methods are rather similar to the ones proposed by us for the Runge-Kutta schemes. Ozawa and Paternoster also introduce linear op-

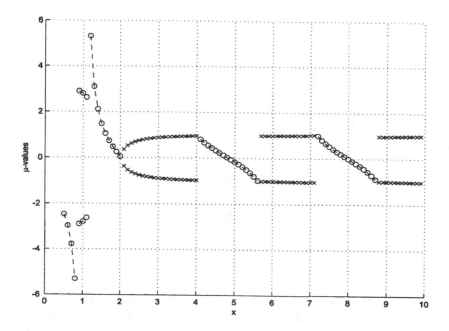

Figure 6.15. Variation with x of the optimal μ in the first component of (6.118).The regions of x where only one curve is shown represent the regions where A1 was chosen while when two curves appear (these are always symmetric) it means that A2 was activated and the two curves give the values of the two μ's. In the latter case, the circles indicate that the two μ's are real while the crosses indicate that they are imaginary.

erators associated to the method. Coleman and Duxbury rely on a technique of so-called mixed interpolation, previously introduced by one of the authors [8] and extended by Coleman [6]. In our presentation we shall follow in first instance Paternoster's presentation, which is closely related to our exponential fitting approach. In all the presentations one always assumes that a good estimate of the frequency is available in advance.

RKN methods are developed for second order ODEs, in which the first derivative does not appear explicitly (for a more detailed description we refer to [12, 13]); in other words one considers

$$y'' = f(t, y), \quad y(x_0) = y_0, \quad y'(x_0) = y_0' .$$

A general s-stage RKN method has the following form:

$$y_{n+1} = y_n + h\, y_n' + h^2 \sum_{j=1}^{s} \bar{b}_j f(x_n + c_j\, h, Y_j),$$

$$y'_{n+1} = y'_n + h \sum_{j=1}^{s} b_j f(x_n + c_j h, Y_j),$$

$$Y_j = y_n + c_j h y'_n + h^2 \sum_{k=1}^{s} a_{jk} f(x_n + c_k h, Y_k), \quad j = 1, 2, \ldots, s.$$

As in the case of Runge-Kutta methods, this can be represented in the Butcher array:

$$
\begin{array}{c|cccc}
c_1 & a_{11} & a_{12} & \cdots & a_{1s} \\
c_2 & a_{21} & a_{22} & \cdots & a_{2s} \\
    &   & \cdots &   &   \\
c_s & a_{s1} & a_{s2} & \cdots & a_{ss} \\
\hline
    & \bar{b}_1 & \bar{b}_2 & \cdots & \bar{b}_s \\
\hline
    & b_1 & b_2 & \cdots & b_s \\
\end{array}
$$

The three research groups use the concept of the *trigonometric order*, which was first introduced by Gautschi [11] for linear multistep methods. For Runge-Kutta-Nyström methods, the trigonometric order is defined in an obvious manner analogous to that for the linear multistep methods.

DEFINITION 6.7 *An s-stage Runge-Kutta-Nyström method is said to be of trigonometric order q relative to the frequency μ if all the associated linear difference operators:*

$$\mathcal{L}_i[h, \mathbf{a}]y(x) = y(x + c_i h) - y(x) - c_i h \, y'(x) - h^2 \sum_{j=1}^{s} a_{ij} y''(x + c_j h),$$

$$i = 1, 2, \ldots, s$$

$$\bar{\mathcal{L}}[h, \bar{\mathbf{b}}] = y(x + h) - y(x) - h \, y'(x) - h^2 \sum_{i=1}^{s} \bar{b}_i y''(x + c_i h),$$

$$\bar{\mathcal{L}}'[h, \mathbf{b}] = h y'(x + h) - h y'(x) - h^2 \sum_{i=1}^{s} b_i y''(x + c_i h).$$

satisfy the follwing sets of conditions:

$$\mathcal{L}_i[h, \mathbf{a}]1 = \bar{\mathcal{L}}[h, \bar{\mathbf{b}}]1 = \bar{\mathcal{L}}'[h, \mathbf{b}]1 = 0, \quad i = 1, \ldots, s,$$

$$\mathcal{L}_i[h, \mathbf{a}]x = \bar{\mathcal{L}}[h, \bar{\mathbf{b}}]x = \bar{\mathcal{L}}'[h, \mathbf{b}]x = 0, \quad i = 1, \ldots, s,$$

$$\mathcal{L}_i[h, \mathbf{a}] \cos(r \mu x) = \mathcal{L}_i[h, \mathbf{a}] \sin(r \mu x) = 0, \quad i = 1, \ldots, s, q = 1 \ldots r,$$

$$\bar{\mathcal{L}}[h, \bar{\mathbf{b}}] \cos(r \mu x) = \bar{\mathcal{L}}[h, \bar{\mathbf{b}}] \sin(r \mu x) = 0, \quad r = 1, \ldots, q,$$

$$\bar{\mathcal{L}}'[h, \mathbf{b}] \cos(r \mu x) = \bar{\mathcal{L}}'[h, \mathbf{b}] \sin(r \mu x) = 0, \quad r = 1, \ldots, q.$$

It is clear from this definition that more attention is paid to the oscillatory-like solutions than to the exponential-like ones. Although the Definition 6.7 is quite general, in all papers published so far as only methods of trigonometric order 1 have been studied.

It is easy to verify that an RKN method has trigonometric order q, according to Definition 6.7 if its parameters satisfy the following systems:

$$
\begin{cases}
\sum_{j=1}^{s} a_{ij} \cos(k\, c_j\, z) = \dfrac{1 - \cos(k\, c_i z)}{k^2\, z^2}, & i = 1, \ldots, s, \\[2mm]
\sum_{j=1}^{s} a_{ij} \sin(k\, c_j\, z) = \dfrac{k c_i z - \sin(k\, c_i z)}{k^2\, z^2}, & i = 1, \ldots, s, \\[2mm]
\sum_{i=1}^{s} \bar{b}_i \cos(k c_i\, z) = \dfrac{1 - \cos(kz)}{k^2\, z^2}, \\[2mm]
\sum_{i=1}^{s} \bar{b}_i \sin(k\, c_i\, z) = \dfrac{k z - \sin(k\, z)}{k^2\, z^2}, \\[2mm]
\sum_{i=1}^{s} b_i \cos(k\, c_i\, z) = \dfrac{\sin(k\, z)}{k\, z}, \\[2mm]
\sum_{i=1}^{s} b_i \sin(k\, c_i z) = \dfrac{1 - \cos(k\, z)}{k\, z},
\end{cases}
$$

for $k = 1, \ldots, q$ and where as usual $z = \mu\, h$. As an example, a two-stage RKN method has trigonometric order 1 when its parameters are given by the following expressions [22]:

$$
P(z) = \cos(c_1\, z)\sin(c_2\, z) - \sin(c_1\, z)\cos(c_2\, z),
$$

$$
a_{i1} = \left(\frac{1 - \cos(c_1\, z)}{z^2} \sin(c_2\, z) - \cos(c_2 z)\frac{c_i\, z - \sin(c_i\, z)}{z^2} \right) / P(z),
$$

$$
a_{i2} = \left(\cos(c_1 z)\frac{c_i z - \sin(c_i z)}{z^2} - \sin(c_1 z)\frac{1 - \cos(c_i\, z)}{z^2} \right) / P(z),
$$

$$
\bar{b}_1 = \left(\frac{1 - \cos(z)}{z^2} \sin(c_2\, z) - \cos(c_2\, z)\frac{z - \sin(z)}{z^2} \right) P(z),
$$

$$
\bar{b}_2 = \left(\cos(c_1\, z)\frac{z - \sin(z)}{z^2} - \sin(c_1\, z)\frac{1 - \cos(z)}{z^2} \right) / P(z),
$$

$$
b_1 = \left(\frac{\sin(z)}{z} \sin(c_2\, z) - \cos(c_2\, z)\frac{1 - \cos(z)}{z} \right) / P(z),
$$

$$
b_2 = \left(\cos(c_1\, z)\frac{1 - \cos(z)}{z} - \sin(c_1\, z)\frac{\sin(z)}{z} \right) / P(z),
$$

with c_1 and c_2 free parameters. Note that the above results can also be expressed in terms of η_{-1}- and η_0- functions. We have preferred to maintain the original notation of the above cited papers. In all investigations the set $\{c_i\}$ is kept frequency-independent.

DEFINITION 6.8 *The (algebraic) order of accuracy of the RKN-method is defined to be* $p = \min(p_1, p_2)$ *for the integers* p_1 *and* p_2 *satifying*

$$y(t_{n+1}) - y_{n+1} = \mathcal{O}(h^{p_1+1}), \quad y'(t_{n+1}) - y'_{n+1} = \mathcal{O}(h^{p_2+1}) .$$

Taking this definition into account all two-stage RKN-methods have order 2, irrespective of the values of c_1 and c_2. Order 3 is achieved whenever

$$c_1 = \frac{2 - 3c_2}{3(1 - 2c_2)} ,$$

and order 4, the maximum possible, is attained when the collocation parameters have the values $1/2 \pm \sqrt{3}/6$. In a similar way Ozawa [21] derived four-stage RKN methods of trigonometric order 1 and algebraic order 4. Coleman and Duxbury [7] used the treatment of mixed interpolation to construct mixed collocation methods with one, two and three collocation points. They show that their methods are trigonometric RKN-methods in the terminology of Ozawa [21]. They also studied the stability properties of their methods and the so-called order of dispersion.

References

[1] Albrecht P. (1987). The extension of the theory of A-methods to RK methods, in: K. Strehmel, ed. *Numerical Treatment of Differential Equations*, Proc. 4th Seminar NUMDIFF-4, Tuebner-Texte zur Mathematik (Tuebner, Leipzig):8–18.

[2] Albrecht, P. (1987) A new theoretical approach to RK methods. *SIAM J. Numer. Anal.*, 24:391–406.

[3] Butcher, J. C. (1964). On Runge-Kutta processes of high order. *Math. Comput.*, 18:50–64.

[4] Butcher, J. C. (1987). *The Numerical Analysis of Ordinary Differential Equations, Runge-Kutta and General Linear Methods*. Chichester John Wiley & Sons.

[5] Butcher, J. C. (2003). *Numerical Methods for Ordinary Differential Equations*. Chichester John Wiley & Sons.

[6] Coleman, J. P. (1998). Mixed interpolation methods with arbitrary nodes. *J. Comp. Appl. Math.*, 92:69–83.

[7] Coleman, J. P. and Duxbury, S. C. (2000). Mixed collocation methods for $y'' = f(x, y)$. *J. Comp. Appl. Math.*, 126: 47–75.

[8] De Meyer, H. , Vanthournout, J. and Vanden Berghe, G. (1990). On a new type of mixed interpolation. *J. Comp. Appl. Math.*, 30: 55–69.

[9] England, R. (1969). Error estimates for Runge-Kutta type solutions to systems of ordinary differential equations. *Comput. J.*, 12: 166–170.

[10] Franco, J. M. (2002). An embedded pair of exponentially fitted explicit Runge-Kutta methods. *J. Comp. Appl. Math.*, 149: 407–414.

[11] Gautschi, W. (1962). Numerical integration of ordinary differential equations based on trigonometric polynomials. *Numer. Math.*, 3: 381–397.

[12] Hairer, E. , Nørsett, S. P. and Wanner G. (1993). *Solving Ordinary Differential Equations I, Nonstiff Problems*. Berlin Springer-Verlag.

[13] Henrici, P. (1962). *Discrete Variable Methods in Ordinary Differential Equations*. John Wiley & Sons,Inc., New-York - London.

[14] Ixaru, L. Gr. (1997). Operations on oscillatory functions. *Comput. Phys. Comm.*, 105:1–19.

[15] Ixaru, L. Gr. , De Meyer, H. and Vanden Berghe, G. (2002). Frequency evaluation in exponential fitting multistep algorithms. *J. Comp. Appl. Math.*, 140: 423–433.

[16] Ixaru, L. Gr. and Paternoster, B. (2001). A Gauss quadrature rule for oscillatory integrands. *Comput. Phys. Comm.*, 133: 177 – 188.

[17] Kutta, W. (1901). Beitrag zur näherungsweisen Integration totaler Differentialgleichungen. *Zeitschr. für Math. u. Phys.*, 46: 435–453.

[18] Lambert, J. D. (1991). *Numerical Methods for Ordinary Differential Systems, The Initial Value Problem*. Chichester John Wiley & Sons.

[19] Lyche, T. (1972). Chebyshevian multistep methods for ordinary differential equations. *Numer. Math.*, 19:65–75.

[20] Oliver, J. (1975). A Curiosity of Low-Order Explicit Runge-Kutta Methods. *Math. Comp.*, 29: 1032–1036.

[21] Ozawa, K. (1999). A Four-stage Implicit Runge-Kutta-Nyström Methods with Variable Coefficients for Solving Periodic Initial Value Problems. *Japan Journal of Industrial and Applied Mathematics*, 16: 25–46.

[22] Paternoster, B. (1998). Runge-Kutta(-Nyström) methods for ODEs with periodic solutions based in trigonometric polynomials, *Appl. Num. Math.*, 28: 401–412.

[23] Runge, C. and König, H. (1924). *Vorlesungen über numerisches Rechnen*, Grundlehren XI, Springer Verlag.

[24] Runge, C. (1895). Über die numerische Auflösung von Differentialgleichungen. *Math. Ann.*, 46: 167–192.

[25] Simos, T. E. , Dimas, E. and Sideridis,A. B. (1994). A Runge-Kutta-Nyström method for the numerical integration of special second-order periodic initial-value problems, *J. Comp. Appl. Math.*, 51 : 317–326.

[26] Simos, T. E. (1998). An exponentially-fitted Runge-Kutta method for the numerical integration of initial-value problems with periodic or oscillating solutions. *Comput. Phys. Comm.*, 115: 1–8.

[27] Simos, T. E. (2001). A fourth algebraic order exponentially-fitted Runge-Kutta method for the numerical solution of the Schrödinger equation. *IMA Journ. of Numerical Analysis*, 21: 919 –931.

[28] Vanden Berghe,G. , De Meyer, H. , Van Daele, M. and Van Hecke, T. (1999). Exponentially-fitted explicit Runge-Kutta methods. *Computer Phys. Comm.*, 123: 7–15.

[29] Vanden Berghe, G. , De Meyer, H. , Van Daele, M. and Van Hecke, T. (2000). Exponentially-fitted Runge-Kutta methods. *J. Comp. Appl. Math.*, 125: 107–115.

[30] Vanden Berghe, G. , Ixaru, L. Gr. and Van Daele, M. (2001). Optimal implicit exponentially-fitted Runge-Kutta methods. *Comp. Phys. Commun.*, 140: 346–357.

[31] Vanden Berghe, G. , Ixaru, L. Gr. and De Meyer, H. (2001). Frequency determination and step–length control for exponentially-fitted Runge–Kutta methods. *J. Comp. Appl. Math.*, 132: 95–105.

[32] Vanden Berghe, G. , Van Daele, M. and Vande Vyver, H. (2003). Exponential-fitted Runge–Kutta methods of collocation type: fixed or variable knot points? *J. Comp. Appl. Math.*, 159: 217–239.

[33] Van der Houwen, P. J. and Sommeijer, B. P. (1987). Explicit Runge-Kutta(-Nyström) methods with reduced phase errors for computing oscillating solution. *SIAM J. Numer. Anal.*, 24: 595–617.

[34] Van der Houwen, P. J. and Sommeijer, B. P. (1987). Phase-lag analysis of implicit Runge-Kutta methods. *SIAM J. Numer. Anal.*, 26: 214–228.

[35] Van der Houwen,P. J. , Sommeijer, B. P. , Strehmel, K. and Weiner, R. (1986). On the numerical integration of second order initial value problems with a periodic forcing force. *Computing*, 37 : 195–218.

[36] Wright, K. (1970). Some relationships between implicit Runge-Kutta, collocation and Lanczos τ-methods, and their stability properties. *BIT*, 10:217–227.

CD contents

The attached CD contains the file `eflib.for` which is a program library for use in various applications of the ef-based methods. The file collects both the directly accessible subroutines described in the book and the ones which are called internally by these. All subroutines were written in FORTRAN–95 and commented.

The subroutines described in the book (in alphabetic order), their purpose and the page number where they are presented are as it follows:

The following subroutines (also in alphabetic order) are called internally by the previous ones: CINFF1, CINFF2, CINFG1, CREGLIS, DCLUDEC, DCLUSBS, DCREGBL, GSYST, INFFENC1, INFFENC2, INFFFOUR, INFFNC1, INFFNC2, INFGENC1, INFGFOUR, INFGNC1, INFGNC2, LUDEC, LUSBS, REGBLOCK, REGLIS.

Topic Index